Hormone and Biofilm Interplay

This book comprehensively reviews the interplay between hormones and biofilms, uncovering the complex ways in which hormonal signals impact microbial communities and the formation of biofilms. It begins with an introductory overview of hormones and reveals the dynamic nature of biofilm formation and its regulation, emphasizing the adaptability of biofilms to hormonal influences. The book also explores the role of hormones in quorum sensing among both Gram-positive and Gram-negative bacteria, illustrating how bacterial communities harness hormone signals to coordinate their behavior and form biofilms. It explores various types of hormones, including peptide hormones, stress hormones, phytohormones, calcium and sugar-regulating hormones, sex hormones, enteric hormones, catecholamines, and melatonin, and their specific roles in biofilm formation and regulation. Each chapter provides insights into the regulatory factors behind biofilm formation, offering a detailed look at the biochemical pathways involved and their potential implications for understanding bacterial behavior and resistance mechanisms. Additionally, the book discusses the implications of hormonal drugs on biofilms, opening new avenues for biomedical research and therapeutic strategies. This book is an important resource for researchers, academicians, and professionals in microbiology, endocrinology, biotechnology, and environmental sciences.

Key Features:

- Explores the complex interplay between hormones and microbiota composition, emphasizing the influence of hormonal levels on microbial diversity and biofilm dynamics
- Highlights the dynamic and adaptable nature of biofilm formation and its regulation by hormonal signals
- Details the significant role of hormones in quorum sensing among bacteria, illustrating how microbial communities utilize hormonal signals for coordinated behavior
- Uncovers regulatory factors behind biofilm formation, providing a deep dive into the biochemical pathways that facilitate biofilm development
- Discusses implications of hormonal drugs on biofilms, highlighting potential targets for combating biofilm-associated infections and their applications

Hormone and Biofilm Interplay

Mechanisms and Applications

Edited by
Rina Rani Ray

CRC Press
Taylor & Francis Group
Boca Raton London New York

CRC Press is an imprint of the
Taylor & Francis Group, an **informa** business

First edition published 2026
by CRC Press
2385 NW Executive Center Drive, Suite 320, Boca Raton FL 33431

and by CRC Press
4 Park Square, Milton Park, Abingdon, Oxon, OX14 4RN

CRC Press is an imprint of Taylor & Francis Group, LLC

ISBN: 978-1-032-85618-6 (hbk)
ISBN: 978-1-032-85624-7 (pbk)
ISBN: 978-1-003-51904-1 (ebk)

DOI: 10.1201/9781003519041

Typeset in Times
by SPi Technologies India Pvt Ltd (Straive)

To all my teachers who taught me Endocrinology and Microbiology

Contents

Preface

Microbiology is a subject that deals with various aspects of microorganisms, such as structure, function, phylogeny, economic importance, and pathogenicity of bacteria, fungi, protozoa, and viruses. In contrast, endocrinology deals with the chemical and functional attributes of different hormones secreted in higher vertebrates. The relationship between these two distant subjects is now being explored, leading to the coining of a new term, "microbial endocrinology."

Research has shown that a few phytohormones and neurotransmitters are produced by the microbes, and that often gut microbiota can produce new steroid hormones and even metabolize a few sex steroids. On the other hand, the host hormones can act as the signaling molecules, and the microbes can respond to regulate their expression levels. Hence, microbes are active participants in the hormonal scenery of their host. Microbes, in the host body, often remain as biofilm-embedded sessile cells, within a biofilm matrix, resistant to antibiotics and host immune reactions. The recalcitrant biofilm-bound cells lead to various chronic diseases of a "hard to treat" nature. The biofilm formed on implants and prosthetics results in the failure of surgical outcomes by increasing the risk of infection and reducing the chance of wound healing. Hence the factors promoting or inhibiting biofilm formation are crucial for determining the remedial success. Although common physiological factors, such as body temperature, pH, availability of food, and the host's immune reactions play a pivotal role in the formation of biofilm, researchers are trying to find out the specific role of a hormone on bacterial biofilm formation. Since, except for some sex steroids and sugar-regulating hormones, relatively little work has been done and there is a dearth of knowledge regarding the role of various hormones on different bacteria and fungi. A thorough appraisal is required regarding the impact of hormones on biofilm formation.

The present book attempts to showcase the experimental observations of almost all mammalian hormones, based on their functional attributes and their impact on biofilm formation or eradication efficiencies of different bacteria and fungi. The hormones secreted from different endocrine glands, namely pituitary, thyroid, gut, pineal, hypothalamus, adrenal, ovary, and testis, and various local hormones are studied to evaluate their effect on microbial biofilm. In addition to the natural hormones, it also highlights the effects of synthetic hormones, used as hormonal drugs, on biofilm formation and thereby tries to give a "bird's eye view" on the interplay of hormones and microbial biofilms.

Rina Rani Ray
Editor
May, 2025

About the Editor

Rina Rani Ray, Associate Professor in the Department of Biotechnology, is currently acting as Head of the Department of Life Science in Maulana Abul Kalam Azad University of Technology, formerly known as West Bengal University of Technology. She obtained her post-graduate degree from the University of Calcutta in Zoology with a specialization in **Endocrinology**. She did her PhD in **Microbiology** from Bose Institute, Kolkata. She has thirty-three years of research experience and her topic of research is microbial biotechnology, which includes microbial enzymology, bioremediation, biofilm research and nanobiotechnology. She has published more than one hundred fifty-odd research papers in different international journals with high impact factors and has edited a few books from renowned publishing houses. She has conducted several extramural research projects successfully and guided several post-graduate projects. A good number of scholars are being guided by her, out of which several students have been awarded their PhD under her sole supervision. She regularly acts as a resource person and chairperson in various conferences, acting as an editorial board member in various reputed international journals and as a reviewer of various journals of repute. She is a regular reviewer of various international research project proposals and adjudicator of a number of doctoral theses. Dr. Ray is a fellow member of various scientific associations and keen to step into the multidisciplinary arena of Life Sciences.

1 Introduction to Hormones

Ayoushmoti Chakraborty, Anjali Rao, and Rina Rani Ray
Department of Biotechnology, Maulana Abul Kalam Azad University of Technology, Kolkata, India

1.1 INTRODUCTION

One of the most important processes to control all sorts of behavior is hormones. This is because it has functions for growth and development. The body is made up of a network of glands that compose the endocrine system, which is responsible for the production and secretion of hormones present in the body. These chemicals, which are all secreted by the endocrine glands, have effects on a wide range of biological activities, such as growth and development, metabolism, sexual function, reproduction, and mood. Hormones also facilitate bodily responses and act to ensure healthy growth, metabolism, and homeostasis. They work without any disturbances which automatically steer an abundance of body processes. Apart from this, they all play a great role in controlling the metabolism wherein excess sources and reserves are stored, and, in times of deficiency, certain nutrients are broken down to produce sufficient amounts of essential energy. The initiation and termination of anabolism and catabolism is also controlled by the system. Two further hormones, insulin and glucagon, play an important part in signalling blood sugar concentrations; the former decreases them, while the latter increases them (Hamid et al., 2018).

In addition to this, respiration, digestion, and circulation involves multiple homeostatic processes that need to be integrated. ADH (antidiuretic hormone) works against dehydration by retaining body fluids and thus increasing blood volume simultaneously. Diabetes mellitus, a common condition of the present, involves the excretion of large amounts of fluid and the inability to retain some of it. This disease is known for decreasing an individual's internal as well as external stability. Therefore, the preservation of homeostasis, along with control of electrolyte imbalance, protects the body. Cortisol and adrenaline are two other hormones which also play a role in relaxing the body for the proper manipulation of stress as well as a balanced cardiovascular function. The transformation of an individual from a child to an adult, resulting in cognitive and physical changes, is governed by a set of different hormones to ensure a good level of growth and development. These are the thyroid hormones which help the brain to grow and influence the control of metabolism to some degree during early life. In addition, it initiates the growth hormone, which enables development of bones and muscles, ultimately driving the development of the skeletal system. Two other hormones, the sex hormones estrogen and testosterone, are responsible for breast development in females and facial hairs in males. They also control and sustain secondary sex characteristics. Along with this, they also control general reproductive health and sexual desire.

By the secretion of melatonin, the pineal gland controls the body's sleep–wake cycle. Secretion of melatonin is also necessary for the preservation of circadian rhythms. Light, however, controls the amount of melatonin that is produced. The level of study about the endocrine system has improved tremendously recently. It is for this reason that hormones that used to be viewed as purely modulators of stress response and emotional control like serotonin and dopamine, are now seen as being part of the active pathology of mood disorders. Any type of hormonal imbalance can further worsen these disorders (Csaba, 2014).

DOI: 10.1201/9781003519041-1

This chapter deals with different types of hormones, and the importance of hormones in maintaining homeostasis in the body. Based on this explanation, it makes sense that imbalances in hormones can have multifaceted repercussions such as the deterioration of physical and mental health, infertility, or problems in metabolism. The topics covered in this chapter will offer an essential perspective on the systemic nature of hormones.

1.2 CLASSIFICATION OF HORMONES

Hormones can be classified based on their function and origin, and how they are chemically structured and processed. This section of the chapter will highlight how hormones perform different roles related to growth, metabolism, reproduction, and the maintenance of metabolic balance, or homeostasis.

1.2.1 Classification Based on Source

1.2.1.1 Phytohormones (Plant Hormones)

To begin with, it is important to know that while animals produce hormones in special glands, plants also produce their own hormones, which are called phytohormones. These are present in different plant tissues. They then travel to other target cells in different tissues to perform functions such as stress reactions, flowering, fruiting, and germination. Phytohormones are also used in the control of physiological functions such as growth and development in reaction to external stimuli. Among these we find:

Auxins (Indole-3-acetic acid): These allow for the elongation of cells and the development of roots.

Gibberellins: These accomplish flowering, along with stem elongation and seed germination.

Cytokinins: These are concerned with the division of cells and the slowing down of the aging of tissues in plants.

Abscisic Acid: This triggers the inhibition of growth as well as controlling how plants respond to stress.

Ethylene: This controls the ripening of fruits and separation of leaves from the stems.

1.2.1.2 Animal Hormones

Glands in an animal's body secrete chemical substances called animal hormones. These are basically released into the circulatory system and their main focus is to regulate growth, reproduction, metabolism, and other bodily functions. These hormones alter specific target organs and tissues using receptor-binding processes and subsequent biochemical cascades. Furthermore, animal hormones play a quite important role in maintaining homeostasis, responding to environmental challenges, and facilitating the healthy development and functioning of organs. Examples include:

Insulin–Manages blood sugar concentration;

Adrenaline–Triggers the body for the impending stress response;

Thyroxine–Regulates metabolic activity and growth;

Estrogen and Testosterone–Manage reproductive organs and functions (Bharati et al., 2023)

1.2.2 Classification based on Chemical Structure

1.2.2.1 Peptide Hormones

Because they are typically water-soluble and comprised of chains of amino acids, peptide hormones have difficulty passing through cell membranes. Rather, they attach to receptors on the cell surface, which sets off intracellular signaling cascades that result in physiological reactions. Before being converted into active hormones and kept in vesicles for controlled release, peptide hormones are first created as larger precursors. Examples include:

Growth Hormone (GH)–Stimulates growth and cell regeneration (Zhang et al., 2025).
Oxytocin–Promotes uterine contractions and lactation.

1.2.2.2 Steroid Hormones

Cholesterol is the source of lipid-soluble steroid hormones. They can readily penetrate cell membranes and attach to intracellular receptors because they are hydrophobic, which has a direct impact on gene expression. The adrenal glands and gonads are the main organs that produce steroid hormones, which are essential for metabolism, immunological response, and reproduction, among other physiological processes. Examples include:

Cortisol–Regulates metabolism and immune response;
Aldosterone–Controls sodium and water balance;
Testosterone–Supports male reproductive functions.

1.2.2.3 Thyroid Hormones

The thyroid gland produces thyroid hormones, which control growth, development, and metabolism. Iodine is a component of these hormones and is necessary for their synthesis. By raising metabolic activity and oxygen consumption, thyroid hormones affect almost every organ system, including the neurological, skeletal, and cardiovascular systems. These are:

Thyroxine (T4)–Controls the body's metabolic rate;
Triiodothyronine (T3)–Active form of thyroid hormone with similar metabolic functions (Shahid et al., 2023).

1.2.2.4 Catecholamines

Tyrosine is the source of catecholamines, which are both hormones and neurotransmitters. Stress causes the release of these hormones, which aid in controlling metabolic, neurological, and cardiovascular processes. They play a vital part in the body's "fight or flight" response and act swiftly with transient effects. Examples include:

Epinephrine (Adrenaline)–Increases heart rate and energy availability during stress;
Norepinephrine–Enhances alertness and blood pressure regulation;
Dopamine–Functions in mood regulation and neurotransmission (Goldstein, 2010).

1.2.3 Classification based on their Effects

1.2.3.1 Autocrine Hormones

The cell that secretes autocrine hormones is also the cell that is affected by them. These hormones play a key role in immunological responses, cancer progression, and local cell signaling. By reacting to the signals they produce, they enable cells to control their own activity. These include: Insulin, which can act on the cells of its source itself.

1.2.3.2 Paracrine Hormones

Paracrine hormones do not enter the bloodstream; instead, they affect neighboring cells. They have an impact on processes such as inflammation, wound healing, and cell differentiation, and are essential for both local tissue responses and intercellular communication. An example includes: Prostaglandins, which act locally on a closely situated cell to regulate pain and inflammation.

1.2.3.3 Intracrine Hormones

Intracrine hormones are synthesized and act within the same cell. They play important roles in intracellular signaling and gene regulation, but, unlike other hormones, they are not secreted and instead affect cellular processes from within. Examples include steroid hormones, as they act through nuclear receptors and can therefore be considered intracrine hormones.

1.2.3.4 Endocrine Hormones

Distant target organs are impacted by endocrine hormones that are released into the bloodstream. These hormones preserve systemic homeostasis and control a variety of physiological processes. One of the examples is Thyroxine, which, being secreted from the thyroid gland, regulates a number of metabolic functions (Hull & Harvey, 2014).

1.2.4 Local Hormones and Trophic Hormones

Local hormones, such as histamine, which mediates inflammatory responses, act in particular tissues where they are produced.

Trophic hormones stimulate other endocrine glands (ductless gland) to secrete their hormones and are controlled by positive and negative feedback systems. As per example, Thyroid-stimulating hormone (TSH), releasing from anterior pituitary, stimulates the thyroid gland to produce T3 and T4 hormones, which, in turn, impact a number of metabolic processes (Wuttke et al., 2010).

1.3 CLASSIFICATION BASED ON FUNCTIONS

We can also categorize hormones according to the physiological functions that they perform in the body:

- **Regulatory Hormones**: Hormones that regulate the release of other hormones. They include various trophic hormones from the pituitary gland, releasing hormones from the hypothalamus and controlling the stimulation or inhibition of other hormones.
- **Reproductive Hormones**: These hormones, such as estrogen, testosterone, and progesterone, control reproductive processes.
- **Metabolic Hormones**: This includes hormones that control various metabolic functions such as insulin, glucagon, and thyroxine.
- **Growth Hormones**: These hormones, such as insulin-like growth factor-1 (IGF-1), stimulate growth and development.
- **Stress Hormones**: Cortisol, adrenaline, and noradrenaline assist the body's reaction to stress.
- **Digestive Hormones**: These regulate the digestive process and include gastrointestinal hormones such as secretin, cholecystokinin, and gastrin.
- **Circadian Rhythm Hormones**: Melatonin controls the cycles of sleep and wakefulness. (Rosol et al., 2024)

1.4 HORMONE-SECRETING GLANDS IN ANIMALS

Hormones are secreted by specific glands, known as endocrine glands, which are then sent directly into circulation. Along with the other glands, they also help to maintain homeostasis, manage growth and development, regulate metabolism, and propagate reproduction. Unlike exocrine glands, which emit chemicals through ducts, endocrine glands are ductless and rely on the circulatory system for their distribution. Endocrinology describes the principal hormone-secreting glands in animals, their secretions for preserving physiological balance, and their functions. Coordination amongst different

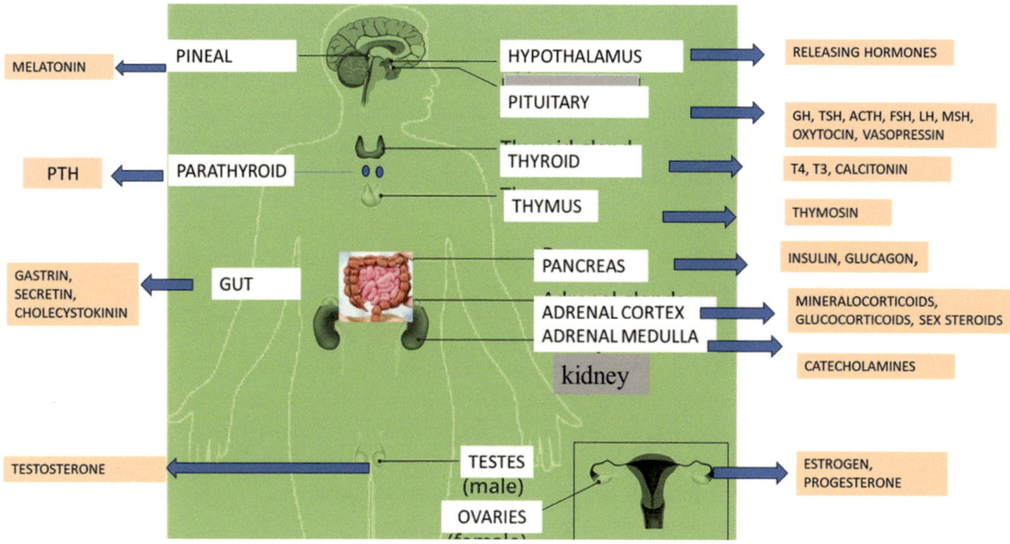

FIGURE 1.1 Anatomical location of human endocrine glands.

endocrine glands (Figure 1.1) is an excellent example of the perfect orchestration of various physiological systems for the maintenance of homeostasis.

1.4.1 Hypothalamus

The hypothalamus is the most important part of the brain, sitting between the nervous and endocrine systems. It maintains homeostasis by controlling appetite, thirst, and emotion-driven temperature regulation. The hypothalamus produces multiple hormones that act on the pituitary gland, such as the thyrotropin-releasing hormone, the gonadotropin-releasing hormone, and the corticotropin-releasing hormone. These hormones help the organism to maintain set homeostatic levels while responding correctly to surroundings providing communication between the brain and endocrine glands. The oxytocin and antidiuretic hormone, which are produced by the hypothalamus, are stored and released by the posterior pituitary (Xie & Dorsky, 2017).

1.4.2 Pituitary Gland

Located at the base of the brain, the pituitary gland is the control center of the body's endocrine system. It coordinates a number of body activities and secretes hormones. The anterior pituitary produces three primary hormones: Growth hormone (GH), which stimulates cell division and growth; thyroid-stimulating hormone (TSH), which maintains the function of the thyroid gland and adrenocorticotropic hormone (ACTH), which triggers the release of cortisol from the adrenal glands. In addition, it produces prolactin (PRL), follicle-stimulating hormone (FSH), and luteinizing hormone (LH), which control reproductive processes. In contrast, the posterior pituitary does not produce hormones. Instead, it stores oxytocin, a hormone associated with childbirth, breastfeeding as well as nursing, and antidiuretic hormone (ADH), which manages the body's water retention through kidney functions (Lechan & Toni, 2016).

1.4.3 Thyroid Gland

The thyroid gland is positioned at the front of the neck. It plays a critical role in controlling energy production, metabolism, and general growth and development. Its two important hormones, thyroxine (T4) and triiodothyronine (T3), control metabolic functions by directly impacting the amount of oxygen and heat produced by body cells. T3 and T4 influence the heart, brain, And muscles, as well as nearly every organ in the body. In addition, another thyroid hormone, calcitonin, assists in calcium blood level management to help prevent excessive bone loss. Among the most vital endocrine glands, the thyroid is critical for normal growth, brain development, and energy homeostasis (Mense & Boorman, 2017).

1.4.4 Parathyroid Gland

There are four tiny parathyroid glands, which are located at the back of the thyroid gland. Their size, however, does not impede their quantity of hormone production; the parathyroid glands produce parathyroid hormone (PTH), which plays a vital role in maintaining calcium homeostasis. Principally, PTH maintains constant levels of calcium in the blood by stimulating bone resorption, increasing intestinal calcium absorption, and decreasing calcium excretion by the kidneys. This balance is important for muscle contraction, neuron functioning, and bone health. To avoid osteoporosis and calcium deficiency, one must have a precise balance of parathyroid hormone and calcitonin. This, in turn, keeps the calcium level in the body within an ideal range (Chen & Goodman, 2004).

1.4.5 Pancreas

The pancreas is a distinct organ since it is also an exocrine and an endocrine gland. The endocrine gland is responsible for the secretion of somatostatin, glucagon, and insulin, which are mandatory in regulating the blood sugar level. While insulin enhances the uptake of glucose into cells and reduces blood glucose levels, glucagon increases blood sugar levels by fueling the conversion of glycogen to glucose in the liver. Somatostatin, as an inhibitory hormone, acts on the secretion of glucagon and insulin in order to maintain glucose homeostasis. The pancreas plays a role in maintaining energy balance, and its malfunctioning endocrine hormones can lead to diabetes mellitus, which disrupts blood sugar regulation (Freeman et al., 2022).

1.4.6 Adrenal Glands

The adrenal gland is a triangle organ, made up of an outer cortex and an inner medulla, located above the kidneys. Steroid hormones like cortisol, which control immunological responses, metabolism, and stress adaptation, are produced by the adrenal cortex. The gland also secretes aldosterone that controls blood pressure and hydration status by maintaining sodium and potassium levels. The hormones released from three distinct layers of the cortex are mineralocorticoid, glucocorticoid, and sex steroids, respectively. In contrast, catecholamines produced by the adrenal medulla, like norepinephrine (noradrenaline) and epinephrine (adrenaline), accelerate the heart rate, enlarge the respiratory passages, and redirect blood to the muscles in preparation of a "fight or flight" response. These glands promote survival in extreme conditions by regulating metabolism, fluid balance, and coping with stress (Berger et al., 2019).

1.4.7 Gonads (Ovaries and Testes)

Sex hormones that control secondary sexual traits and reproductive processes are produced by the gonads, which consist of the testes in males and the ovaries in females. The ovaries produce estrogen and progesterone, which govern the menstrual cycle, pregnancy, and processes such as breast development and body fat distribution. The testes secrete testosterone, which is essential for the development of muscle bulk, spermatogenesis, as well as other male secondary sexual characteristics such as beard growth and the lowering of voice. These hormones are essential for puberty, reproduction, and sexual health in general. Developmental problems and infertility can result from any imbalance in the production of gonadal hormones. (Zhou et al., 2023)

1.4.8 Pineal Gland

The function of the pineal gland, which is the size of a pea, is to regulate sleep–wake cycles along with circadian rhythms. This small gland is responsible for producing melatonin, which helps the body internal clock to maintain the day–night cycle. However, light affects the secretion of melatonin, with greater amounts being released at night to encourage sleep. In certain animal species, this gland modifies melatonin levels during different periods. This helps sustain seasonal reproductive cycles. The preservation of sleep cycles and the overall health of the body is taken care of by the pineal gland. Although it is not as large as the others, it still makes it possible to work in such a way, maintaining general health (Gheban et al., 2019).

1.4.9 Thymus Gland

The thymus gland is located behind the sternum and within the upper chest in the superior mediastinum. It is known for producing the hormone thymosin, which helps in the formation and maturation of T-cells. T-cells are a subdivision of white blood cells that play an important role in immune response. With age, the thymus shrinks and deteriorates, resulting in lower T-cell production over time. The thymus plays a critical role in the performance of the immune system, which is used in fighting infection and fostering immunity against pathogens (Miller, 2020).

1.5 DIFFERENCE BETWEEN PLANT HORMONES AND ANIMAL HORMONES

Hormones are necessary for many biological functions in animals (Table 1.1), as was previously mentioned. Along with vital metabolic processes like sleep, digestion, appetite, heart rhythm, blood pressure, and respiration, these also encompass growth, puberty, and reproduction. Furthermore, fulfillment, affection, and joy, as well as other psychological states, are greatly influenced by hormones. Specialized ductless glands, known as endocrine glands, secrete these hormones within the animal's body. Here the bloodstream acts as the means of transport. The pituitary, adrenal, thyroid, parathyroid, testes, ovaries, pineal gland, islets of Langerhans, and thymus are the important endocrine glands in mammals. A vital component of the brain, the hypothalamus, produces a number of hormones that modulate the release of pituitary hormones by acting as both releasing and inhibiting agents on the pituitary gland. The pituitary gland, which is termed the "master" gland, arranges the activity of other glands by secreting trophic hormones that instruct these glands to release their hormones. (Lakna, 2019).

Remarkably, the endocrine functions of numerous organs and tissues, such as the gut, thymus, adipose tissue, kidneys, and heart, were not first understood. They are now recognized as essential parts of the endocrine system, though. When a gland produces too much or too little of a hormone, or when a gland is impacted by tumors or lesions, endocrine diseases result. The purpose of the body's endocrine feedback system is to control blood hormone levels. However, if this feedback system is unable to maintain the right hormone levels, imbalances may result. The significance of the endocrine system in preserving general physiological and psychological well-being is highlighted by the fact that these imbalances can result in a number of health problems (Tilbrook & Ralph, 2017).

On the other hand, plant hormones—also known as phytohormones or plant growth regulators—are essential for controlling the growth and development of plants. Numerous processes, including cell division, germination, bud formation, root expansion, and fruit development, are stimulated by these hormones. They are also essential for the periodical shedding of leaves and the development

TABLE 1.1
Plant and Animal Hormones and Their Functions

Category	Plant Hormones	Function in Plants	Animal Hormones	Function in Animals
Growth & Development	Auxins	Cell elongation, root initiation, apical dominance.	Growth Hormone (GH)	Stimulates growth, cell division, protein synthesis.
	Gibberellins	Stem elongation, seed germination, flowering.	Thyroxine (T4) & Triiodothyronine (T3)	Regulate metabolic rate, energy use, growth.
	Cytokinins	Stimulates cell division, delays senescence	Testosterone	Develops male reproductive tissues, muscle/bone growth.
Stress Response & Adaptation	Abscisic Acid (ABA)	Regulates water stress, seed dormancy.	Cortisol	Aids in stress response, regulates metabolism, reduces inflammation
	Jasmonates	Defense against herbivores, wound healing.	Adrenaline (Epinephrine)	Triggers fight-or-flight response, increases heart rate, oxygen supply.
Reproduction & Maturation	Brassino ster oids	Enhances stress resilience, pollen tube growth.	Estrogen	Controls female reproductive system, bone density, menstrual cycle.
	Salicylic Acid	Defense against pathogens, flowering regulation.	Prolactin	Stimulates milk production, regulates metabolism, immune system.
Energy Regulation	Ethylene	Fruit ripening, leaf abscission, stress response.	Insulin	Lowers blood glucose, promotes glucose uptake and storage.
			Glucagon	Raises blood glucose, promotes glycogen breakdown.

of fruits. In plants, hormone synthesis takes place in different amounts across the plant, with feedback mechanisms functioning inside the same cells or adjacent tissues, rather than in specialized glands. Gibberellin, ethylene, auxin, cytokinin, and abscisic acid are the five traditional plant hormones. These hormones are easily transported through intercellular gaps because of their small molecular size (Tarkowská, 2019). In recent years, other families of plant hormones, such as strigolactones, salicylic acid, jasmonic acid, brassinosteroids, and different tiny proteins or peptides, have been identified. Plant hormones are typically divided into two groups according to how they work: growth inhibitors (like abscisic acid) and growth stimulants (like auxins, gibberellins, and cytokinins). Ethylene is special because, depending on the surroundings, it may serve either as an inhibitor or as a growth booster. Phytohormones efficiently regulate the entire plant development process through intricate interactions called cross-talk, guaranteeing healthy growth, reproduction, and environmental adaptation (Bari & Jones, 2008).

Despite this, the way in which plant and animal hormones interact within the body varies, even though they are both effective at low quantities. While plant hormones function more gradually and are incorporated into the plant's overall developmental cycle, animal hormones can have quick and noticeable impacts. These essential distinctions show the various ways in which plants and animals control growth, development, and environmental adaptability. They also represent the different evolutionary adaptations and life strategies of these two groups (Table 1.2).

TABLE 1.2
Difference between Plant Hormone and Animal Hormone

Characteristics	Plant Hormones	Animal Hormones
Origin	Produced by plants	Produced by animals
Type	Auxins, Gibberellins, Cytokinins, Ethylene, Abscisic Acid, Brassinosteroids	Peptides (e.g., insulin), Steroids (e.g., estrogen), Amines (e.g., adrenaline)
Production sites	Various tissues and organs such as roots, stems, leaves, and flowers.	Specific glands such as pituitary, thyroid, adrenal glands
Transport	Through xylem, phloem, or cell diffusion	Through the bloodstream
Mechanism of action	Typically work through altering gene expression or enzyme activity	Typically bind to specific receptors on target cells
Function	Regulate growth, development, and responses to environmental stimuli	Regulate various physiological processes such as metabolism, growth, reproduction
Concentration	Effective at very low concentrations	Effective at low concentrations, but generally higher than plant hormones
Response time	Generally slower and longer-lasting	Can be rapid or slow, depending on the hormone
Feedback mechanism	Less complex, often not involving feedback loops	Complex feedback mechanisms, often involving negative feedback loops
Synthetic forms	Used in agriculture (e.g., rooting powders)	Used in medicine (e.g., hormone replacement therapy)

1.6 SIMILARITIES BETWEEN PLANT HORMONES AND ANIMAL HORMONES

Regardless of the different biological contexts, there are a number of similarities between plant and animal hormones. Both forms of life have a variety of essential biochemical message systems that interrelate numerous physiological processes which are vital for development, growth, and adaptation to the environment. Each hormone is synthesized in certain blood organs or tissues and thereafter transported to target areas where certain physiological actions are enlisted. Both types of organisms have lower doses of hormones, but exert powerful control over their systems. Hormones from both plants and animals can interact in either opposing or complementary ways, forming a dynamic and integrated network of signals that regulate the overall functioning of an organism. Plants and animal hormones can affect each other in opposing or cooperative ways, creating a complex network of signals that integrate the functioning of the organism. For example, auxins and cytokinins have a synergistic interaction in plants for the control of cell division and differentiation; in animals, in contrast, insulin and glucagon have interdependent roles in the control of blood glucose concentration. These intricate interrelationships provide an insight into the complexity of hormonal regulation in plants and animals and show the importance of the hormones in maintaining the constancy and dynamicity in living organisms (Nürnberger et al., 2004).

1.7 GENERAL MECHANISM OF HORMONE ACTION

There are signaling pathways that guarantee appropriate communication between endocrine glands and target tissues following the hormone action mechanism. Different kinds of receptors allow hormones to act, triggering physiological and biochemical reactions that support growth, metabolism, reproduction, and homeostasis. Hormone synthesis and secretion, transportation, receptor binding, signal transduction, and physiological response are the main categories into which the general mechanism of hormone activity can be divided (Lyte, 2014).

1.7.1 Hormone Synthesis and Secretion

Feedback systems involving the hypothalamus, pituitary gland, and target organs control the synthesis of hormones. Once they are created, hormones are either released straight into the bloodstream (for steroid hormones) or retained in secretory vesicles (for peptide hormones). Hormone secretion is influenced by a number of variables, including metabolic alterations, brain impulses, and other hormones.

1.7.2 Hormone Transport in the Bloodstream

To get to their target cells, hormones must travel through the bloodstream. Lipid-soluble hormones, like thyroid and steroid hormones, need carrier proteins to move around, while water-soluble hormones, such as peptide hormones, dissolve in plasma and move freely. These carrier proteins guarantee the controlled release of hormones at their intended locations and shield them from deterioration (Puzianowska-Kuznicka et al., 2013).

1.7.3 Hormone Binding to Target Cell Receptors

By attaching to particular receptors on or within target cells, hormones produce their effects. Because these receptors are so specific, a particular hormone will only cause the right cells to react. A hormone's method of action is determined by the kind of receptor it binds to.

Hormone receptors come in two main varieties:

1.7.3.1 Cell Surface Receptors (for Water-Soluble Hormones)

Water-soluble hormones, including catecholamine and peptide hormones, are unable to penetrate target cells' lipid membranes. Rather, they initiate intracellular signaling cascades by binding to cell surface receptors. Usually, these receptors are enzyme-linked receptors that start second-messenger pathways or G-protein-coupled receptors (GPCRs).

1.7.3.2 Intracellular Receptors (For Lipid-Soluble Hormones)

Steroid and thyroid hormones are examples of lipid-soluble hormones that readily pass through cell membranes and attach to receptors found in the nucleus or cytoplasm. Long-term cellular alterations result from these hormone-receptor complexes' direct interactions with DNA to control gene expression (Zedalis & Eggebrecht, 2018)

1.7.4 Signal Transduction Pathways

Depending on the type of hormone and receptor involved, signal transduction is the series of molecular events that begins when a hormone attaches to its receptor. Second-messenger systems are necessary for water-soluble hormones, including catecholamine and peptide hormones, to transmit signals from the cell surface to intracellular destinations. One frequent mechanism is the cyclic adenosine monophosphate (cAMP) pathway, in which adenylyl cyclase converts ATP to cAMP after hormones such as glucagon and adrenaline activate G-protein-coupled receptors. Protein kinase A (PKA) is subsequently activated by cAMP, phosphorylating target proteins to initiate biological reactions.

A different mechanism involves IP3 (inositol triphosphate) and DAG (diacylglycerol), for which phospholipase C is stimulated by hormones such as oxytocin, which cleaves phospholipids into IP3 and DAG. While the latter promotes calcium efflux from the endoplasmic reticulum, the former acts on and stimulates protein kinase C (PKC) to propagate the signal, thus enhancing it further. For example, development factors and insulin are hormones that associate with the so-called tyrosine kinase receptors and these receptors become auto-phosphorylated and activate signal transduction cascades that modulate development and metabolism. Steroid and thyroid hormones, for example, are lipid-soluble hormones that can cross cell membranes and attach to intracellular receptors to create hormone–receptor complexes that go into the nucleus. These complexes influence gene transcription and trigger the production of proteins that mediate the effects of hormones by binding to particular DNA sequences called hormone response elements (HREs). Long-lasting physiological changes are the outcome of this slower mechanism (Balla & Hunyady, 2020).

1.7.5 Physiological Response and Feedback Regulation

Following the transmission of the hormone signal, the target cell produces a particular physiological response that may include the regulation of cell division, growth, and differentiation; modifications to membrane permeability and ion transport; activation or inhibition of metabolism-related enzymes; and modulation of gene expression leading to protein synthesis. Feedback mechanisms carefully regulate hormone action in order to preserve homeostasis. Positive and negative feedback are the two main categories of feedback regulation. Negative feedback loops govern the majority of hormone effects; as the hormone's influence increases, more hormone secretion is inhibited. To stop excessive hormone synthesis, for instance, high thyroid hormone levels inhibit the pituitary gland's release of thyroid-stimulating hormone (TSH). On the other hand, by encouraging further hormone secretion, positive feedback systems intensify the effects of hormones. One important instance is during childbirth, when uterine contractions brought on by oxytocin cause further oxytocin production, which intensifies and reinforces labor contractions. These regulatory systems guarantee that hormone levels stay in check and that bodily functions run as efficiently as possible (Themes, 2017).

1.8 MICROBIAL ENDOCRINOLOGY

An emerging interdisciplinary subject called "microbial endocrinology" revolves around the reciprocal connections between human neuroendocrine systems and microbes. It combines endocrinology, neurobiology, and microbiology to show how microbial communities affect and react to host hormone signals. This area of study especially deals with human health and illness because it clarifies how bacteria affect behavior, metabolism, and immunological responses. Microbial endocrinology research has important consequences for fields like stress reactions, infection dynamics, gut–brain communication, and treatment approaches.

1.8.1 Microbes as Endocrine Signal Responders

Microorganisms have the ability to sense and respond to neuroendocrine signals emitted by their host. Many bacterial species come along with receptors for hormones such as cortisol, adrenaline, and norepinephrine. These hormones facilitate behavioral modifications in reaction to the host's physiological state. This ability, combined with the hormonal background of the host, allows infections to regulate virulence, motility, biofilm formation, and antibiotic resistance.

Stress hormones can influence the function and disease-causing ability of bacteria, fungi, and viruses. Pathogens such as *Salmonella* and *Escherichia coli* recognize norepinephrine and epinephrine, and they can enhance their growth, adhesion, and toxin production

(Sarkodie et al., 2019). Similar to this, these stress hormones cause *Pseudomonas aeruginosa* and *Staphylococcus aureus* to alter biofilm production and antibiotic resistance, ultimately improving their virulence and survival (Cambronel et al., 2019) *Candida albicans* and similar fungal infections can change their degree of invasiveness based on hormonal signals presented by host factors (Kumwenda et al., 2022).

1.8.2 Microbes as Hormone Producers

Some microbes can synthesize and secrete bioactive compounds resembling human hormones that modify the physiology of the host. Their bioactive endocrine secretions are of great importance to the metabolic balance, as well as for communication between the intestine and the brain. Gut bacteria also produce hormones and neurotransmitters that influence one's mood, behavior, and physiology. A neurotransmitter which is known to affect mood and anxiety, gamma-aminobutyric acid (GABA), is produced by members of the *Lactobacillus* and *Bifidobacterium* species. Some strains of *Escherichia coli* produce serotonin, which plays an important role in regulating mood and bowel movement (Sittipo et al., 2022). Dopamine and norepinephrine, which affect the autonomic nervous system and behavioral reactions, are also produced by certain gut bacteria (Mhanna et al., 2024). Microbes have an impact on metabolic hormones in addition to neurological impacts. Gut bacteria create short-chain fatty acids (SCFAs), which are involved in controlling insulin secretion and preserving glucose homeostasis. Additionally, microbial metabolites influence hunger and energy balance by interacting with hormones that regulate appetite, such as ghrelin and leptin (O'Riordan et al., 2022).

1.8.3 Gut Microbiota and Host Endocrine System

Through microbial metabolites, immunological signals, and direct interactions with host receptors, the gut microbiota significantly influences host endocrine functioning. The gut–brain axis, impacting mood, cognition, and stress reactions, is one of the main ways in which bacteria impact physiological functions. The hypothalamic–pituitary–adrenal (HPA) axis's control is a crucial component of this connection (Freimer et al., 2022). Stress-related problems can be exacerbated after increased cortisol production caused by microbial imbalances, or dysbiosis. However, it has been demonstrated that probiotics like *Lactobacillus* and *Bifidobacterium* lessen these effects by lowering cortisol levels induced by stress (Misiak et al., 2020). In addition to regulating stress, the gut microbiome also affects hormones involved in reproduction. Gut bacteria affect estrogen metabolism through the estrobolome, which affects menopausal symptoms and disorders, including polycystic ovarian syndrome (PCOS). Testosterone and other androgen levels, essential for reproduction and general metabolic health, are impacted by microbial metabolites (D'Afflitto et al., 2022).

1.8.4 Microbial Endocrinology in Infection and Immune Modulation

The endocrine signaling enhances the microbes to engage with the host immune system, influencing inflammatory responses and preserving immunological homeostasis. Stress hormones like norepinephrine can increase the adherence and invasion of bacteria, making a person more vulnerable to illnesses. The dynamics between microbial activity and endocrine signals is demonstrated by the way in which some infections, such as *Helicobacter pylori*, modify their virulence in response to the host's hormonal state. The gut microbiome affects inflammatory and autoimmune disorders in addition to infection susceptibility (Mohammadzadeh et al., 2023). Apart from this, rheumatoid arthritis and inflammatory bowel disease have been connected to dysbiosis, underscoring the significance of microbial balance in immune control. Also, gut bacteria shape immunological and inflammatory responses by influencing immune cell function through metabolites and hormones generated by microbes (Zheng et al., 2020).

1.9 COMMON DISEASES/DISORDERS DUE TO HYPER AND HYPO SECRETION OF MAJOR HORMONES IN MAMMALS

Hormones are known for controlling numerous physiological functions in mammals. Thus, a small amount of hormonal imbalance, be it excessive secretion of hormones known as hypersecretion, or reduced secretion of hormones, known as hyposecretion, can lead to several health issues (Table 1.3).

TABLE 1.3
Diseases due to Hyper and Hypo Secretion of Major Hormones in Mammals

Hormone	Hyper secretion (Excess Production)	Hyposecretion (Deficient Production)
Growth Hormone (GH)	Gigantism (in children), Acromegaly (in adults) – excessive growth of bones and tissues	Pituitary Dwarfism – stunted growth and delayed development
Thyroid Hormones (T3, T4)	Hyperthyroidism (Graves' disease) – rapid heartbeat, Weight loss, nervousness	Hypothyroidism (Hashimoto's disease, Cretinism in infants) – fatigue, weight gain, slow metabolism
Insulin	Hypoglycemia –excessively low blood sugar, causing dizziness and confusion	Diabetes Mellitus (Type 1 and Type 2) – high blood sugar levels, excessive thirst, frequent urination
Cortisol	Cushing's Syndrome – weight gain, high blood pressure, muscle weakness	Addison's Disease – chronic fatigue, low blood pressure, weight loss
Parathyroid Hormone (PTH)	Hyper-parathyroidism weakened bones, kidney stones, excessive calcium in the blood	Hypoparathyroidism –muscle cramps, seizures, low blood calcium
Estrogen and Testosterone	Estrogen excess – risk of breast cancer, blood clots, mood swings; Testosterone excess – aggressive behavior, acne, increased muscle mass	Estrogen deficiency – osteoporosis, irregular menstrual cycles; Testosterone deficiency – reduced libido, muscle loss
Aldosterone	Conn's Syndrome – high blood pressure, excessive sodium retention	Addison's Disease – dehydration, low blood pressure, fatigue
Adrenaline (Epinephrine)	Pheochromocytoma, high blood pressure, rapid heartbeat, sweating	Addison's Disease – low blood pressure, muscle weakness

These issues will eventually cause harm to organ systems. That's why it is important to know more about endocrine issues related to health so that one can get prevented from chronic illness requiring serious medical help.

One of the most common conditions, diabetes mellitus, occurs due to insulin hypo-secretion from pancreas. The major cause for it is hormonal imbalance. Insulin is known for controlling the blood sugar level and a decrease or increase in insulin can cause changes in the blood sugar level. This will eventually lead to diabetes–either Type 1 or Type 2. The former is an autoimmune disease which involves the destruction of beta cells that are produced by insulin. The destruction is caused by the immune system itself. On the other side, Type 2 diabetes is due to the resistance of insulin or lifestyle changes seen mostly in adults. Feeling sluggish, having fatigue, and having an urge to urinate frequently are the common signs of diabetes. One should take this disease seriously because it can cause dangerous side effects such as neuropathy, renal disease, vision issues, and even cardiovascular problems. By contrast, hyperinsulinemia, characterized by the excessive secretion of insulin, causes hypoglycemia, which is a threat to an individual with low blood sugar levels. This leads to development of symptoms such as lightheadedness, disorientation, and even the loss of consciousness (Sapra & Bhandari, 2023).

Another problem, known as thyroid dysfunction, involves both hypothyroidism and hyperthyroidism. It is also considered a serious hormonal condition because of the overproduction of thyroid hormones from the thyroid gland. These hormones include the T3 and T4 hormones, which are involved in hyperthyroidism, and may even lead to autoimmune diseases such as Graves' disease. Rapid weight loss, anxiety, excess perspiration, increase in heat sensitivity and a rapid heart rate are among the side effects of this condition. In some cases, it is seen that the situation becomes serious, leading to thyroid storm. This is a type of fatal illness that causes high fever, delusion, and, in rare cases, also heart failure (Pokhrel & Bhusal, 2023). Hashimoto's thyroiditis is another autoimmune disease which attacks the thyroid gland, causing hypothyroidism. Its signs and symptoms include fatigue, weight gain, dry skin, depression, and a slower metabolism (Chaker et al., 2017).

The gland that has the most important function is the pituitary gland. The excessive secretion of growth hormone (GH) results in acromegaly in adults and gigantism in children. Acromegaly causes joint pain, the enlargement of internal organs, and excessive growth of the bones in the hands, feet, and face. In contrast, children suffering from GH deficiency exhibit pituitary dwarfism, which is identified by stunted growth and retarded development. Deficiency of growth hormone in adults causes fatigue, reduced strength, and an increase in body weight (Ramírez-Rentería and Hernández-Ramírez, 2024). Pituitary disorders also include Cushing syndrome, which is caused by the excess secretion of adrenocorticotropic hormone (ACTH). This leads to the elevated production of cortisol. Symptoms of this condition include an increase in weight, rounder face, stria, muscular atrophy, and increased blood pressure (Savas et al., 2022). Addison's disease is caused by the adrenal glands as it does not produce sufficient amounts of cortisol and aldosterone, resulting in lower blood pressure, a loss of weight, darker skin, and constant fatigue. If there is no urgent medical assistance, Addison's disease can also lead to a life-threatening condition called adrenal crisis (Charmandari et al., 2014).

Another type of health problem associated with reproductive hormonal imbalance is polycystic ovarian syndrome (PCOS). This is a condition where women suffer irregularity in their menstrual periods. People with this condition develop excessive hair growth, acne, and, in major cases, infertility. This happens all because of the hypersecretion of the hormone estrogen (Tay et al., 2023). Osteoporosis, hot flashes, and mood swings can be caused by estrogen insufficiency, which is frequently seen in postmenopausal women. All these conditions are quite common at present (Panahi et al., 2021). Similarly, low testosterone levels result in decreased libido, exhaustion, and low mood. But excessive testosterone release in males, often due to adrenal tumors or steroid abuse, can result in aggressive behavior, acne, and increased muscular mass (Corona & Maggi, 2022).

Similarly, parathyroid dysfunction, which is another endocrine condition, affects calcium metabolism. Its excessive secretion (PTH) results in hyperparathyroidism, developing hypercalcemia, the weakening of muscle and bones, kidney stones and, on occasion, lethal abnormalities. Its opposite, hypoparathyroidism, involves low levels of calcium, which can lead to tetany, muscle pain and seizures (Mannstadt et al., 2017).

Finally, fluctuations in hormone levels can also affect the functioning of the pineal gland, which is responsible for melatonin secretion and the regulation of the sleep–wake cycle. While the underproduction of melatonin is connected to insomnia, and seasonal affective disorder (SAD), hypersecretion leads to excessive sleepiness as well as mood disorders (Srinivasan et al., 2013).

1.10 CONCLUSION

Hormones, being chemical messengers produced by plants and animals, perform various vital metabolic functions. Phytohormones regulate esvery aspect of plant growth, from embryogenesis to reproductive development. In animals, hormones are produced either in ductless or endocrine glands or in any cell as local hormones. They play a pivotal role in maintaining the steady state in the body by regulating proper sugar and calcium titre in blood, combatting stress and other adverse conditions. The pituitary, as the bandmaster of the endocrine orchestra, controls various organs to synthesize and secrete hormones. Although many of the functions of theses hormones are known, their relationship with microbes (present as normal microbiota or as pathogens) are still unclear. Hence, the interplay between the microbiota and the endocrine system becomes the need of the hour.

REFERENCES

Balla, T., & Hunyady, L. (2020). Editorial: Hormone action and signal transduction in endocrine physiology and disease. *Frontiers in Endocrinology*, *11*. https://doi.org/10.3389/fendo.2020.00589

Bari, R., & Jones, J. D. G. (2008). Role of plant hormones in plant defence responses. *Plant Molecular Biology*, *69*(4), 473–488. https://doi.org/10.1007/s11103-008-9435-0

Berger, I., Werdermann, M., Bornstein, S. R., & Steenblock, C. (2019). The adrenal gland in stress – Adaptation on a cellular level. *The Journal of Steroid Biochemistry and Molecular Biology*, *190*, 198–206. https://doi.org/10.1016/j.jsbmb.2019.04.006

Bharati, A. C., Prasad, B., Mallick, S., Masram, D. S., Kumar, A., & Saxena, G. K. (2023). Animal and plant hormone. In *Elsevier eBooks* (pp. 151–175). https://doi.org/10.1016/b978-0-323-91684-4.00028-1

Cambronel, M., Tortuel, D., Biaggini, K., Maillot, O., Taupin, L., Réhel, K., Rincé, I., Muller, C., Hardouin, J., Feuilloley, M., Rodrigues, S., & Connil, N. (2019). Epinephrine affects motility, and increases adhesion, biofilm and virulence of *Pseudomonas aeruginosa* H103. *Scientific Reports*, *9*(1). https://doi.org/10.1038/s41598-019-56666-7

Chaker, L., Bianco, A. C., Jonklaas, J., & Peeters, R. P. (2017). Hypothyroidism. *The Lancet*, *390*(10101), 1550–1562. https://doi.org/10.1016/s0140-6736(17)30703-1

Charmandari, E., Nicolaides, N. C., & Chrousos, G. P. (2014). Adrenal insufficiency. *The Lancet*, *383*(9935), 2152–2167. https://doi.org/10.1016/s0140-6736(13)61684-0

Chen, R. A., & Goodman, W. G. (2004). Role of the calcium-sensing receptor in parathyroid gland physiology. *AJP Renal Physiology*, *286*(6), F1005–F1011. https://doi.org/10.1152/ajprenal.00013.2004

Corona, G., & Maggi, M. (2022). The role of testosterone in male sexual function. *Reviews in Endocrine and Metabolic Disorders*, *23*(6), 1159–1172. https://doi.org/10.1007/s11154-022-09748-3

Csaba, G. (2014). Hormones in the immune system and their possible role. A critical review. *Acta Microbiologica Et Immunologica Hungarica*, *61*(3), 241–260. https://doi.org/10.1556/amicr.61.2014.3.1

D'Afflitto, M., Upadhyaya, A., Green, A., & Peiris, M. (2022). Association between sex hormone levels and gut microbiota composition and diversity—A systematic review. *Journal of Clinical Gastroenterology*, *56*(5), 384–392. https://doi.org/10.1097/mcg.0000000000001676

Freeman, S. C., Malik, A., & Basit, H. (2022, September 26). *Physiology, exocrine gland*. StatPearls – NCBI Bookshelf. https://www.ncbi.nlm.nih.gov/books/NBK542322/

Freimer, D., Yang, T. T., Ho, T. C., Tymofiyeva, O., & Leung, C. (2022). The gut microbiota, HPA axis, and brain in adolescent-onset depression: Probiotics as a novel treatment. *Brain Behavior & Immunity –Health*, *26*, 100541. https://doi.org/10.1016/j.bbih.2022.100541

Gheban, B. A., Rosca, I. A., & Crisan, M. (2019). The morphological and functional characteristics of the pineal gland. *Medicine and Pharmacy Reports*. https://doi.org/10.15386/mpr-1235

Goldstein, D. S. (2010). Catecholamines 101. *Clinical Autonomic Research*, *20*(6), 331–352. https://doi.org/10.1007/s10286-010-0065-7

Hamid, A. A., Issa, M. B., & Nizar, N. N. A. (2018). Hormones. In *Elsevier eBooks* (pp. 253–277). https://doi.org/10.1016/b978-0-08-101892-7.00013-4

Hull, K. L., & Harvey, S. (2014). Growth hormone and reproduction: A review of endocrine and autocrine/paracrine interactions. *International Journal of Endocrinology*, *2014*, 1–24. https://doi.org/10.1155/2014/234014

Kumwenda, P., Cottier, F., Hendry, A. C., Kneafsey, D., Keevan, B., Gallagher, H., Tsai, H., & Hall, R. A. (2022). Estrogen promotes innate immune evasion of *Candida albicans* through inactivation of the alternative complement system. *Cell Reports*, *38*(1), 110183. https://doi.org/10.1016/j.celrep.2021.110183

Lakna. (2019, July 31). *What is the Difference Between Animal and Plant Hormones*. https://pediaa.com/. https://pediaa.com/what-is-the-difference-between-animal-and-plant-hormones/

Lechan, R. M., & Toni, R. (2016, November 28). *Functional anatomy of the hypothalamus and pituitary*. Endotext – NCBI Bookshelf. https://www.ncbi.nlm.nih.gov/books/NBK279126/

Lyte, M. (2014). Microbial endocrinology. *Gut Microbes*, *5*(3), 381–389. https://doi.org/10.4161/gmic.28682

Mannstadt, M., Bilezikian, J. P., Thakker, R. V., Hannan, F. M., Clarke, B. L., Rejnmark, L., Mitchell, D. M., Vokes, T. J., Winer, K. K., & Shoback, D. M. (2017). Hypoparathyroidism. *Nature Reviews Disease Primers*, *3*(1). https://doi.org/10.1038/nrdp.2017.55

Mense, M. G., & Boorman, G. A. (2017). Thyroid gland. In *Elsevier eBooks* (pp. 669–686). https://doi.org/10.1016/b978-0-12-391448-4.00034-4

Mhanna, A., Martini, N., Hmaydoosh, G., Hamwi, G., Jarjanazi, M., Zaifah, G., Kazzazo, R., Mohamad, A. H., & Alshehabi, Z. (2024). The correlation between gut microbiota and both neurotransmitters and mental disorders: A narrative review. *Medicine*, *103*(5), e37114. https://doi.org/10.1097/md.0000000000037114

Miller, J. F. A. P. (2020). The function of the thymus and its impact on modern medicine. *Science*, *369*(6503). https://doi.org/10.1126/science.aba2429

Misiak, B., Łoniewski, I., Marlicz, W., Frydecka, D., Szulc, A., Rudzki, L., & Samochowiec, J. (2020). The HPA axis dysregulation in severe mental illness: Can we shift the blame to gut microbiota? *Progress in Neuro-Psychopharmacology and Biological Psychiatry*, *102*, 109951. https://doi.org/10.1016/j.pnpbp.2020.109951

Mohammadzadeh, R., Menbari, S., Pishdadian, A., & Farsiani, H. (2023). Helicobacter pylorivirulence factors: subversion of host immune system and development of various clinical outcomes. *Expert Reviews in Molecular Medicine*, *25*. https://doi.org/10.1017/erm.2023.17

Nürnberger, T., Brunner, F., Kemmerling, B., & Piater, L. (2004). Innate immunity in plants and animals: striking similarities and obvious differences. *Immunological Reviews*, *198*(1), 249–266. https://doi.org/10.1111/j.0105-2896.2004.0119.x

O'Riordan, K. J., Collins, M. K., Moloney, G. M., Knox, E. G., Aburto, M. R., Fülling, C., Morley, S. J., Clarke, G., Schellekens, H., & Cryan, J. F. (2022). Short chain fatty acids: Microbial metabolites for gut–brain axis signalling. *Molecular and Cellular Endocrinology*, *546*, 111572. https://doi.org/10.1016/j.mce.2022.111572

Panahi, N., Ostovar, A., Fahimfar, N., Gharibzadeh, S., Shafiee, G., Heshmat, R., Raeisi, A., Nabipour, I., Larijani, B., & Ghasem-Zadeh, A. (2021). Grand multiparity associations with low bone mineral density and degraded trabecular bone pattern. *Bone Reports*, *14*, 101071. https://doi.org/10.1016/j.bonr.2021.101071

Pokhrel, B., & Bhusal, K. (2023, June 20). *Graves disease*. StatPearls – NCBI Bookshelf. https://www.ncbi.nlm.nih.gov/books/NBK448195/

Puzianowska-Kuznicka, M., Pawlik-Pachucka, E., Owczarz, M., Budzińska, M., & Polosak, J. (2013). Small-Molecule Hormones: Molecular mechanisms of action. *International Journal of Endocrinology*, *2013*, 1–21. https://doi.org/10.1155/2013/601246

Ramírez-Rentería, C., & Hernández-Ramírez, L. C. (2024). Genetic diagnosis in acromegaly and gigantism: From research to clinical practice. *Best Practice & Research Clinical Endocrinology & Metabolism*, *38*(3), 101892. https://doi.org/10.1016/j.beem.2024.101892

Rosol, T. J., Brändli-Baiocco, A., Hoenerhoff, M. J., & Vahle, J. L. (2024). Endocrine system. In *Elsevier eBooks* (pp. 517–631). https://doi.org/10.1016/b978-0-12-821046-8.00002-5

Sapra, A., & Bhandari, P. (2023, June 21). *Diabetes*. StatPearls - NCBI Bookshelf. https://www.ncbi.nlm.nih.gov/books/NBK551501/

Sarkodie, E. K., Zhou, S., Baidoo, S. A., & Chu, W. (2019). Influences of stress hormones on microbial infections. *Microbial Pathogenesis*, *131*, 270–276. https://doi.org/10.1016/j.micpath.2019.04.013

Savas, M., Mehta, S., Agrawal, N., Van Rossum, E. F. C., & Feelders, R. A. (2022). Approach to the patient: Diagnosis of cushing syndrome. *The Journal of Clinical Endocrinology & Metabolism*, *107*(11), 3162–3174. https://doi.org/10.1210/clinem/dgac492

Shahid, M. A., Ashraf, M. A., & Sharma, S. (2023), June 5. *Physiology, thyroid hormone*. StatPearls – NCBI Bookshelf. https://www.ncbi.nlm.nih.gov/books/NBK500006/

Sittipo, P., Choi, J., Lee, S., & Lee, Y. K. (2022). The function of gut microbiota in immune-related neurological disorders: A review. *Journal of Neuroinflammation*, *19*(1). https://doi.org/10.1186/s12974-022-02510-1

Srinivasan, V., Lauterbach, E. C., Shillcutt, S. D., De Berardis, D., Kuppuswamy, P. S., Zakaria, R., Othman, Z., Acuña-Castroviejo, D., & Partonen, T. (2013). Melatonin in mood disorders and agomelatine's antidepressant efficacy. In *Springer eBooks* (pp. 281–295). https://doi.org/10.1007/978-81-322-0825-9_20

Tarkowská, D. (2019). Plants are capable of synthesizing animal steroid hormones. *Molecules*, *24*(14), 2585. https://doi.org/10.3390/molecules24142585

Tay, C. T., Garrad, R., Mousa, A., Bahri, M., Joham, A., & Teede, H. (2023). Polycystic ovary syndrome (PCOS): international collaboration to translate evidence and guide future research. *Journal of Endocrinology*, *257*(3). https://doi.org/10.1530/joe-22-0232

Themes, U. (2017, February 17). *Hormone action & signal transduction*. Basicmedical Key. https://basicmedicalkey.com/hormone-action-signal-transduction/

Tilbrook, A. J., & Ralph, C. R. (2017). Hormones, stress and the welfare of animals. *Animal Production Science*, *58*(3), 408. https://doi.org/10.1071/an16808

Wuttke, W., Jarry, H., & Seidlova-Wuttke, D. (2010). Definition, classification and mechanism of action of endocrine disrupting chemicals. *Hormones*, *9*(1), 9–15. https://doi.org/10.1007/bf03401276

Xie, Y., & Dorsky, R. I. (2017). Development of the hypothalamus: conservation, modification and innovation. *Development*, *144*(9), 1588–1599. https://doi.org/10.1242/dev.139055

Zedalis, J., & Eggebrecht, J. (2018, March 8). *28.2 how hormones work – Biology for AP® Courses | OpenStax*. https://openstax.org/books/biology-ap-courses/pages/28-2-how-hormones-work

Zhang, Z., Han, H., Zhao, J., Liu, Z., Deng, L., Wu, L., Niu, J., Guo, Y., Wang, G., Gou, X., Li, C., Li, C., & Liu, C. (2025). Peptide hormones in plants. *Molecular Horticulture*, *5*(1). https://doi.org/10.1186/s43897-024-00134-y

Zheng, D., Liwinski, T., & Elinav, E. (2020). Interaction between microbiota and immunity in health and disease. *Cell Research*, *30*(6), 492–506. https://doi.org/10.1038/s41422-020-0332-7

Zhou, X., Ma, J., Shen, Y., Xie, X., & Ren, W. (2023). Effects of growth hormone on adult human gonads: Action on reproduction and sexual function. *International Journal of Endocrinology*, *2023*, 1–16. https://doi.org/10.1155/2023/7492696

2 Hormones and Microbiota Composition

Anjali Rao, Ayoushmoti Chakraborty, and Rina Rani Ray
Department of Biotechnology, Maulana Abul Kalam Azad University of Technology, Kokata, India

2.1 INTRODUCTION

The smooth balance between hormonal functioning and microbiota composition is paramount for one's health and physiological homeostasis. A human's microbiota is composed of an extensive and intricate array of microorganisms, viruses, fungi, and even archaea that reside within various parts of a person's body, mainly the intestines. These microorganism colonies perform essential functions such as food digestion, metabolism, immune response, and protection from disease. Different people possess varying proportions of the microorganisms that constitute the microbiota which are influenced by many factors, including dietary habits, genetics, personal lifestyle, usage of antibiotics, and other allergens. Gut microbiota is one of the many types of microbiotas. It has attracted the greatest amount of research, particularly with regard to its profound relationship with the endocrine system (Pires et al., 2024).

A wide range of biological activities are regulated by hormones, which are chemical messengers secreted by the endocrine glands and which also impact microbiota composition. Cortisol, insulin, sex and thyroid hormones can influence the composition, number and metabolic activity of microbes. At the same time, however, microbiota can biochemically modify the levels of hormones, through biosynthesis, metabolism and receptor binding, which results in an active form of the hormone. There is a reciprocal relation within some physiological and pathophysiological conditions such as obesity, diabetes, polycystic ovarian syndrome, stress disorders, and autoimmune disorders.

Normal flora, also known as resident microbiota, is the term given to the populations of microorganisms that maintain a stable relationship with the host. These resident microbiotas prevent the invasion of harmful pathogens and help to maintain immune balance. Nonetheless, some external factors, such as diet patterns, stress, infection, or hormonal alterations, can result in changes in the microbiota composition. This can result in dysbiosis, an imbalance of the bacterial populations which can lead to many health problems. The microbiota composition is also impacted by age, method of birth, the use of antibiotics, and even the region one lives in. Hormones interact with the microbiota in a number of different ways, with stress hormones, for example, stimulating some pathogenic bacteria and sex hormones increasing diversity to other parts of the body such as the gut and vagina. In addition, certain metabolites produced by microbiomes such as short-chain fatty acids (SCFAs), secondary bile acids, and some neurotransmitters are shown to influence hormonal, metabolic, and signaling pathways. This interplay requires further exploration to find out potential targets for interventions regarding the microbiome and hormone axis (Neuman et al., 2015).

The present chapter mostly revolves around a general description of microbiotas and their composition and moves into the various factors affecting their compositions. In addition, a detailed discussion of the different hormones such as the sex hormones and the steroid hormones has been made which covers their effects on the microbiota present in a mammalian body. The topics covered in this chapter will offer an essential perspective on hormones and its relationship with the microbiome.

 DOI: 10.1201/9781003519041-2

2.2 MICROBIOTA

Microbiota refers to the different populations of microorganisms, such as bacteria, viruses, fungi, and protozoa, which are capable of living in a wide array of environments in humans, animals, and even plants (Table 2.1). It is well known that every living organism carries microorganisms, humans included. Besides the gut microbiota, which is the best known, humans also have skin, oral, and lung microbiota. Additionally, other organisms also host certain microbiota, referred to as plant microbiota, which are found in external plant parts like stems and leaves, and in internal plant tissues such as roots and the rhizosphere and endosphere (Valdes et al., 2018).

The microbiota is essential to the preservation of health and also to any maladies that may hinder a person's wellbeing. These microorganisms aid in the digestion and nutrition of the human body while also helping to prevent disease. Also, the immune system uses microbiota to create immune cells, immunoglobulins, and to modulate responses to infections. These microorganisms also help to transform complex human carbohydrates, proteins, and fats that cannot be used by humans into SCFAs that are crucial for the colon and the general metabolism. Furthermore, microbiota perform an antimicrobial function, enhance the immune responsiveness to infection, and antagonize

TABLE 2.1
Generic Composition of Microbiota based on Origin

Origin	Dominant Microbial Groups	Common Genera	Key Functions
Animal Microbiota			
Gut Microbiota (GI Tract)	Bacteria (Firmicutes, Bacteroidetes, Actinobacteria, Proteobacteria)	*Bacteroides, Lactobacillus, Bifidobacterium, Escherichia*	Digestion, immune modulation, vitamin production
Oral Microbiota	Bacteria (Firmicutes, Proteobacteria, Bacteroidetes, Actinobacteria, Fusobacteria)	*Streptococcus, Porphyromonas, Veillonella, Fusobacterium*	Oral health, pathogen defense, nitric oxide metabolism
Skin Microbiota	Bacteria (Actinobacteria, Firmicutes, Proteobacteria)	*Staphylococcus, Corynebacterium, Cutibacterium*	Barrier function, immune interaction, odor production
Microbiota in Conjunctiva	Bacteria (Proteobacteria, Firmicutes, Actinobacteria)	*Staphylococcus, Corynebacterium, Propionibacterium*	Ocular defense, immune modulation
Microbiota of Nasopharynx	Bacteria (Firmicutes, Proteobacteria, Bacteroidetes, Actinobacteria)	*Streptococcus, Haemophilus, Moraxella*	Respiratory defense, pathogen resistance
Microbiota of Urogenital Tract	Bacteria (Firmicutes, Actinobacteria, Proteobacteria)	*Lactobacillus, Gardnerella, Atopobium*	pH balance, pathogen inhibition
Vaginal Microbiota	Bacteria (Firmicutes, Actinobacteria, Bacteroidetes)	*Lactobacillus, Gardnerella, Atopobium*	pH maintenance, pathogen inhibition, reproductive health
Ear Microbiota	Bacteria (Firmicutes, Actinobacteria, Proteobacteria)	*Staphylococcus, Corynebacterium, Pseudomonas*	Defense against infections
Plant Microbiota			
Rhizosphere	Bacteria (Proteobacteria, Actinobacteria, Firmicutes)	*Pseudomonas, Bacillus, Streptomyces*	Nitrogen fixation, plant growth promotion
Phyllosphere	Bacteria (Proteobacteria, Firmicutes, Bacteroidetes)	*Pseudomonas, Methylobacterium, Bacillus*	Photosynthesis enhancement, disease prevention

pathogenic microbes by competing for nutrients and space. Differences in parts of the host, age, sex, ethnicity, or even diet suggest diversification in macrobiotic synthesis and functionality. The host is colonized by commensal microbes soon after birth. This allows the gradual development of the host, enabling them to develop a completely different biological system at maturity.

As microorganisms cycle and solubilize nutrients such as nitrogen and phosphorus, microbiota improves a plant's capability of nutrient acquisition. Other microbes promote plant growth by either producing or stimulating plant growth regulators. In addition, they protect the plants from diseases by down-regulating aggressive microbes, producing antimicrobial agents, and activating plant defenses. Additionally, plant-associated microbes enhance the tolerance of abiotic factors such as drought, salinity, and heavy metals by refining the plant's stress response (Trivedi et al., 2020).

2.3 DIFFERENT TYPES OF MICROBIOTAS

2.3.1 Plant Microbiota

2.3.1.1 Underground Plant Microbiota

Microorganisms in the soil that surrounds the plant roots, or rhizosphere, have a symbiotic interaction with underground plant microbiota, who play a central part in meriting plant health, stress resistance, and nutrient acquisition. Plants pull in microbial communities from the soil, seeds, and other plant parts, creating a complex biosystem. Within the soil, an incredibly complex ecosystem of microorganisms is found with *Acidobacteria, Verrucomicrobia, Bacteroidetes, Proteobacteria, Planctomycetes*, and *Actinobacteria* being the prominent colonizers of the rhizospheric and the root tissues of the plants through horizontal gene transfer. Moreover, seeds act as vertical transmission vehicles as they introduce new microbes to the emerging roots (Bulgarelli et al., 2013). Certain bacteria termed as endophytes infiltrate the internal root tissues (root endosphere) by colonizing into the root through the root fractures or points where lateral roots emerge, or by some degree of active means enabling the planting of mutualistic associations assisting in the acquisition of nutrients, resistance to diseases, and tolerance to stress. Research has indicated that roots of the grapevine are inhabited mainly by species of *Proteobacteria, Acidobacteria, Actinobacteria, Bacteroidetes, Firmicutes*, whereas in the roots of rice, the predominant families of bacteria include *Rhizobiaceae*, *Comamonadaceae*, *Streptomycetaceae*, and *Bradyrhizobiaceae*. Likewise, in the roots of maize the major bacterial phyla are some *Proteobacteria*, *Firmicutes*, *Bacteroidetes*, and these are affected by the history of soil cultivation (Marasco et al., 2018).

2.3.1.2 Surface-Ground Plant Microbiota

Above-ground plant tissues, such as leaves, stems, and flowers, offer unique environments for microbial populations which include epiphytes and endophytes. Microbes, on the other hand, have differing compositions based on the environmental characteristics they come from. Many endophytes spread through xylem vessels to leaves, stems, and fruits, while some penetrate through aerial parts of the plant such as flowers and fruits. The surface of soil, seeds, and the atmosphere are primary sources of phyllosphere microbiota, but local conditions such as humidity, temperature, and ultraviolet radiation also play a major role in determining their composition (Peng et al., 2024). Microbial community composition at genera and species levels differs between the phyllosphere and the endosphere. In grapevines, for example, the *Pseudomonas, Sphingomonas, Bacillus, Acinetobacter, Enterobacter*, and *Methylobacterium* are constituents of the phyllosphere microbiota, whereas *Ralstonia, Burkholderia, Staphylococcus, Mesorhizobium*, and *Propionibacterium* are constituents of grape berry endophytes (Campisano et al., 2014). In the same manner, seed microbiota are an intermediate source between the soil, flower, and fruit microbiota of a plant, creating continuity between different plant generations. Above-ground microbiota, which have important functions in plant health, growth, and defense against diseases and other stressors, are affected by farm management practices, plant species, and environmental factors.

2.3.2 Animal Microbiota

2.3.2.1 Skin Microbiota

The skin microbiota is composed of a variety of microorganisms living in different parts of the skin. The bacterial composition differs based on microenvironments; sebaceous areas are rich in lipophilic species of *Propionibacterium*, while moist areas are populated by *Staphylococcus* and *Corynebacterium*. In contrast, the fungal element is fairly consistent across locations, with *Malassezia* at the central body and arm locations and a more diverse fungus population at the foot locations. There are more bacteria in such systems, even though there is a lower amount of fungi. The relationship of acne vulgaris and *P. acnes*, the skin bacterium, is clear to all sufferers from the condition; however, there are also individuals who harbor this bacterium and do not develop acne. Genetics, immune response, and sebaceous gland activity are among the factors that influence its development. It is clear that the *P. acnes* found in follicles, along with biofilm formation, contributes to the pathology of acne. This type of *P. acnes* is associated with this condition type 1A1 (Byrd et al., 2018).

2.3.2.2 Microbiota in Conjunctiva

These microorganisms are part of the normal flora, meaning they do not cause disease under normal conditions and help maintain ocular health. Abundant microbes found in the conjunctiva are *Staphylococcus epidermidis* and *Propionibacterium acnes*. Other species, such as *Staphylococcus aureus, Streptococcus spp., Actinobacteria, Lactobacillus, Clostridium, Helicobacter, Sphingomonas, Haemophilus spp.*, and *Neisseria spp.*, may also be present, but in smaller amounts. The glands found underneath the eyes are known as lacrimal glands. These secrete tears, which are formed and flow over the eye surface. Because tears contain lysozyme and other antimicrobial agents, they serve to maintain a homeostatic microbial environment.

Thus, tears serve the dual purpose of washing away potentially harmful particulates, as well as inhibiting bacterial growth, thereby preventing dangerous infections (Ren et al., 2022). Various eyelid functions that aid in facilitating moisture and cleanliness for the eye are termed mechanical eyelid functions. These protect the eye surface to some degree. The conjunctival microbiota is a generally beneficial addition to the human body ecosystem. However, like anything in the world, there can always be a problem in the case of dysregulation which can contribute to inflammation or infection in the eye.

2.3.2.3 Microbiota of Nasopharynx

The nasopharynx is a sterile region at birth, but it acquires microbial flora within days. It acts as a natural reservoir for benign and potentially harmful bacteria that play a significant role in respiratory diseases. The upper respiratory tract (nose and throat) contains a mixture of normal flora and opportunistic pathogens, such as *Streptococcus pneumoniae, Haemophilus influenzae*, and *Moraxella spp.* which can result in infection in immunocompromised individuals. In contrast, the lower respiratory tract (lungs and bronchi) is mostly sterile because of ciliated activity clearing microbes and mucus membranes. The nasopharyngeal microbiota is an important component of immune defense, but it can also modify the likelihood of a host developing pneumonia or sinusitis. Viral infections, immunosuppression, and antibiotic use are examples of external factors that disturb this balance and increase the likelihood of infection. To secure respiratory health and disease prevention, a stable nasopharyngeal microbiota must be maintained (Henares et al., 2021).

2.3.2.4 Microbiota of the Oral Cavity

Aerobic and anaerobic bacteria, fungi, viruses, protozoa, and other microorganisms exist within the oral cavity, which hosts one of the most diverse microbial communities. For example, *Streptococcus*, which is the inhabitant of normal flora, forms a part of this community. These microorganisms, in balance, are essential not only for oral health, but also for general well-being. Unhygienic dental practices permit the proliferation of bacteria, resulting in the development of dental caries (tooth decay) and gingivitis (gum disease). The presence of oral infections in more serious cases is thought to worsen periodontitis, a condition that is known to be associated with other diseases such as cardiovascular diseases and diabetes. Following dental work or surgery, there is a possibility of bacteria being released into the circulation, which could, for example, infect the heart, resulting in endocarditis. The saliva plays an important role in the oral microbiome as it contains antimicrobial proteins, and helps to neutralize acids as well as wash away food residues (Santacroce et al., 2023). Factors such as diet, smoking, severe medical conditions, and antibiotic treatment can all contribute to this balance disruption, thereby leading to infection.

2.3.2.5 Microbiota of the Gastrointestinal Tract

Microbiota residing within the gastrointestinal (GI) tract play an important role in digestion, metabolism, immune response and protection from disease. Its composition is changed geographically because of the variation in pH, oxygen, and nutrients. The stomach contains the lowest number of microbes because the acidic environment restricts microbial presence. However, some acidophilic species, such as *Lactobacillus* and *Helicobacter pylori*, do manage to survive. The small intestine is inhabited by moderate numbers of microbes which include *Lactobacillus, Enterococcus, E. coli*, and *Klebsiella*, and these tend to be more abundant in the lower parts of the intestine. The highest number of microbes and the greatest diversity is found in the colon, which contains 99% of anaerobic bacteria consisting mainly of *Bacteroides spp.* and some facultative anaerobes like *E. coli, Klebsiella* and *Proteus* (Hasan & Yang, 2019). Combinations of these microbes perform the functions of digesting fiber, synthesizing vitamins such as K, and B12, immune system regulation, and the maintenance of gut health. The balance of gut microbiota is altered by diet, antibiotics, stress, age, infections, and overall health, and can cause an imbalance (dysbiosis) associated with inflammatory bowel disease (IBD), inflammatory bowel disease (IBS), obesity, and lowered levels of immunity (Iliev et al., 2025).

2.3.2.6 Microbiota of Urogenital Tract

The urogenital microbiota is comprised of the microorganisms that reside in the vagina and the urine bladder, which are fundamental to the health of people and tor the promotion of infection prevention. The urinary microbiota was previously considered sterile, but some researchers have found *Lactobacillus, Staphylococcus, Corynebacterium*, and *Streptococcus*. In adult females, *Lactobacillus spp.* dominates the vaginal microbiota due to its production of lactic acid, which maintains a low pH of 3.5 to 4.5, preventing pathogen proliferation. The susceptibility to vaginal infection such as Candida vaginitis infection caused by *Candida albicans* increases with a decline in *Lactobacilli* levels and elevated pH during the prepubertal and postmenopausal periods of low estrogen levels. Hormonal shifts, antibiotic intake, sexual behavior, personal cleanliness, and food are among the underlying reasons that impact urogenital microbiota (Čeprnja et al., 2023). Certain shifts can cause bacterial dysbiosis and urinary tract infections and, therefore, proper urogenital hygiene, nutrition behavior, and probiotics supplements all remain crucial for urogenital wellbeing.

2.4 FACTORS AFFECTING MICROBIOTA COMPOSITION

Specific and nonspecific factors are used by the hosts to cultivate their personal microbiota. The host promotes the survival of certain microbes while excreting other ones. The composition of microbiota is shaped at a young age; however, it may be further modified by numerous factors that influence its development and variety.

2.4.1 Age and Delivery Pattern

The process of microbial colonization of the intestine starts with the microbiota from the amniotic fluid and placenta during gestation. The infant microbiota is relatively simple at birth, but becomes more complex as the infant interacts both with the mother and the environment. The beneficial bacteria *Bifidobacterium* and *Lactobacillus*, which are critical for the maintenance of gut health, are present in large numbers in breastfed infants (Saturio et al., 2021). With increasing age among children and diversification in their diet, the composition of microbes becomes more stable and a balanced microbiome develops in adults, predominantly composed of *Firmicutes* and *Bacteroidetes*. In advanced age, however, diminished dietary diversity, lowered immunity, and high levels of consumption of medication leads to significant decreases in microbiota diversity, which comes with an enhanced probability of infection, inflammation, and metabolic issues (Ma et al., 2020).

The delivery of a baby also significantly channels the immediate microbiome of an infant (Figure 2.1). For instance, infants who are born through the vaginal route will have preliminary gut microbiota dominated (*Lactobacillus* and *Prevotella*) by those in the mother's vagina, while infants born via Cesarean section derive their initial gut microbiota from the skin, resulting in the dominance of *Streptococcus, Corynebacterium* and *Propionibacterium*, a factor which is often associated with lower levels of beneficial microbes (Zhang et al., 2021).

2.4.2 Effect of Diet

The diet during infancy is important for establishing the gut microbiota and breastfeeding gives infants a *Lactobacillus*- and *Bifidobacterium*-dominated microbiota because of the oligosaccharides in the breast milk. These are beneficial bacteria which, along with the mother's immune system, helps to further develop the infant's immune system. On the other hand, formula-fed infants have a more diverse and less stable microbiota comprised of *Enterococcus, Enterobacteria, Bacteroides, Clostridia*, and *Streptococcus*, and hence tend to respond in an immunologically different manner.

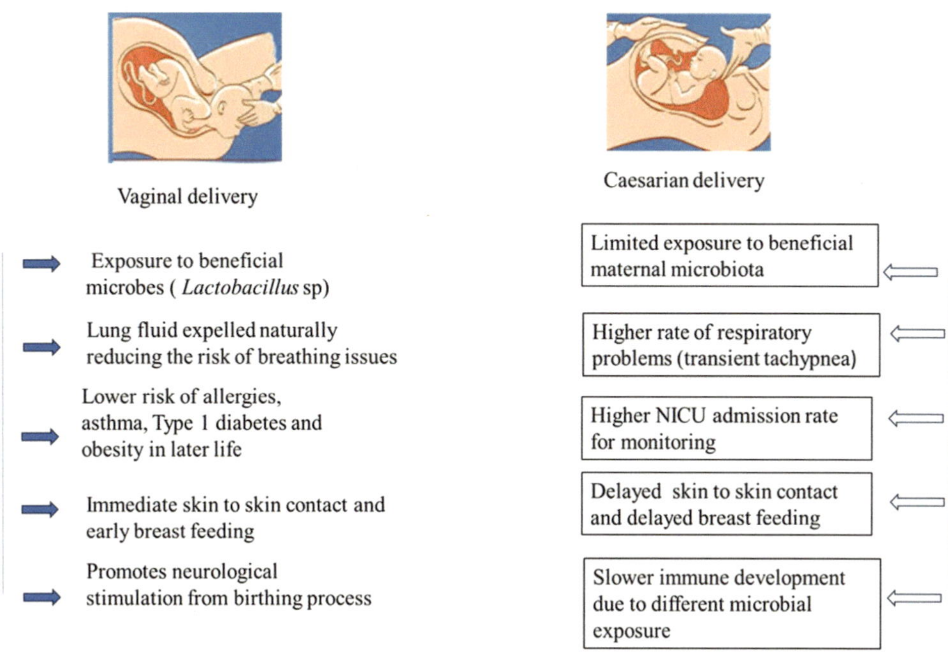

FIGURE 2.1 Impact of normal delivery and Cesarean delivery on a newborn.

The infant who benefits from breastfeeding tends to have a stable gut microbiota and greater mucosal immune response. The diet that we follow over infancy also has an impact on the gut microbiota. Supporting a vegetarian diet is correlated with a healthy diverse microbiota in bacteria such as *Ruminococcus*, *Roseburia*, and *Eubacterium*, because they help metabolize insoluble carbohydrates. In contrast, a Western diet laden with animal proteins and fats reduces *Firmicutes* and raises *Bacteroides*, which can prove harmful by causing super-metabolic products to be formed (Tomova et al., 2019).

2.4.3 Use of Antibiotics

Antibiotics have a profound impact on gut microbiota, as they eliminate both harmful and beneficial microbes, leading to dysbiosis and the overgrowth of unwanted pathogens such as *Clostridium difficile*. Studies have shown that antibiotic use disrupts essential microbial functions, including bile acid and serotonin metabolism, which can delay intestinal motility. Additionally, antibiotics weaken the microbiota's ability to exclude pathogens, increasing the risk of infections (Buffie et al., 2011).

Certain antibiotics have long-lasting effects on microbiota composition. Clindamycin alters gut microbiota for up to two years, reducing *Bacteroides* diversity. Clarithromycin, used for *Helicobacter pylori* treatment, decreases Actinobacteria, while Ciprofloxacin lowers *Ruminococcus* levels. Vancomycin, although effective for *C. difficile* infections, depletes beneficial microbes such as *Bacteroidetes* and *Faecalibacterium*, while increasing *Proteobacteria*, contributing to recurrent infections. The extent of microbiota disruption and recovery time varies among individuals (Haak et al., 2018).

In addition to treating sickness, antibiotics are commonly used in farming, especially in raising chickens and cattle, in order to increase their weight and size. Research suggests that even small amounts of antibiotics fed to animals for meat may cause obesity in people. Even though herbicides and other substances are standard in agricultural production, their effect on the gut microbiome is still uncertain owing to insufficient data (Manyi-Loh et al., 2018).

2.4.4 Lifestyle

Individual lifestyle choices have a substantial impact on diet, health and fitness levels, all of which are interlinked with gut microbiome composition. The absence of physical activity contributed to reduced bacterial diversity and the abundance of important microbes such as *Akkermansia muciniphila*, which plays an important role in metabolic and anti-inflammatory activity. Premature sedentarism, on the other hand, can lead to gut dysbiosis along with associated metabolic disorders (Marlicz & Loniewski, 2014). Microbiota differ according to hygiene practices. Cleanliness and hygiene limit infection and disease, but an overly clean environment during the formative years may limit microbial contact and, in combination with poor gut microbiome diversity, increases the rate of immune maladies such as allergies. Mental factors and stress have been shown to influence gut microbiome composition via the gut–brain axis. Long-term stress can shift a microbiota's balance towards higher levels of detrimental bacteria such as *Escherichia coli* and lower levels of helpful bacteria like *Lactobacillus* and *Bifidobacterium*, causing potential harm to the digestive and immune systems (Madison & Kiecolt-Glaser, 2019). Sleep is just as important. The relationship between sleep and the microbiome is a complex one; insufficient sleep results in the disruption of the circadian rhythm, which can lead to an imbalance of gut microbiota and other metabolic diseases. However, a consistent sleep schedule can nurture a healthy microbiota. Finally, substance abuse such as smoking and excessive alcohol intake, as well as drug use, has an adverse effect on gut health. Alcohol consumption, together with cigarette smoking, encourages the expansion of harmful microflora like *Proteobacteria* and suppresses useful bacteria such as *Firmicutes*, resulting in increased gut inflammation and permeability (Cicchinelli et al., 2023).

2.4.5 Disease

Microbiota composition may be changed through several diseases, leading to microbial imbalances and complications in health. Gastrointestinal diseases such as IBS, IBD, and Crohn's disease have been linked with diminished microbial diversity and impaired gut homeostasis. Dysbiotic intestines are associated with obesity, diabetes, cardiovascular diseases, and other metabolic disorders, as these conditions affect the gut, which, in turn, affects their metabolism, inflammation, and insulin resistance (Hou et al., 2022). Other examples of autoimmune disorders, like rheumatoid arthritis, multiple sclerosis, and lupus, develop because of moderating microbiota dysbiosis and immune perturbation (De Luca & Shoenfeld, 2018). Likewise, the link between the gut and the brain emphasizes the growing tendency of gut involvement in neurological disorders, such as Parkinson's disease, Alzheimer's disease, and autism spectrum disorders (Ullah et al., 2023).

Hence, all these aforementioned factors have a contribution in the development of diseases due to dysbiosis of microbiota, which may be exemplified by gut microbiota dysbiosis (Figure 2.2).

2.5 EFFECTS OF HORMONES ON THE MICROBIOTA

Different hormones affect the microbiota in organs such as the gut, skin, mouth, and even the reproductive organs through immune system responses, metabolism, and nutrient supply. For instance, sex hormones, particularly estrogen and testosterone, affect the makeup of gut and skin microbiota, where estrogen enhances microbial diversity and mitigates inflammation, while androgens stimulate sebum production, encouraging the proliferation of acne-associated bacteria. Cortisol, a stress hormone, can compromise microbial balance by suppressing immunity, resulting in the predominance of pathogenic microbes, whereas insulin changes influence skin and gut microbiota, resulting in acne and other metabolic disorders. The changes cause changes in the microbiota composition, which alters the health risk for disease, and immunity. In this section we will consider the various effects of hormones on different microbiota in detail.

2.5.1 Effect of Hormones on Oral Microbiota

It is primarily sex hormones, such as estrogen and progesterone, that have a definitive impact on the oral microbiome, which also affects oral hygiene. Shifts in the composition of the oral environment that favor specific bacteria occur in line with hormonal changes during the menstrual cycle, pregnancy, and menopause.

2.5.1.1 Role of Major Hormones

Changes in estrogen levels have a noteworthy effect on the fluctuations of the oral microbiome. Changes in estrogen, for instance, around puberty, pregnancy, or menopause, often results in a decrease in microbial diversity. Increased levels of estrogen tend to correlate with higher levels of certain bacterial species responsible for gingivitis and periodontitis. This is, however, the opposite of what happens during menopause where there is a drop in estrogen levels. During this stage, the levels of helpful bacteria also decline, which can worsen the oral environment, resulting in either inflammation or infection. Another of the four major sex hormones, progesterone, also impacts the oral microbiota. When pregnant, elevated levels of this hormone can cause inflammation in the gums and pregnancy is associated with gingivitis as a result of increased bacterial growth. In addition, this hormone has also been shown to increase blood flow to the gums, making them sensitive and prone to swelling, thereby, in turn, shifting the balance of microbes even further (Tramice et al., 2022).

2.5.1.2 Role of Other Hormones

Stress causes the production of cortisol, which is known to suppress the immune response. In this case, the body is less able to manage bacteria in the mouth. In addition, stress can dry the mouth,

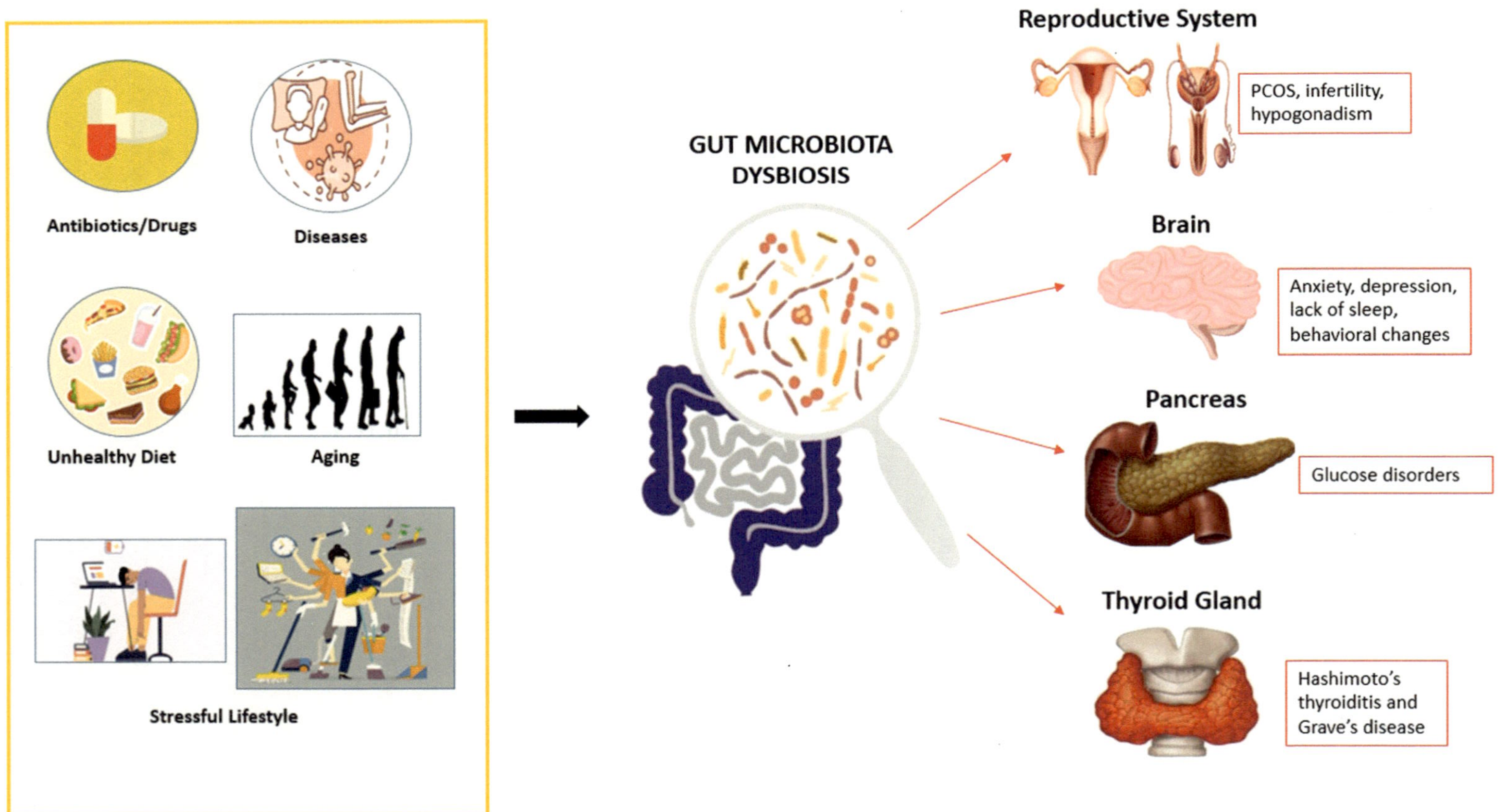

FIGURE 2.2 Gut microbiota dysbiosis due to different factors affecting microbiota composition.

enabling the growth of pathogenic bacteria, which increases the chances of cavities, bad breath, and gingivitis. Thyroid glands, like other endocrine glands, have a direct effect on the microbial population in the mouth. Both hypothyroidism and hyperthyroidism change the quantity and acidity of saliva along with the bacterial composition of the mouth, leading to an increased incidence of dry mouth, unpleasant smell from the mouth, and increased chances of falling ill. Similarly, people who suffer from diabetes usually have some degree of insulin resistance, which tends to be associated with increased oral bacterial changes that are linked to gum infection and impaired healing of the tissues.

2.5.2 Effect of Hormones on Gut Microbiota

Certain hypotheses suggest that the gut microbiota can affect the levels of sex hormones due to the involvement of microbial metabolites, the immune system, chronic inflammation, and even neuroendocrine systems such as the gut–brain axis. Furthermore, there is evidence of two-way communication between the microbiome and the endocrine system, and research is underway to understand the details of these interactions.

2.5.2.1 Role of Major Hormones

Estrogen facilitates the development of helpful bacteria such as *Lactobacillus* and *Bifidobacterium*, which are useful in preserving gut balance and minimizing inflammation. Diminishing estrogen, as seen during menopause, correlates with lower microbial diversity and higher gut leakage, which may lead to metabolic as well as inflammatory diseases. During pregnancy, progesterone impacts microbes in order to protect the mother and fetus. Gut microbiota composition is subject to change by testosterone, and thus contributes to differences in gut bacterial composition among the sexes. An abundance of *Bacteroides* is associated with high levels of testosterone, while low levels of testosterone may induce more chances of metabolic disorders (Santos-Marcos et al., 2023).

2.5.2.2 Role of Other Hormones

Elevated cortisol levels are associated with a negative shift in microbes such as *Lactobacillus* and *Bifidobacterium* while fostering the growth of detrimental pathogens. This undermines the gut barrier and triggers inflammation alongside illnesses such as IBS and leaky gut syndrome. Rises and falls in adrenaline further gut motility and microbial growth, which is usually associated with stress-related digestive problems. In addition, the bacteria that reside in the gut act in a self-governing manner and coordinate directly with insulin sensitivity and glucose metabolism, and shifts away from the balance relative to the insulin, obesity, and type 2diabetes. Good bacteria are of great importance in blood sugar management as they feel dietary fibers to SCFAs, consequently increasing insulin activities. Thyroid hormones (T3 and T4) change the gut motility along with its microflora. Many cases of hypothyroidism showed slow digestion which causes an imbalance of microbes and inflammation to the gut while hyperthyroidism accelerated digestion to disrupt the absorption of nutrients and the stability of microbes (Marwaha et al., 2025).

2.5.3 Effect of Hormones on Skin Microbiota

By controlling sebaceous and sweat gland activity, hormones greatly influence the skin microbiota through regulating both nutrient availability and growth conditions for microbes. Across the lifespan, changes in hormones affect the physiology of the skin, the maturation of the immune system, and the skeletonization of the body's microbial community.

2.5.3.1 Role of Major Hormones

Androgens, such as testosterone and dihydrotestosterone (DHT), enhance the functionality of the sebaceous gland, resulting in the overproduction of sebum and the proliferation of lipophilic bacteria such as *Corynebacterium*, *Cutibacterium*, and *Malassezia*. During puberty, higher levels of androgens, especially testosterone, are produced, which creates suitable conditions for the *C. acnes* acne bacteria to multiply, resulting in acne and inflammation. Skin microbiome disruption as a result of hormonal imbalance, as in polycystic ovary syndrome, leads to greater risk of acne and seborrheic dermatitis (Park et al., 2021).

Estrogens protect by controlling hydration, modulating immunity, and overseeing skin barrier functions. As in the case of pregnancy, higher levels of estrogen support a normal microbiome and less inflammation, preventing bacterial overgrowth. On the other hand, during the menopause phase, lower skin hydration and elasticity from reduced estrogen cause microbial composition changes that increase the chances of skin infections and dryness in combination with other skin issues (Pagac et al., 2024).

2.5.3.2 Role of Other Hormones

The skin also interfaces with the neuroendocrine system as an organ capable of expressing major components of the cytokine-signaling pathway, the hypothalamic–pituitary–adrenal axis (HPA). While psychological stress aids in releasing cortisol and catecholamines, it can trigger disruption in the immune function, resulting in increased inflammation and, consequently, impaired wound healing (Holmes et al., 2014). This imbalance caused by stress reduces the ability of the skin to repair itself and maintain microbial balance. This also aids in the overpopulation of pathogenic bacteria such as *Staphylococcus aureus*, which, in turn, aggravate eczema and psoriasis (Duarte et al., 2024).

In addition to controlling blood sugar levels, insulin affects the skin microbiome. Insulin resistance coupled with high levels of insulin stimulates sebaceous glands, thereby increasing the risk of acne. Moreover, insulin dysregulation also contributes to excessive fungal growth, which increases the risk of developing dandruff and seborrheic dermatitis (Niedźwiedzka et al., 2024).

2.5.4 Effect of Hormones on Vaginal Microbiota

Alongside stressful scenarios such as lifestyle acts and hormones, the vaginal microbiome can be altered to a worrying extent. This results in a dip in *Lactobacillus* (LAB) counts, a lower incidence of microbes, and a heightened likelihood of dysbiosis. Generally, this imbalance is linked with bacterial vaginosis, yeast infection, sexually transmitted disease, urinary tract infection, or even infertility. The periods that are defined by sharp increases in hormones like puberty, pregnancy, and menopause are linked with a greater degree of diversity of microbes within the vaginal ecosystem. Additional studies indicate that the vaginal microbiota shifts from being dominated by anaerobic microbes to one softer with *Lactobacillus* species during these tumultuous hormonal transitions. The excessive colonization of *Lactobacillus* species for microbial dominance is encouraged by estrogen replacement therapy among post-menopausal women. In these women, *Lactobacillus* species take over and restore balance, which is pro-dysbiotic. Such observations endorse the idea of a 'hormone-'driven' model of the vaginal flora diversity wherein the levels of the gonadal hormones determine the type of the vaginal microbiota during various periods of the woman's lifespan and even during the post-reproductive years (Park et al., 2023).

2.5.4.1 Role of Major Hormones

Both estrogen and progesterone are critical to vaginal microbiota maintenance and reproduction. Estrogen facilitates the expansion of *Lactobacillus* species capable of lactic acid fermentation and acidogenesis, thereby preserving an acidic pH within the vagina (~3.5–4.5), which is hostile to disease-causing bacteria and fungal infections. The lowering of estrogen during menopause leads to a reduction in dominance of *Lactobacillus*, which increases the chances of bacterial vaginosis, vaginal atrophy, and infections. Moreover, progesterone in its various forms, especially in the case of pregnancy and during menstruation, affects microbial content. Increased levels of estrogen during pregnancy stimulates the multiplication of *Lactobacillus* and augments the immune response in the vagina. However, during progesterone-dominant phases, such as the luteal phase, the overgrowth of *Candida* can be stimulated, thereby increasing the risk of vaginal candidiasis (Kaur et al., 2020).

2.5.4.2 Role of Other Hormones

Constant stress combined with high cortisol hormones may disturb the balance of microbes, thereby increasing the chances of infections. Fungal infections and bacterial imbalances become more likely due to metabolic disorders, including diabetes and thyroid problems that alter the composition of vaginal microbes (Amabebe & Anumba, 2018).

2.5.4.3 Role of Hormonal Contraceptives

The use of hormonal contraceptives, such as oral pills, injections, and IUDs that contain estrogen and/or progesterone, can change the composition of the vaginal microbiota. Some researchers think these contraceptives may disrupt the balance of *Lactobacillus* species, which are essential for a healthy vagina. While some contraceptive methods, for instance, combination oral contraceptives, may help sustain *Lactobacillus* dominance and lower bacterial vaginosis (BV) risk, others, particularly progestin-only contraceptives, have been linked to a reduction in *Lactobacillus* and an increase in anaerobic bacteria associated with BV and dysbiosis. In addition, hormonal contraception has shown to change vaginal pH and the immune response and may therefore alter infection risk (Song et al., 2020).

2.5.5 Effect of Hormones on Ocular Microbiota

Androgens and estrogens, as well as other hormones, have a profound impact on the eye's microbiome as well as the overall health of the eye's surface, which affects the quality of tears and meibomian gland activity, and may lead to inflammation as well as various eye disorders such as dry eye syndrome.

2.5.5.1 Role of Major Hormones

Research indicates that estrogen contributes to immune responses in the eye by modulating tear composition as well as the growth of both beneficial and pathogenic bacteria. After the woman's gradual transition into post-menopause, which is often accompanied by a decline in estrogen hormones, they seem to shift their ocular microbiome, which furthers her chances of developing dry eyes and increasing the risk of infections. Also known for its immune system modulation effects, progesterone can inhibit inflammation and modulate bacterial populations within the conjunctiva. The hormonal changes that a woman undergoes during her pregnancy due to amplified levels of progesterone can shift the ocular microbiome, thereby predisposing certain individuals to infections. Moreover, testosterone, which is responsible for controlling the action of the sebaceous glands, has an effect on the lipid composition of the secretion from the meibomian glands. This may alter the bacterial population of the glands and could lead to meibomian gland dysfunction combined with blepharitis (McKay et al., 2022).

2.5.5.2 Role of Other Hormones

Cortisol has been shown to increase permissive opportunistic pathogen colonization, leading to diseases such as conjunctivitis and keratitis. In the same way, fluctuations in adrenaline can interfere with the integrity of the ocular surface and tear production, which may change normal bacterial flora and make some individuals more susceptible to infections. Metabolic diseases such as diabetes and thyroid disease are found to correlate closely with changes in the ocular microbiome. Insulin resistance among diagnosed diabetes patients causes elevated sugar levels in the tear fluid which favors the growth of microbes. This increases the likelihood of ocular infections, especially bacterial keratitis and fungal infections (Zysset-Burri et al., 2022). T3 and T4 thyroid hormones can affect the immune system and mucocutaneous integrity, which has an effect on the ocular surface microbiome. Dysfunctions in the thyroid, including underactive and overactive thyroids, have been associated with changes in microbial diversity, which may have consequences for the eyes. There is also an impact from melatonin, which is produced and secreted from the pineal gland and aids in the function of retina, skin, and other underlying organs (Nuzzi & Caselgrandi, 2022).

2.5.6 Effect of Hormones on Ear Microbiota

Bacteria, fungi, and viruses, renowned as ear microbiota, populate the human ear, especially the external auditory canal region. Administration of antibiotics may disrupt this microbial ecosystem. It's notable that an external or an internal set of factors could impact the composition of the ear microbiota, with hormone regulation being attributed to a more internal factor.

2.5.6.1 Role of the Major Hormones

Estrogen has an impact on the composition and function of the ear microbiota. One of the life events that is associated with a drop in estrogen level is menopause, which has been linked to increased hearing loss among women. In addition, estrogen affects the inflammatory response in the middle ear, setting the stage for more severe and/or frequent infections. There is evidence suggesting that the potent form of estrogen, 17β-estradiol, decreases inflammation in response to infections from nontypeable *Haemophilus influenzae*, a common infective organism in middle ear infections. In addition, estrogen also plays a role in the bone metabolism within the otic capsule needed for hearing and balance. Disturbances in this process may cause hearing loss and vestibular disorders. This shows how estrogen influences the auditory system (Hirano et al., 2023).

Progesterone is another crucial sex hormone which assists the auditory system, especially when utilized alongside estrogen in Hormone Replacement Therapy (HRT). Although HRT is principally employed to alleviate menopausal symptoms, there are studies that show mixed estrogen and progesterone therapy could be detrimental to hearing health. There are receptor sites for progesterone in the cochlea, which means that changes in hormone content could lead to a lessening of blood flow to the cochlea or a potassium imbalance, both of which may prove detrimental to hearing. Additionally, progesterone is known to increase the chances of otosclerosis and some other disorders associated with the inner ear which affect the sense of balance and orientation in space.

The hormone testosterone, normally secreted by males, also impacts the microflora of the external ear. Increased testosterone levels are correlated with heightened occurrence of some skin bacteria, for example, *Cutibacterium* and *Staphylococcus*, which might disrupt the macrobiotic equilibrium. Moreover, the associated increase in sebum caused by testosterone might lead to microbial settlement, which could alter the ear microbiota. The combination of all these hormonal factors are responsible for the differences in the health of ears and the composition of microbiota, reflecting the intricate connection between the endocrine system and the functions of ears (D'Afflitto et al., 2022).

2.5.6.2 Role of Other Hormones

Like cortisol and adrenaline, stress hormones significantly influence the ear microbiome by changing immune and microbial responses. Long-term stress elicits inflammation through elevated cortisol levels, which, in turn, disrupts microbial homeostasis. This imbalance is associated with a higher susceptibility to infections such as otitis externa, and fungal overgrowth. Shifts in adrenaline also influence the permeability of ear tissues, enabling opportunistic pathogens to infect easily. In addition to the above, ear microbiota is also affected by metabolic hormones such as insulin and thyroid hormones. In diabetics, insulin resistance leads to hyperglycemia, which supports the growth of pathogenic bacteria and fungi and thereby increasing the chances of fungal otitis externa. An imbalance in thyroid hormones, whether hypothyroid or hyperthyroid, changes immune response and microbial profile to disturb homeostasis, which leads to a greater infection rate of *Staphylococcus aureus* and *Candida* (Beurel, 2024).

2.5.6.3 Microbiome, Hormones and Specific Microbiota Changes

Microbiome interacts with hormones, which is important in the context of the ear. Some gut bacteria have the potential to metabolize systemic hormones, thereby affecting their overall concentration within the body. This also means that probiotics, which alter the composition of gut microbiota, might also change the hormonal equilibrium, which, in turn, affects the health of the ear. Some bacterial genera, such as *Ruminococcus*, *Bacteroides* and *Prevotella*, have been associated with the level of estrogen, which shows the relation of microbiota and hormones. Also, it has been established that ear microbiota differs between males and females, probably because of the action of hormones. Beneficial ear bacteria, such as *Staphylococcus capitis* and *S. capitis/caprae*, are typically found in healthy ears, whereas diseased and unbalanced ear microbiomes are populated with pathogenic bacterial and fungal species, thereby stressing the significance of microbiota in ear health (Sjövall et al., 2020).

2.6 EFFECT OF PLANT HORMONES ON ROOT MICROBIOTA

The shape and behavior of plant roots, the secretion of nutrients, and immune activities are determined by plant hormones, which together decide the microbes' selection and settlement. These processes are guided by plant hormones that are first-level signals. Among them, rhizobacteria like indole-3-acetic acid (IAA) are the strongest exudates known to increase root elongation and lateral root formation through auxin. In all these processes, exogenous auxins produced by many rhizobacteria, like *Rhizobium* and *Pseudomonas*, increase the root microbe interactions. In addition, root exudation augmented by auxins influences the abundance of certain microbes through carbon oxide check modification. Auxiliary exudation inhibition in legumes modifies the supplanting microbial compost rhythms. These cytokinins that influence cell division and differentiation also modify the legume's symbiont interactions with the nitrogen-fixing bacteria which has two sides: favourable and destructive. Growth-promoting gibberellins (GAs) have a key role in assortment of root exudates such as organic acids and sugars which increase the levels of recruiting microbes. Certain conditions suppress arbuscular mycorrhizal (AM) fungal colonization by changing symbiotic associations (Gonin et al., 2023).

As a significant stress hormone, abscisic acid (ABA) alters root exudates during periods of water shortage, which affects microbiota composition and promotes the recruitment of drought-tolerant microbes such as *Bacillus* and *Pseudomonas*. ABA also makes some contribution toward systemic resistance through the suppression of soilborne pathogens in a more indirect manner. Ethylene, which controls the formation of root hairs and responses to stress, has mixed effects on microbes; it assists with fungal benefactors but inhibits rhizobial nodule formation. Furthermore, plant-growth-promoting rhizobacteria (PGPR) that are attracted by ethylene-modulated exudates help the plant to defend itself against root pathogens. Key defense hormones, jasmonic acid (JA) and salicylic acid (SA), control the immune responses which in turn define the root microbiota composition. SA relieves hoof pathogenic microbes and allows colonization by beneficial ones, while JA does the opposite; it promotes mycorrhizal interactions and PGPR recruitment while inhibiting pathogenic fungi. Microbial diversity, in turn, is governed by the plant's trade-off between defense and growth as the balance is struck by signaling between SA and JA. Together, these hormones shape the recruited microbiota in root-associated regions which, in turn, affects the health, stress tolerance, and productivity of the plant, guiding toward better outcomes (Salas-González et al., 2020).

2.7 CONCLUSION

Microbiota, being a regulator of physical and mental homoeostasis, maintains an intimate relationship with hormones. Gut microbiota is often considered to be an endocrine organ, as they surpass the biochemical complexity of traditional endocrine organs, generating neurotransmitters such as GABA, dopamine, and serotonin. Despite the challenges in culturing gut bacteria, advances in research methodologies have elucidated their role in behaviour, metabolism, appetite, and insulin resistance, which reveal their hormone-like activities. Interestingly, plant hormones also influence human health and even modulate animal–microbiota relationships. As microbial endocrinology continues to evolve, further exploration of the intricate connections between hormones and the microbiome are anticipated, highlighting hormones' pivotal role in the dynamic host–microbiota relationship.

REFERENCES

Amabebe, E., & Anumba, D. O. C. (2018). Psychosocial stress, cortisol levels, and maintenance of vaginal health. *Frontiers in Endocrinology*, *9*. https://doi.org/10.3389/fendo.2018.00568

Beurel, E. (2024). Stress in the microbiome-immune crosstalk. *Gut Microbes*, *16*(1). https://doi.org/10.1080/19490976.2024.2327409

Buffie, C. G., Jarchum, I., Equinda, M., Lipuma, L., Gobourne, A., Viale, A., Ubeda, C., Xavier, J., & Pamer, E. G. (2011). Profound alterations of intestinal microbiota following a single dose of clindamycin results in sustained susceptibility to clostridium difficile-induced colitis. *Infection and Immunity*, *80*(1), 62–73. https://doi.org/10.1128/iai.05496-11

Bulgarelli, D., Schlaeppi, K., Spaepen, S., Van Themaat, E. V. L., & Schulze-Lefert, P. (2013). Structure and functions of the bacterial microbiota of plants. *Annual Review of Plant Biology*, *64*(1), 807–838. https://doi.org/10.1146/annurev-arplant-050312-120106

Byrd, A. L., Belkaid, Y., & Segre, J. A. (2018). The human skin microbiome. *Nature Reviews Microbiology*, *16*(3), 143–155. https://doi.org/10.1038/nrmicro.2017.157

Campisano, A., Antonielli, L., Pancher, M., Yousaf, S., Pindo, M., & Pertot, I. (2014). Bacterial endophytic communities in the grapevine depend on pest management. *PLoS ONE*, *9*(11), e112763. https://doi.org/10.1371/journal.pone.0112763

Čeprnja, M., Hadžić, E., Oros, D., Melvan, E., Starcevic, A., & Zucko, J. (2023). Current viewpoint on female urogenital microbiome—The cause or the consequence? *Microorganisms*, *11*(5), 1207. https://doi.org/10.3390/microorganisms11051207

Cicchinelli, S., Rosa, F., Manca, F., Zanza, C., Ojetti, V., Covino, M., Candelli, M., Gasbarrini, A., Franceschi, F., & Piccioni, A. (2023). The impact of smoking on microbiota: A narrative review. *Biomedicines*, *11*(4), 1144. https://doi.org/10.3390/biomedicines11041144

D'Afflitto, M., Upadhyaya, A., Green, A., & Peiris, M. (2022). Association between sex hormone levels and gut microbiota composition and diversity—A systematic review. *Journal of Clinical Gastroenterology*, *56*(5), 384–392. https://doi.org/10.1097/mcg.0000000000001676

De Luca, F., & Shoenfeld, Y. (2018). The microbiome in autoimmune diseases. *Clinical & Experimental Immunology*, *195*(1), 74–85. https://doi.org/10.1111/cei.13158

Duarte, M., Pedrosa, S. S., Khusial, P. R., & Madureira, A. R. (2024). Exploring the interplay between stress mediators and skin microbiota in shaping age-related hallmarks: A review. *Mechanisms of Ageing and Development*, *220*, 111956. https://doi.org/10.1016/j.mad.2024.111956

Gonin, M., Salas-González, I., Gopaulchan, D., Frene, J. P., Roden, S., Van De Poel, B., Salt, D. E., & Castrillo, G. (2023). Plant microbiota controls an alternative root branching regulatory mechanism in plants. *Proceedings of the National Academy of Sciences*, *120*(15). https://doi.org/10.1073/pnas.2301054120

Haak, B. W., Lankelma, J. M., Hugenholtz, F., Belzer, C., De Vos, W. M., & Wiersinga, W. J. (2018). Long-term impact of oral vancomycin, ciprofloxacin and metronidazole on the gut microbiota in healthy humans. *Journal of Antimicrobial Chemotherapy*, *74*(3), 782–786. https://doi.org/10.1093/jac/dky471

Hasan, N., & Yang, H. (2019). Factors affecting the composition of the gut microbiota, and its modulation. *PeerJ*, *7*, e7502. https://doi.org/10.7717/peerj.7502

Henares, D., Brotons, P., De Sevilla, M. F., Fernandez-Lopez, A., Hernandez-Bou, S., Perez-Argüello, A., Mira, A., Muñoz-Almagro, C., & Cabrera-Rubio, R. (2021). Differential nasopharyngeal microbiota composition in children according to respiratory health status. *Microbial Genomics*, *7*(10). https://doi.org/10.1099/mgen.0.000661

Hirano, T., Kawano, T., Kadowaki, Y., Moriyama, M., Umemoto, S., Yoshinaga, K., Matsunaga, T., & Suzuki, M. (2023). Impact of IL-17-producing γδ T cells on chronic otitis media induced by nontypeable Haemophilus influenzae in a mouse model. *Pathogens and Disease*, *81*. https://doi.org/10.1093/femspd/ftad029

Holmes, C. J., Plichta, J. K., Gamelli, R. L., & Radek, K. A. (2014). Dynamic role of host stress responses in modulating the cutaneous microbiome: Implications for wound healing and infection. *Advances in Wound Care*, *4*(1), 24–37. https://doi.org/10.1089/wound.2014.0546

Hou, K., Wu, Z., Chen, X., Wang, J., Zhang, D., Xiao, C., Zhu, D., Koya, J. B., Wei, L., Li, J., & Chen, Z. (2022). Microbiota in health and diseases. *Signal Transduction and Targeted Therapy*, *7*(1). https://doi.org/10.1038/s41392-022-00974-4

Iliev, I. D., Ananthakrishnan, A. N., & Guo, C. (2025). Microbiota in inflammatory bowel disease: mechanisms of disease and therapeutic opportunities. *Nature Reviews Microbiology*. https://doi.org/10.1038/s41579-025-01163-0

Kaur, H., Merchant, M., Haque, M. M., & Mande, S. S. (2020). Crosstalk between female gonadal hormones and vaginal microbiota across various phases of women's gynecological lifecycle. *Frontiers in Microbiology*, *11*. https://doi.org/10.3389/fmicb.2020.00551

Ma, J., Li, Z., Zhang, W., Zhang, C., Zhang, Y., Mei, H., Zhuo, N., Wang, H., Wang, L., & Wu, D. (2020). Comparison of gut microbiota in exclusively breast-fed and formula-fed babies: a study of 91 term infants. *Scientific Reports*, *10*(1). https://doi.org/10.1038/s41598-020-72635-x

Madison, A., & Kiecolt-Glaser, J. K. (2019). Stress, depression, diet, and the gut microbiota: Human–bacteria interactions at the core of psychoneuroimmunology and nutrition. *Current Opinion in Behavioral Sciences*, *28*, 105–110. https://doi.org/10.1016/j.cobeha.2019.01.011

Manyi-Loh, C., Mamphweli, S., Meyer, E., & Okoh, A. (2018). Antibiotic use in agriculture and its consequential resistance in environmental sources: potential public health implications. *Molecules*, *23*(4), 795. https://doi.org/10.3390/molecules23040795

Marasco, R., Rolli, E., Fusi, M., Michoud, G., & Daffonchio, D. (2018). Grapevine rootstocks shape underground bacterial microbiome and networking but not potential functionality. *Microbiome*, *6*(1). https://doi.org/10.1186/s40168-017-0391-2

Marlicz, W., & Loniewski, I. (2014). The effect of exercise and diet on gut microbial diversity. *Gut*, *64*(3), 519–520. https://doi.org/10.1136/gutjnl-2014-307909

Marwaha, K., Cain, R., Asmis, K., Czaplinski, K., Holland, N., Mayer, D. C. G., & Chacon, J. (2025). Exploring the complex relationship between psychosocial stress and the gut microbiome: implications for inflammation and immune modulation. *Journal of Applied Physiology*. https://doi.org/10.1152/japplphysiol.00652.2024

McKay, T. B., Priyadarsini, S., & Karamichos, D. (2022). Sex hormones, growth hormone, and the cornea. *Cells*, *11*(2), 224. https://doi.org/10.3390/cells11020224

Neuman, H., Debelius, J. W., Knight, R., & Koren, O. (2015). Microbial endocrinology: the interplay between the microbiota and the endocrine system. *FEMS Microbiology Reviews*, *39*(4), 509–521. https://doi.org/10.1093/femsre/fuu010

Niedźwiedzka, A., Micallef, M. P., Biazzo, M., & Podrini, C. (2024). The role of the skin microbiome in acne: Challenges and future therapeutic opportunities. *International Journal of Molecular Sciences*, *25*(21), 11422. https://doi.org/10.3390/ijms252111422

Nuzzi, R., & Caselgrandi, P. (2022). Sex hormones and their effects on ocular disorders and pathophysiology: current aspects and our experience. *International Journal of Molecular Sciences*, *23*(6), 3269. https://doi.org/10.3390/ijms23063269

Pagac, M. P., Stalder, M., & Campiche, R. (2024). Menopause and facial skin microbiomes: a pilot study revealing novel insights into their relationship. *Frontiers in Aging*, *5*. https://doi.org/10.3389/fragi.2024.1353082

Park, J., Schwardt, N. H., Jo, J., Zhang, Z., Pillai, V., Phang, S., Brady, S. M., Portillo, J. A., MacGibeny, M. A., Liang, H., Pensler, M., Soldin, S. J., Yanovski, J. A., Segre, J. A., & Kong, H. H. (2021). Shifts in the skin bacterial and fungal communities of healthy children transitioning through puberty. *Journal of Investigative Dermatology*, *142*(1), 212–219. https://doi.org/10.1016/j.jid.2021.04.034

Park, M. G., Cho, S., & Oh, M. M. (2023). Menopausal changes in the microbiome—A review focused on the genitourinary microbiome. *Diagnostics*, *13*(6), 1193. https://doi.org/10.3390/diagnostics13061193

Peng, D., Wang, Z., Tian, J., Wang, W., Guo, S., Dai, X., Yin, H., & Li, L. (2024). Phyllosphere bacterial community dynamics in response to bacterial wildfire disease: Succession and interaction patterns. *Frontiers in Plant Science*, *15*. https://doi.org/10.3389/fpls.2024.1331443

Pires, L., Gonzalez-Paramás, A. M., Heleno, S. A., & Calhelha, R. C. (2024). Gut microbiota as an endocrine organ: Unveiling its role in human physiology and health. *Applied Sciences*, *14*(20), 9383. https://doi.org/10.3390/app14209383

Ren, Z., Li, W., Liu, Q., Dong, Y., & Huang, Y. (2022). Profiling of the conjunctival bacterial microbiota reveals the feasibility of utilizing a microbiome-based machine learning model to differentially diagnose microbial keratitis and the core components of the conjunctival bacterial interaction network. *Frontiers in Cellular and Infection Microbiology*, *12*. https://doi.org/10.3389/fcimb.2022.860370

Salas-González, I., Reyt, G., Flis, P., Custódio, V., Gopaulchan, D., Bakhoum, N., Dew, T. P., Suresh, K., Franke, R. B., Dangl, J. L., Salt, D. E., & Castrillo, G. (2020). Coordination between microbiota and root endodermis supports plant mineral nutrient homeostasis. *Science*, *371*(6525). https://doi.org/10.1126/science.abd0695

Santacroce, L., Passarelli, P. C., Azzolino, D., Bottalico, L., Charitos, I. A., Cazzolla, A. P., Colella, M., Topi, S., Godoy, F. G., & D'Addona, A. (2023). Oral microbiota in human health and disease: A perspective. *Experimental Biology and Medicine*, *248*(15), 1288–1301. https://doi.org/10.1177/15353702231187645

Santos-Marcos, J. A., Mora-Ortiz, M., Tena-Sempere, M., Lopez-Miranda, J., & Camargo, A. (2023). Interaction between gut microbiota and sex hormones and their relation to sexual dimorphism in metabolic diseases. *Biology of Sex Differences*, *14*(1). https://doi.org/10.1186/s13293-023-00490-2

Saturio, S., Nogacka, A. M., Alvarado-Jasso, G. M., Salazar, N., De Los Reyes-Gavilán, C. G., Gueimonde, M., & Arboleya, S. (2021). Role of bifidobacteria on infant health. *Microorganisms*, *9*(12), 2415. https://doi.org/10.3390/microorganisms9122415

Sjövall, A., Aho, V. T., Hyyrynen, T., Kinnari, T. J., Auvinen, P., Silvola, J., Aarnisalo, A., & Laulajainen-Hongisto, A. (2020). Microbiome of the healthy external auditory canal. *Otology & Neurotology*, *42*(5), e609–e614. https://doi.org/10.1097/mao.0000000000003031

Song, S. D., Acharya, K. D., Zhu, J. E., Deveney, C. M., Walther-Antonio, M. R. S., Tetel, M. J., & Chia, N. (2020). Daily vaginal microbiota fluctuations associated with natural hormonal cycle, contraceptives, diet, and exercise. *mSphere*, *5*(4). https://doi.org/10.1128/msphere.00593-20

Tomova, A., Bukovsky, I., Rembert, E., Yonas, W., Alwarith, J., Barnard, N. D., & Kahleova, H. (2019). The effects of vegetarian and vegan diets on gut microbiota. *Frontiers in Nutrition*, *6*. https://doi.org/10.3389/fnut.2019.00047

Tramice, A., Paris, D., Manca, A., Agudelo, F. A. G., Petrosino, S., Siracusa, L., Carbone, M., Melck, D., Raymond, F., & Piscitelli, F. (2022). Analysis of the oral microbiome during hormonal cycle and its alterations in menopausal women: The "AMICA" project. *Scientific Reports*, *12*(1). https://doi.org/10.1038/s41598-022-26528-w

Trivedi, P., Leach, J. E., Tringe, S. G., Sa, T., & Singh, B. K. (2020). Plant–microbiome interactions: From community assembly to plant health. *Nature Reviews Microbiology*, *18*(11), 607–621. https://doi.org/10.1038/s41579-020-0412-1

Ullah, H., Arbab, S., Tian, Y., Liu, C., Chen, Y., Qijie, L., Khan, M. I. U., Hassan, I. U., & Li, K. (2023). The gut microbiota–brain axis in neurological disorder. *Frontiers in Neuroscience*, *17*. https://doi.org/10.3389/fnins.2023.1225875

Valdes, A. M., Walter, J., Segal, E., & Spector, T. D. (2018). Role of the gut microbiota in nutrition and health. *BMJ*, k2179. https://doi.org/10.1136/bmj.k2179

Zhang, C., Li, L., Jin, B., Xu, X., Zuo, X., Li, Y., & Li, Z. (2021). The effects of delivery mode on the gut microbiota and health: state of art. *Frontiers in Microbiology*, *12*. https://doi.org/10.3389/fmicb.2021.724449

Zysset-Burri, D. C., Morandi, S., Herzog, E. L., Berger, L. E., & Zinkernagel, M. S. (2022). The role of the gut microbiome in eye diseases. *Progress in Retinal and Eye Research*, *92*, 101117. https://doi.org/10.1016/j.preteyeres.2022.101117

3 Dynamic Nature of Biofilm Formation and Regulation

Rina Rani Ray
Department of Biotechnology, Maulana Abul Kalam Azad University of Technology, Kolkata, India

3.1 INTRODUCTION

The term dynamic nature refers to a system that undergoes constant changes. These changes may be in composition, attributes, dimension, architecture, and other features. Dynamism leads to the emancipation from an unchanged and static mode. Biofilm, being a conglomeration of various microbes on a biotic or abiotic surface, is an ever-changing biological model, representing the dynamic lifestyle of complex microbial communities. The dynamic nature is evident from its complex interactions with its members and also with the environment, which confers adaptations with an assurance of longevity. As a dynamic system, it becomes fine-tuned with adaptive attributes such as tolerance to stress, desiccation, a shortage of food, anti-microbials and the immunological reactions of the host. The biofilms are not just a natural glue but also a complex biological community, which may be formed by a single type or multiple types of microbial species, establishing themselves into a harmonized functional community. This community remains in a constantly changing mode and experiences both vertical and horizontal growth over the course of time. Although it is difficult to measure its dynamism experimentally with any degree of accuracy, the change in architecture, gradual expansion, interplay between microbial signalling molecules, genetic regulation, and overall interaction with the environment clearly reveal the dynamic nature of the biofilm. Previously dynamics of biofilm was measured either by simply measuring the biofilm growth kinetics, by determining the gradual increase in the amount of biofilm matrix components or by studying the activities of component microorganisms (Characklis et al., 1982). At present, however, the analysis of spaciotemporal dynamics (Vidakovic et al., 2018) reveals the growth and evolution of the biofilm in space and time, the change in matrix patterns, microbial processes, and cell-to-cell interactions, including quorum sensing and interactions with environment at a specific location and over a certain period of time. In order to achieve better understanding of the dynamic nature of biofilm, scientists took advantage of 3D bioprinting, where they used a customizable compatible platform to visualize all the phases of biofilm formation, from surface attachment right up to maturation, including the regulatory parameters (Dhall et al., 2022).

The present chapter deals with the architecture and dynamic nature of biofilms formed in various parts of the body, and also their regulations.

3.2 GENERAL ARCHITECTURE OF BIOFILM

Bacteria are known to be the pioneer microbe which first colonize to form biofilm. They exist in two forms: free-floating planktonic forms; and sessile forms adhered to some surface. As a consequence of certain gene expressions, planktonic forms transform into transform into sessile forms and develop a biofilm through the production of an exopolysaccharide matrix. The biofilm-bound microbes remain protected from various environmental stresses, such as antibiotic treatment, host immune reaction or phage attack.

DOI: 10.1201/9781003519041-3

Hence, biofilms are an immovable complex structure made up of either a monoculture or many species of microbes, host cells, and cellular by-products, irreversibly attached to the substratum, remaining protected by a biofilm matrix made up of extracellular polymeric substance. Structurally, biofilm is a three-dimensional construction made of genetically similar, although functionally different microbial cells that develop into distinct cell types and facilitate biofilm formation, maintenance, maturation and dispersal (Saxena et al., 2019). Biofilm-bound microbes can interact with the changing environment, and various environmental stresses by constantly developing and dispersing resistance to the adjacent cells. Such cell-to-cell communication among neighboring cells is accomplished by a specialized mechanism known as quorum sensing.

Generally, a biofilm contains microbial mass (10%) in water (90%) (Donlan, 2002). Among the matrix components, a major part is contributed by polysaccharides (50–90%), which form chain-like structures. The hydroxyl groups attached to these polysaccharides supply mechanical strength. The biofilm construction can have positively charged ions, such as Ca^{2+} or Mg^{2+}, which form strong cross-bridges between polymers so that the biofilm can grow up to a thickness of about 300 μm. The polysaccharides in biofilms can be neutral, polyanionic, or cationic in nature, depending on the specific components present in the matrix. The various charges and ions present in the biofilm contribute to the structural stability of EPS, allowing biofilms to endure environments with extreme shear forces, such as at points of waterfall impact (Sharma et al., 2023). Therefore, the physical structure of a biofilm can be characterized as an intertwined polymer network that develops and adapts in response to gradients of nutrients, cellular differentiation, quorum sensing, bacterial movement, and environmental interactions

3.3 INSIGHTS OF BIOFILM DYNAMICS

An analysis of biofilm dynamics requires comprehension of how microbial communities, surrounded by a matrix produce themselves, develop, thrive, and spread. This is essential for understanding their functions in different ecosystems, health contexts, and industrial activities. The growth and demographic changes of biofilms are shaped by multiple influencing factors, including biochemical signals, competition, cooperation, cell mortality, and mechanical interactions. Notably, research has identified that interactions among neighbouring cells, and also between cells and surfaces, play a significant role in the formation of new variants, as demonstrated by both experimental and theoretical investigations. A study by Postek et al. (2024) revealed that the dynamics of bacterial populations within the biofilm can be regulated through alterations in surface geometry.

After the bacteria were grown in a biocompatible 3D-printed microfluidic device, an analysis was made of the complex dynamics of bacterial flow, cell adhesion to the surface, and their aggregation for biofilm formation (Ashish et al., 2024). Computational Fluid Dynamics (CFD), a numerical simulation technique that uses mathematical equations, is increasingly being used to study biofilm dynamics by simulating fluid flow both around and within biofilms, allowing for the analysis of shear stress, shear rate, and biofilm detachment, which are all key factors in biofilm growth and behaviour (Wang et al., 2022). Intricate dynamics of biofilm formation and maturation on a transparent substrate can be studied by using digital holographic tomography, which specifically highlights the quantitative analysis of biofilm dynamics (Buzalewicz et al., 2024). Today, the technique of digital in-line holographic microscopy (DIHM) is being used, for the quantitative measurement of microbe–cell–surface interactions, monitor the three-dimensional (3D) orientation of microbes, and check the dynamics of biofilm formation (Kim and Lee, 2024).

3.4 VERTICAL GROWTH DYNAMICS OF BIOFILMS

Biofilm formation involves a sequential episode of attachment to a surface and growth, leading to vertical and horizontal expansion. Vertical growth in biofilms, although less often studied than horizontal expansion, is very important since it reflects the nature of microbial interaction, inter-community interactions, and growth of the conglomeration (Bravo et al., 2023) and it also has an influence on horizontal expansion (Pokhrel et al., 2024).

In some environments, the prevailing conditions can allow only vertical growth instead of both vertical and horizontal growth. According to Bravo et al. (2023), the vertical growth rate is depicted by the change in height per unit time and it varies with time as in the case of typical logistic growth. Just after inoculation, the biofilm growth is very slow, in order to adapt to the new environment. This is followed by a relatively rapid growth up to the achievement of a climax. It is therefore termed "two regime growth", consisting of a linear increase in the vertical growth rate up to a maximum height; this is then followed by a linear decrease in the vertical growth rate. Such two-phase growth is found to be governed by a multiple parameter.

Nutrient, being a major parameter for biofilm development, is found to enter the biofilms through the biofilm–air and the biofilm–substrate interfaces and then to percolate into the colony in order to be consumed by the microbial members. This restricted infiltration of food prepares a number of microenvironments and its differential uptake restricts the thickness of the expanding zone within the colony.

To understand the dynamics of biofilm growth, many models are suggested, including: (1) cellular automata-based models, i.e., computational models that simulate complex systems by representing them as a grid of cells, each with a state that evolves based on the states of its neighbours and a set of rules; (2) individual-based models, which focus on the behaviour and interactions of individual microbes within the biofilm, allowing for the emergence of complex population-level dynamics; and (3) active fluid models, where the microbes are essentially growing as energy-consuming particles.

Finally, it was suggested that a simple population dynamics model should be used to explain vertical growth dynamics. According to the heuristic model of vertical growth, the colony is regarded as a material with a height. Its vertical growth initially increases at a fixed rate up to a maximum height. This height is fixed in accordance with nutrient consumption and limited nutrient availability. The growth then declines at a different fixed rate. In the case of biofilm, the growth is the combined effect of cell doubling and extracellular matrix production, while decay is caused by an incomplete reabsorption of EPS.

Time is also an important parameter, since the achievement of maximum height of biofilm growth may occur after a different time of inoculation, and this varies from species to species.

Relative fitness of the biofilm determined by activities of various genes is also an important condition to either restrict or encourage the biofilm growth.

To understand how biofilms develop their three-dimensional structure, various models are proposed. The model of the vertical growth dynamics plays a very important role in explaining the original biophysical process of biofilm formation. Bravo et al. (2023) asserted that there exists a universality in following this model by the microorganisms.

Paula et al. (2020) proposed that the formation of biofilms shares certain spatial and structural characteristics with urban development, in which growth in population and expansion can be affected by the types of settlers, neighbouring cells, community integration, and scaffolding processes occurring at multiple levels.

Copeland and Yunker (2025) presented a theoretical basis for a straightforward mathematical model of vertical growth, allowing for a better comprehension of how biological and biophysical elements interact during the development of biofilms.

3.5 HORIZONTAL GROWTH DYNAMICS OF BIOFILMS

Biofilm's horizontal growth, or range expansion, involves bacteria attaching to a surface, reproducing, and forming communities that expand outwards, often at a constant rate. This expansion can be influenced by factors like nutrient availability, cell-to-cell interactions, and the surrounding environment.

In nature, horizontal biofilm growth, or expansion across a surface, is a key aspect of biofilm development, often involving factors like physical interactions, environmental conditions, and inter-strain competition, and bringing about a complex community structure.

Biofilms formed horizontally can sometimes prove fragile, whereas vertical and horizontally expanded biofilms are highly resistant to detachment. Irrespective of strain, the highest number of CFU were detected on a Petri plate after rinsing, indicating both horizontal and vertical growth of the biofilm. Jha et al. (2022) suggested that biofilms formed on vertical surfaces were less dense than those formed on horizontal surfaces, although the later are often more brittle, and, hence, failed to resist even very low shear forces.

3.6 ARCHITECTURE AND DYNAMICS OF THE BIOFILMS OF MAJOR HUMAN MICROBIOMES

3.6.1 Architecture and Dynamics of the Oral Biofilm

The oral ecosystem consists of complex, three-dimensional communities of microorganisms, including bacteria, remaining embedded as a polymicrobial biofilm. They are found on various surfaces, including dental surfaces, on subgingival and supragingival pockets, on the tongue, and in saliva. Despite its constant irrigation through the neutralization of saliva, the pH of the oral environment remains constantly acidic, a fact which is further aggravated by the activities of the acid-producing bacteria dwelling in oral cavities.

The composition and activity of oral microbiomes can change significantly over time and in different environments, and they are developmentally dynamic in relation to the age and physiology of their host (Ray, 2024).

Oral microbiomes can experience considerable shifts in their composition and function over time and across different environments, and they are developmentally responsive to the host. A variety of factors contribute to these intricate and variable dynamics, such as dietary habits, pH sensitivity, interactions among microbes, and, over time, genetic mutations and horizontal gene transfer, which can impart new traits to bacterial strains. The oral cavity serves as a dynamic microbial ecosystem that changes over the course of a person's life. In the first few months of life, the mouth primarily consists of mucosal areas that are conducive to microbial colonization. However, with the appearance of teeth, a new hard and non-shedding surface is established, promoting increased bacterial growth in the form of biofilm (dental plaque). The gingival crevicular fluid, which is produced in the gum sulcus, provides nutrients for microorganisms in the subgingival region. Over time, the oral ecosystem adjusts in response to factors such as the eruption of teeth, extractions, the placement of dentures, the installation of orthodontic bands, and various dental procedures, including scaling and repairs. Temporary fluctuations in the stability of the oral environment may also be influenced by the frequency and type of dietary intake (Ali and Tanwir, 2012).

As a person grows older, the scenario of the oral microbiota changes with the change in life style and the change in the types of food consumed. The entry of pathogens leads to the dysbiosis of oral biofilm which leads to a fluctuation in saliva flow, and the development of diseases. Moreover, dental operative measures such as tooth extraction and the insertion of dentures and, above all, antibiotic treatment may lead to a change in the dynamics of oral biofilm.

The members of the normal oral microbiota include different microbial species belonging to phyla of bacteria, archaea, fungi, protozoa, and viruses. These remain both in planktonic, i.e. free-floating, form and as biofilm-bound sessile forms present on oral surfaces (Samaranayake and Matsubara, 2017).

Saliva contains a diverse array of bacterial species, such as *Streptococcus mitis*, *S. salivarius*, *Granulicatella adiacens*, *Neisseria flavescens*, *Rothia mucilaginosa*, and *Prevotella melaninogenica* (Ray, 2024).

The entrance of pathogenic microbes induces an imbalance in the normal oral microbial rhythm and initiates plaque formation. Such intruding bacteria generally form a Cytophaga–Flavobacterium–Bacteroides (CFB) cluster, one of the most dominant members of which is *Porphyromonas* spp. However, the CFB are found to form microcolonies in the top and middle layers of oral biofilm. Other bacteria found at the top layer are *Prevotella* sp and *Parvimonas micra*. Microorganisms are generally located in groups within oral fluids, creating a diverse community of both aerobic and anaerobic organisms, comprising of bacteria, once classified as early and late colonizers. These microbial clusters rapidly proliferate and form three-dimensional structures, affecting growth rates, spatial organization, and community dynamics. The majority of these clusters gradually transform into organized biofilm communities (Simon-Soro et al., 2022).

3.6.2 Architecture and Dynamics of the Gut Biofilm

The gastrointestinal (GI) tract, also known as the gut, is the alimentary canal that stretches from the mouth to the anus. It is the area of the human body with the highest concentration of microorganisms, which exhibit a variety of lifestyles. Within the gut, the microbiome comprises diverse communities of viruses, bacteria, fungi, and eukaryotes that coexist in the mucus layers. The colonization, diversity, and density of microbes can differ throughout the length of the GI tract.

The gut microbiota is typically comprised of six main phyla, Firmicutes, Bacteroidetes, Actinobacteria, Proteobacteria, Fusobacteria, and Verrucomicrobia, with Firmicutes and Bacteroidetes being the most prevalent types. The fungi most commonly studied within the gut mycobiota include *Candida, Saccharomyces, Malassezia*, and *Cladosporium*. In addition to bacteria and fungi, the human gut microbiota also contains viruses, phages, and archaea, with *Methanobrevibacter smithii* being the predominant representative (Hou et al., 2022).

Microbial, specifically bacterial, density increases along the GI tract, peaking in the colon (10^9–10^{11} bacteria/m). The interaction between bacteria, the surrounding environment, and the host influences the composition and characteristics of the microbiota. Many gut microbes exist as free-floating cells in the lumen, while others form structured groups known as biofilms (Jandl et al., 2024). Scientists found site-specific difference in the nature of biofilm formation, as the scattered, poorly developed type of biofilm, formed by a scanty number of microbes are found in the upper part of gut, including the stomach, whereas the rich and compact types of microbial biofilms are found in large intestinal mucosa (Motta et al., 2021). Such difference is attributed by the existences of well-established distinctness in their composition, genetic makeup, environmental preferences, and microbial characteristics. In certain situations, sessile microbes might detach from these gut biofilms and act as free-swimming planktonic microorganisms (Buret et al., 2019). Bacterial commensals inhabit the structurally intricate environment of the GI tract as closely-knit mixed communities. While the majority of species typically compete for resources and space, they can also form beneficial relationships that enhance diversity. Numerous bacteria produce substances such as nutrient chelators, digestive enzymes, and signalling molecules that support the survival and growth of nearby cells, whether they belong to the same or a different taxonomic group. The arrangement of various species within biofilms affects the dynamics of these competitive or cooperative interactions, ultimately influencing the composition and functioning of the bacterial community (Wong et al., 2023).

Biofilms formed in the gut are composed of complex microbial communities of dynamic nature, remaining enclosed by an extra polymeric substance made biofilm matrix. They play a pivotal role in the maintenance of health or the onset of diseases, The dynamicity of the biofilm is influenced by the food, the immunity of the host, and microbial types and interactions. Food particles remaining attached to various regions of the intestine promote biofilm formation in the lumen or on the gut mucosal surface. The functional attributes of the gut biofilm forming microbiota are thought to play a significant role in metabolism, pathogenesis, and immunity. In polymicrobial biofilm, the dynamic relations among microbes typically involve complicated and multifaceted metabolic networks among several species. In multispecies biofilms, relationships such as neutralism, mutualism, commensalism, and competition can certainly influence the diversity of food digestion as a result of metabolic variances and interspecies cross-feeding. Neutral and competitive relationships encourage seclusion and structural discontinuity, whereas mutualism and commensalism lead to a somewhat more peaceful coexistence, resulting in maximum biofilm growth, maturation, and natural dispersion (Valiei et al., 2024). When a new bacterial species with a competitive attitude invades a mature biofilm of competitive, neutral, or commensal characteristics, it generally prefers to grow rapidly and to flourish at the upper biofilm layer, as overgrowth will be most advantageous for the newcomer in gaining strong foothold establishing domination over the existing communities. After the invasion of new species, those microbes with mutualistic relationships receive a greater benefit, since they can exploit the available food from both the topmost layer and the void areas. Thus, the behavioural pattern and metabolic interactions of the microbial communities, initial abundances, and food concentration contribute to spatiotemporal variation of the biofilm components and its dynamic nature. Bacteria exhibit flexible collective behaviors and create colonies with distinct geometric shapes. The dynamics of these colonies involve the paths of aggregates that grow, migrate, and reproduce at the same time.

The biofilm morphology is found to be dependent on the softness of the substrate to which they adhere and also nutrient availability. Therefore, movement among the bacterial colonies play a strong role in determining the biofilm morphology. Bacteria in colonies can move together over surfaces by swarming to generate dynamic patterns of large swirls and jets. However, swarming, in contrast to collective swimming, involves several cellular processes such as alteration in the levels of key proteins, the involvement of mechanical factors and intercellular chemical communication, leading to a super-diffusive spreading of bacteria (Be'er and Ariel, 2019).

In attempts to understand the dynamics of colonies, various mathematical models have been suggested. Mimura et al. (2000) introduced a model that integrates a reaction–diffusion system for nutrient dynamics, nucleation theory for the creation of aggregates, and individual-based dynamics concerning the movement and growth of these aggregates. This model was regularly validated through experimental reproduction. Subsequently, this model was altered by other researchers. Gerlee and Anderson (2007) presented a hybrid cellular automaton model for colony growth, in which the growth of the colony is influenced by nutrient availability, which limits cell division below a specific threshold. Dong and Klumpp (2018) utilized a cellular automaton model to explore colony pattern formation based on the differential adhesion hypothesis while taking into account cell proliferation.

Farrell et al. (2013) proposed that mechanical interactions between bacterial cells to get hold of a diffusing food particle is responsible for the transition between circular and branching colonies. This model was supported by Ghosh et al. (2015), whose study finds that only the repulsive forces can generate such phase-separated patterns in the growing colony. Although a number of models have been proposed regarding colony growth, the significance of swarming on transport phenomena is yet to be quantitatively described (Ishikawa et al., 2020).

3.6.3 Architecture and Dynamics of the Skin Biofilm

The normal skin microbiome comprises bacteria, fungi, and viruses, with bacteria being the most abundant, and including major species like *Staphylococcus, Corynebacterium*, and *Malassezia*. The skin, as the largest organ in the human body, covers an area of approximately 2 m², which notably expands (by at least 30 m²) for the gut (Gallo, 2017) at the limbs. The skin microbiota may be composed of two types of microbes: short-staying and permanent resident (Cogen et al., 2008). The resident flora consists of organisms that are consistently found across the entire skin surface, including deeper regions such as the ducts of sebaceous and sweat gland follicles. At least 19 different phyla have been identified as contributors to the bacterial skin microbiome. Prominent examples include Actinobacteria, Firmicutes, Proteobacteria, and Bacteroidetes. The most commonly identified genera are Corynebacterium, Propionibacterium, and Staphylococcus. Other groups detected include *Alphaproteobacteria, Bacteroidales, Betaproteobacteria, Clostridiales, Flavobacteriales, Gammaproteobacteria, Lactobacillales*, and *Streptococcaceae*, alongside various yeasts and viruses that are also part of the skin microbiota.

The skin, being a distinct and variable ecosystem, offers numerous niches that support rich populations of diverse microbes (Figure 3.1). These are influenced by varying ecological factors such as humidity, temperature, pH, and the makeup of antimicrobial peptides and lipids. Furthermore, skin structures like hair follicles and sebaceous, eccrine, and apocrine glands create specific niches that host unique microbiota. Investigations into the topographical diversity of microorganisms inhabiting these regions of human skin through 16S rRNA gene phylotyping have shown that these habitats significantly impact microbial composition.

Human skin features four distinct microenvironments: the moist area, which is primarily inhabited by *Staphylococcus, Corynebacteria, and Proteobacteria*; the dry zone, dominated by Flavobacteriales, Proteobacteria, and Corynebacteria; the regions around sebaceous glands featuring *Cutibacterium* and *Staphylococcus*; and other areas within the dermal layers, such as hair follicles and sweat glands, which host a variety of microbes (Grice & Segre, 2011).

The prevalence of each microbial group is largely influenced by the specific characteristics of their respective niches. For instance, sebaceous areas on the face are mainly occupied by Propionibacterium and Staphylococcus species. In moist zones like the axilla, Corynebacterium species are the most commonly found, with Staphylococcus species also present. Conversely, in dry environments, a combination of bacterial species from *Betaproteobacteria* and *Flavobacteriales* constitute the resident microbiota.

According to research, the fungi, namely, the genus *Malassezia*, predominates on healthy human skin and its distribution is dependent on the characteristics of the habitat. *M. globose* was found particularly on the chest and arms and even in the inguinal fold; *M. restricta* predominates on face, scalp and the external auditory canal, retro-auricular crease, and glabella; *M. sympodialis* is frequently found in trunk regions. A variety of fungal communities, such as *Malassezia, Cryptococcus, Aspergillus, Epicoccum, Rhodotorula*, and others, are found to colonize at the foot sites (Findley et al., 2013). *Malassezia* sp. consumes oily sebum and remain associated with sebaceous glands of the skin.

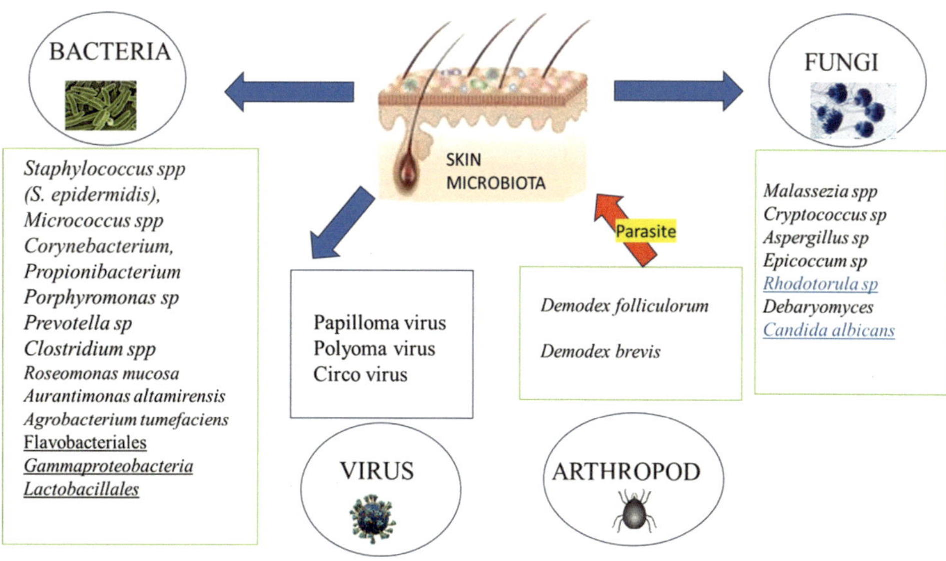

FIGURE 3.1 Natural inhabitants of human skin.

Mites, which are the eukaryotic organism belonging to the phylum Arthropoda, inhabit human skin. Specifically, the *Demodex folliculorum* (larger) and the *Demodex brevis* (smaller) reside, respectively, on the skin surface and sebaceous gland, and on hair follicles.

Metagenomic study reveals the presence of several types of DNA virus on human skin, although cultivational difficulty makes it tough to identify them individually (Foulongne et al., 2012). To date, it is unclear if these viruses are really the part of normal human skin microbiota, though some scientists do actually consider HPV to be a normal member of skin microbiome.

The skin biofilm, like other microbiota, is dynamic in nature and its composition can change based on a number of factors: individual surroundings, age, health, hygiene practices, nutritional state and even certain medical conditions. Environmental parameters such as temperature, humidity, and lipid content impact on the formation of various microhabitats on the skin (Baquero et al., 2021). Apart from protecting from invading pathogens, the skin microbiota produces characteristic and discrete smell of each individual (Lucas-Barbosa et al., 2022).

An analysis of samples collected from approximately 130 males and 110 females indicated that the diversity observed within individual samples varies according to the diversity found between samples from the same environment among different subjects. This suggests that skin demonstrates moderate alpha diversity, but high beta diversity. Due to the differences in microenvironments across various skin areas and the diverse external treatments that individuals apply to their skin, it has proven challenging to create clear links between specific organisms and skin functions. Notably, the distinct microbial makeup for each individual appears to remain stable over time, indicating a level of consistency for both the individual subject and the study (Reuben et al., 2022). SanMiguel and Grice (2014) made a thorough exploration of the connection between various skin conditions and the makeup of skin microbiota, revealing that the interactions between *S. epidermidis*, an opportunistic bacterium, and *P. acnes*, a helpful propionic acid-producing bacterium, shift between mutualistic and antagonistic due to the production of antimicrobial compounds like fatty acids and peptides. In a similar vein, Iwase et al. (2010) found that serine proteinase produced by *S. epidermidis* inhibited biofilm formation by *S. aureus*, thereby reducing the likelihood of infection. Additionally, the cell-free supernatant from cultures of *S. epidermidis* was effective in preventing lipase buildup and biofilm development in yeasts such as *Candida albicans* and *Malassezia* spp., which are among the most common members of the skin microbiota.

3.6.4 Biofilm Architecture and Dynamics of the Vagina

The vaginal cavity of a matured female is enriched with a diverse microbial population and the healthy dynamic vaginal microbiome reflects the status of good reproductive health. The human vagina hosts a distinct collection of microbes that differs from those present in other organ surfaces and mucosal sites (Parks, 2022).

The human vagina, unlike oral and gut microbiota, lacks the extremely high bacterial diversity, because of the existence of a less diverse micro environment. Although anatomically situated at close proximity, the vagina does not harbour rectal microflora, which reveals the existence of some selection parameters in the vaginal ecosystem. This does not make the vaginal ecosystem static, however, since it experiences regular changes like continuous mucus secretion, the periodic shedding of vaginal epithelial cells, a monthly outpouring in iron-enriched endometrial tissue in menstruating women, and the introduction of exogenic microbes with intercourse. Hence, the human vagina is an extremely dynamic and precise ecosystem, where the balanced interactions between the host and her microflora, influenced by both internal and external factors, maintain her vaginal heath (Figure 3.2).

Vaginal pH, being extremely dynamic, fluctuates in response to the physiological, sexual, and reproductive state of the host, which adds to the dynamic nature of the vaginal ecosystem.

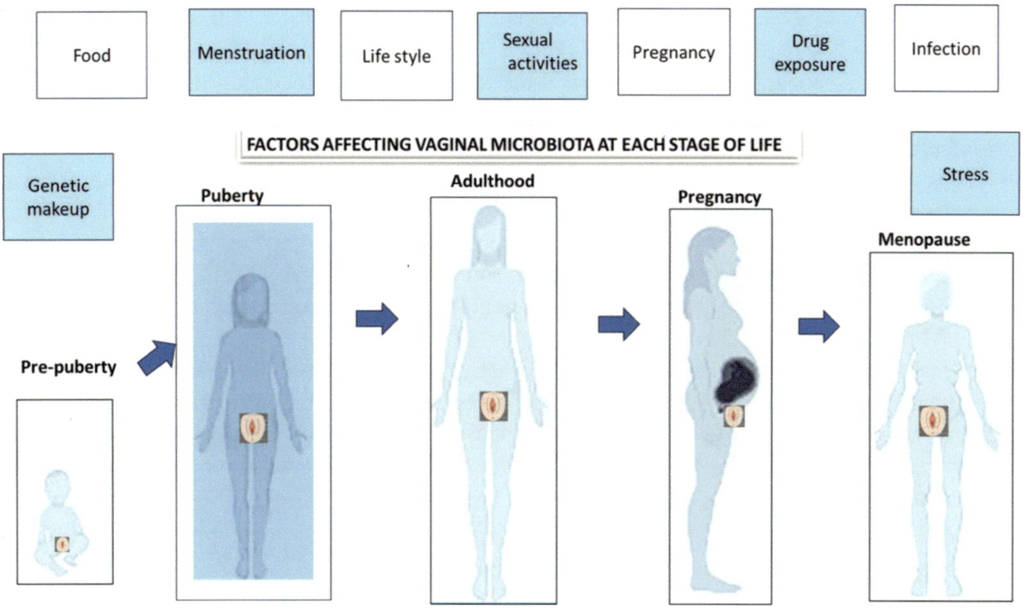

FIGURE 3.2 Various factors impacting on the dynamicity of vaginal biofilm.

The vaginal microbiota changes significantly according the age of the host: in pre-puberty, post-puberty and menopausal woman, and also on the pathogenic condition of the host. The vaginal microbiome is composed of more than two hundred different bacterial species. These are dominated by members from *Lactobacillus* genus (Auriemma et al., 2021), which protect the vagina from potential pathogenic attack. Actually, in the vaginal epithelial cells the bacteria ferment glycogen into lactic acid, which gives rise to the extreme low pH of vagina. Moreover, some antimicrobial compounds are being produced which, through competitive inhibition, checks the entry of the intruding pathogens (Chee et al., 2020).

On the other hand, the patient with bacterial vaginosis (BV) shows the presence of increased proportions of diverse types of other bacteria, like *Atopobium vaginae, Gardnerella vaginalis, Prevotella* spp., *Sneathia* spp., and even some anaerobes and a decrease in lactobacilli concentration (Hardy et al., 2017). Vaginal microbiome sometimes harbours protists, fungi, archaea and viruses, each of which have their own niche requirement (Abou Chacra and Fenollar, 2021).

Hormonal changes play a key role in regulating the reproductive life of the host and influence the vaginal microbiota. One such hormone is estrogen, which brings about remarkable changes in the vaginal microbiota. However, the saga of vaginal biofilm dynamics relies on various factors such as hormonal surges during the menstrual cycle, sexual activities, genetic makeup and the exposure of drugs on the part of the host, along with the environment and other lifestyle-related activities (Abou Chacra and Fenollar, 2021).

3.7 REGULATION OF BIOFILM ARCHITECTURE AND DYNAMICS

Biofilm architecture, the intricate 3D structure of biofilms, is regulated by a complex interplay of factors, including quorum sensing, c-di-GMP signaling, small non-coding RNAs (sRNAs), and environmental cues, all of which impact biofilm formation, matrix composition, and overall structure. The process of biofilm formation is initiated and controlled by quorum sensing, challenging environmental conditions, the availability of nutrients, hydrodynamic factors, cell-to-cell interactions, signaling pathways, and secondary messengers. Cyclic di-GMP (c-di-GMP) plays a crucial role as a regulator that oversees the development and spread of bacterial biofilms. The enzymes diguanylate cyclase (DGC) and phosphodiesterase (PDE), which facilitate the synthesis and degradation of this second messenger, respectively, control the levels of c-di-GMP within the cytoplasm. This molecule manages gene expression, the activity of metabolic enzymes, and biofilm development by binding to its receptors, leading to allosteric modifications that activate its effectors. Elevated levels of c-di-GMP restrict bacterial movement while promoting the production of the extracellular polymeric substance (EPS) matrix, ultimately resulting in biofilm formation (Gong et al., 2024). The development of biofilm growth is shaped by numerous environmental and host-related factors, such as proteases, fatty acids, oxygen levels, nutrients, and iron (Motta et al., 2021). Enteric infections can trigger dormant virulence genes in otherwise benign microbiota and disrupt the structure of the biofilm (Buret & Allain, 2023).

The invasion of pathobionts during the acute phase of protozoan infections cause gastrointestinal hypersensitivity. Communities that exhibit a remarkable array of adaptations to diverse environmental stresses, including antibiotic resistance, nutrient scarcity, and harsh physical conditions. These adaptations are achieved through various mechanisms, including the production of an extracellular matrix, spatial organization, and metabolic shifts.

3.8 CONCLUSION

Biofilm, being a syntrophic microbial association, is not a static entity. The dynamicity of biofilm is accomplished by a regular change in its architecture, composition, and characteristics. This change in the microbial composition and their nature of interaction are influenced by several factors, including genetic makeup, lifestyle, and infections. The positive interactions between normal microbiota and environment denote health and wellbeing. Hence, in order to cope with the changed environment, a regular change in microbiome composition and biofilm structure is unquestionably necessary. The accumulation of such changes, over time, will lead to the new adaptations, new variations, in phenotype and genotype leading to evolution.

REFERENCES

Abou Chacra, L., & Fenollar, F. (2021). Exploring the global vaginal microbiome and its impact on human health. *Microbial Pathogenesis*, *160*, 105172.

Ali, S. M. F., & Tanwir, F. (2012). Oral microbial habitat a dynamic entity. *JOBCR*, *2*(3), 181–187. https://doi.org/10.1016/j.jobcr.2012.07.001

Ashish, A., Biswas, P., Datta, B., Kanike, C., Atta, A., & Dutta, G. (2024). A study on the dynamics of bacterial growth using biocompatible 3D-printed microfluidic device. *Chemical Engineering Journal*, *493*, 152687.

Auriemma, R. S., Scairati, R., Del Vecchio, G., Liccardi, A., Verde, N., Pirchio, R., Pivonello, R., Ercolini, D., & Colao, A. (2021). The vaginal microbiome: A long urogenital colonization throughout woman life. *Frontiers in Cellular and Infection Microbiology*, *11*, 686167.

Baquero, F., Saralegui, C., Marcos-Mencía, D., Ballestero, L., Vañó-Galván, S., Moreno-Arrones, Ó. M., & Del Campo, R. (2021). Epidermis as a platform for bacterial transmission. *Frontiers in Immunology*, *12*, 774018.

Be'er, A., & Ariel, G. (2019). A statistical physics view of swarming bacteria. *Movement Ecology*, *7*, 9. https://doi.org/10.1186/s40462-019-0153-9

Bravo, P., Lung Ng, S., MacGillivray, K. A., Hammer, B. K., & Yunker, P. J. (2023). Vertical growth dynamics of biofilms. *Proceedings of the National Academy of Sciences of the United States of America*, *120*(11), e2214211120. https://doi.org/10.1073/pnas.2214211120

Buret, A. G., & Allain, T. (2023). Gut microbiota biofilms: From regulatory mechanisms to therapeutic targets. *The Journal of Experimental Medicine*, *220*(3), e20221743. https://doi.org/10.1084/jem.20221743

Buret, A. G., Motta, J. P., Allain, T., Ferraz, J., & Wallace, J. L. (2019). Pathobiont release from dysbiotic gut microbiota biofilms in intestinal inflammatory diseases: a role for iron? *Journal of Biomedical Science*, *26*(1), 1.

Buzalewicz, I., Kaczorowska, A., Fijałkowski, W., Pietrowska, A., Matczuk, A. K., Podbielska, H., Wieliczko, A., Witkiewicz, W., & Jędruchniewicz, N. (2024). Quantifying the dynamics of bacterial biofilm formation on the surface of soft contact lens materials using digital holographic tomography to advance biofilm research. *International Journal of Molecular Sciences*, *25*(5), 2653. https://doi.org/10.3390/ijms25052653

Characklis, W. G., Trulear, M. G., Bryers, J. D., & Zelver, N. (1982). Dynamics of biofilm processes: Methods. *Water Research*, *16*(7), 1207–1216.

Chee, W. J. Y., Chew, S. Y., & Than, L. T. L. (2020). Vaginal microbiota and the potential of Lactobacillus derivatives in maintaining vaginal health. *Microbial Cell Factories*, *19*(1), 203.

Cogen, A. L., Nizet, V., & Gallo, R. L. (2008). Skin microbiota: A source of disease or defence? *British Journal of Dermatology*, *58*, 442–455. https://doi.org/10.1111/j.1365-2133.2008.08437.x

Copeland, R., & Yunker, P. J. (2025). Biofilm vertical growth dynamics are captured by an active fluid framework. *bioRxiv*, 2025-03.

Dhall, A., Ramjee, R., Oh, M. J., Tao, K., & Hwang, G. (2022). A 3D-printed customizable platform for multiplex dynamic biofilm studies. *Advanced Materials Technologies*, *7*(7), 2200138.

Dong, J., & Klumpp, S. (2018). Simulation of colony pattern formation under differential adhesion and cell proliferation. *Soft Matter*, *14*, 1908–1916. https://doi.org/10.1039/C7SM02451G

Donlan, R. M. (2002). Biofilms: Microbial life on surfaces. *Emerging Infectious Diseases*, *8*, 881–890.

Farrell, F. D. C., Hallatschek, O., Marenduzzo, D., & Waclaw, B. (2013). Mechanically driven growth of quasi-two-dimensional microbial colonies. *Physical Review Letters*, *111*, 168101.

Findley, K., Oh, J., Yang, J., Conlan, S., Deming, C., Meyer, J. A., Schoenfeld, D., Nomicos, E., Park, M., NIH Intramural Sequencing Center Comparative Sequencing Program, Kong, H. H., & Segre, J. A. (2013). Topographic diversity of fungal and bacterial communities in human skin. *Nature*, *498*(7454), 367–370. https://doi.org/10.1038/nature12171

Foulongne, V., Sauvage, V., Hebert, C., Dereure, O., Cheval, J., Gouilh, M. A., Pariente, K., Segondy, M., Burguière, A., Manuguerra, J. C., Caro, V., & Eloit, M. (2012). Human skin microbiota: high diversity of DNA viruses identified on the human skin by high throughput sequencing. *PloS One*, *7*(6), e38499. https://doi.org/10.1371/journal.pone.0038499

Gallo R. L. (2017). Human skin is the largest epithelial surface for interaction with microbes. *The Journal of Investigative Dermatology*, *137*(6), 1213–1214. https://doi.org/10.1016/j.jid.2016.11.045

Gerlee, P., & Anderson, A. R. A. (2007). Stability analysis of a hybrid cellular automaton model of cell colony growth. *Physical Review E*, *75*, 051911. https://doi.org/10.1103/PhysRevE.75.051911

Ghosh, P., Mondal, J., Ben-Jacob, E., & Levine, H. (2015). Mechanically-driven phase separation in a growing bacterial colony. *Proceedings of the National Academy of Sciences*, *112*, E2166–E2173. https://doi.org/10.1073/pnas.1504948112

Gong, F., Xin, S., Liu, X., He, C., Yu, X., Pan, L., Zhang, S., Gao, H., & Xu, J. (2024). Multiple biological characteristics and functions of intestinal biofilm extracellular polymers: Friend or foe? *Frontiers in Microbiology*, *15*, 1445630.

Grice, E. A., & Segre, J. A. (2011). The skin microbiome. *Nature Reviews. Microbiology*, *9*(4), 244–253. https://doi.org/10.1038/nrmicro2537

Hardy, L., Cerca, N., Jespers, V., Vaneechoutte, M., & Crucitti, T. (2017). Bacterial biofilms in the vagina. *Research in Microbiology*, *168*(9–10), 865–874.

Hou, K., Wu, Z. X., Chen, X. Y. et al. (2022). Microbiota in health and diseases. *Signal Transduction and Targeted Therapy*, *7*, 135. https://doi.org/10.1038/s41392-022-00974-4

Ishikawa, T., Omori, T., & Kikuchi, K. (2020). Bacterial biomechanics—from individual behaviors to biofilm and the gut flora. *APL Bioengineering*, *4*(4), 041504-1–041504-14.

Iwase, T., Uehara, Y., Shinji, H., Tajima, A., Seo, H., Takada, K., Agata, T., & Mizunoe, Y. (2010). Staphylococcus epidermidis Esp inhibits Staphylococcus aureus biofilm formation and nasal colonization. *Nature*, *465*(7296), 346–349. https://doi.org/10.1038/nature09074

Jandl, B., Dighe, S., Baumgartner, M., Makristathis, A., Gasche, C., & Muttenthaler, M. (2024). Gastrointestinal biofilms—Endoscopic detection, disease relevance and therapeutic strategies. *Gastroenterology*, *167*(6), 1098–1112.

Jha, P. K., Dallagi, H., Richard, E., Deleplace, M., Benezech, T., & Faille, C. (2022). Does the vertical vs horizontal positioning of surfaces affect either biofilm formation on different materials or their resistance to detachment? *Food Control*, *133*, 108646.

Kim, J., & Lee, S. J. (2024). Digital in-line holographic microscopy for label-free identification and tracking of biological cells. *Military Medical Research*, *11*(1), 38.

Lucas-Barbosa, D., DeGennaro, M., Mathis, A., & Verhulst, N. O. (2022). Skin bacterial volatiles: Propelling the future of vector control. *Trends in Parasitology*, *38*(1), 15–22.

Mimura, M., Sakaguchi, H., & Matsushita, M. (2000). Reaction–diffusion modelling of bacterial colony patterns. *Physica A*, *282*, 283–303. https://doi.org/10.1016/S0378-4371(00)00085-6

Motta, J. P., Palese, S., Giorgio, C., Chapman, K., Denadai-Souza, A., Rousset, P., Sagnat, D., Guiraud, L., Edir, A., Seguy, C., Alric, L., Bonnet, D., Bournet, B., Buscail, L., Gilletta, C., Buret, A. G., Wallace, J. L., Hollenberg, M. D., Oswald, E., Barocelli, E., Le Grand, S., Le Grand, B., Deraison, C., & Vergnolle, N. (2021). Increased mucosal thrombin is associated with Crohn's disease and causes inflammatory damage through protease-activated receptors activation. *Journal of Crohn's and Colitis*, *15*(5), 787–799.

Parks, D. (2022). The vaginal microbiome. In Dylan Parks (Ed.), *Microbiomes: Health and the Environment*. Mavs Open Press. https://uta.pressbooks.pub/microbiomeshealthandtheenvironment/chapter/the-vaginal-microbiome/

Paula, A. J., Hwang, G., & Koo, H. (2020). Dynamics of bacterial population growth in biofilms resemble spatial and structural aspects of urbanization. *Nature Communications*, *11*, 1354.

Pokhrel, A. R., Steinbach, G., Krueger, A., Day, T. C., Tijani, J., Bravo, P., Ng, S. L., Hammer, B. K., & Yunker, P. J. (2024). The biophysical basis of bacterial colony growth. *Nature Physics*, *20*(9), 1509–1517.

Postek, W., Staśkiewicz, K., Lilja, E., & Wacław, B. (2024). Substrate geometry affects population dynamics in a bacterial biofilm. *Proceedings of the National Academy of Sciences of the United States of America*, *121*(17), e2315361121. https://doi.org/10.1073/pnas.2315361121

Ray R. R. (2024). Biofilm architecture and dynamics of the oral ecosystem. *Biotechnologia*, *105*(4), 395–402. https://doi.org/10.5114/bta.2024.145259

Reuben, R. C., Sarkar, S. L., & Jahid, I. K. (2022). Multi-species biofilms: Friends or foe to the human? In Dijendra Roy (Ed.), *A Complete Guidebook on Biofilm Study* (pp. 327–359). Academic Press.

Samaranayake, L., & Matsubara, V. H. (2017). Normal oral flora and the oral ecosystem. *Dental Clinics*, 61(2), 199–215. https://doi.org/10.1016/j.cden.2016.11.002

SanMiguel, A., & Grice, E. A. (2014). Interactions between host factors and the skin microbiome. *Cellular and Molecular Life Sciences: CMLS*, *72*(8), 1499–1515. https://doi.org/10.1007/s00018-014-1812-z

Saxena, P., Joshi, Y., Rawat, K., & Bisht, R. (2019). Biofilms: Architecture, resistance, quorum sensing and control mechanisms. *Indian Journal of Microbiology*, *59*(1), 3–12. https://doi.org/10.1007/s12088-018-0757-6

Sharma, S., Mohler, J., Mahajan, S. D., Schwartz, S. A., Bruggemann, L., & Aalinkeel, R. (2023). Microbial biofilm: A review on formation, infection, antibiotic resistance, control measures, and innovative treatment. *Microorganisms*, *11*(6), 1614.

Simon-Soro A., Ren Z., Krom B. P., Hoogenkamp M. A., Cabello-Yeves P. J., Daniel S. G., et al. (2022) Polymicrobial aggregates in human saliva build the oral biofilm. *MBio*, *13*(1), e00131-22. https://doi.org/10.1128/mbio.00131-22

Valiei, A., Dickson, A., Aminian-Dehkordi, J. et al. (2024). Metabolic interactions shape emergent biofilm structures in a conceptual model of gut mucosal bacterial communities. *NPJ Biofilms Microbiomes*, *10*, 99. https://doi.org/10.1038/s41522-024-00572-y

Vidakovic, L., Singh, P. K., Hartmann, R., Nadell, C. D., & Drescher, K. (2018). Dynamic biofilm architecture confers individual and collective mechanisms of viral protection. *Nature Microbiology*, *3*(1), 26–31. https://doi.org/10.1038/s41564-017-0050-1

Wang, S., Zhu, H., Zheng, G., Dong, F., & Liu, C. (2022). Dynamic changes in biofilm structures under dynamic flow conditions. *Applied and Environmental Microbiology*, *88*(22), e01072.

Wong, J. P., Fischer-Stettler, M., Zeeman, S. C., Battin, T. J., & Persat, A. (2023). Fluid flow structures gut microbiota biofilm communities by distributing public goods. *Proceedings of the National Academy of Sciences*, *120*(25), e2217577120.

4 Hormones and Quorum Sensing in Gram-Positive and Gram-Negative Bacteria

Bandita Dutta, Sedevino Ltu, and Rina Rani Ray
Department of Biotechnology, Maulana Abul Kalam Azad University of Technology, Kolkata, India

4.1 INTRODUCTION

Bacteria operate through quorum sensing (QS), a communication system driven by molecules called the autoinducers (AI) to synchronize the expression of a gene once a critical population density is reached. QS plays an important role in modulating virulence factor production and biofilm formation, allowing for coordinated pathogenicity and microbial community development (Ziegert et al., 2024). AIs, are signaling molecules that make use of bacteria to dictate a population of bacteria and conduct their behavior in a synchronized manner. These signaling molecules are also referred to as quorum-sensing (QS) pheromones (Yajima, 2016). Recently, in order to modulate bacterial pathogenicity, stress hormones are surfacing as a key modulator to combat infection epidemics in industries such as aquaculture. Among the pool of hormones, catecholamines, which belong to a neuroendocrine hormone family and include norepinephrine, epinephrine, and dopamine, are the most studied in microbial endocrinology. These hormones can exert direct influence on bacterial collective behavior by escalating their growth, pathogenicity factors like adhesins and toxins, biofilm development, and worsening infection progression. Catecholamines, the stress-inducing hormones, also weaken the overall function of the host-immune defense mechanism, which further increases pathogen growth. Catecholamines can also function as a siderophore or influence the iron transport mechanisms, which allows pathogenic bacteria to acquire iron, an essential nutrient crucial for their existence. The dual capability of bacteria in iron acquisition and the activation of pathogenic virulent genes via stress hormones permits bacteria to flourish within host cells under extreme environmental conditions. Such interplay focuses attention on the duality of catecholamines in responding to host stress while, simultaneously, fueling resilience and infectivity in bacteria (Sarkodie et al., 2019).

4.2 BIOFILM AND QUORUM SENSING

Biofilms are polymicrobial structured communities that are capable of adhering to surfaces and interfaces, encased within a self-produced extracellular matrix. Within this enclosed matrix, bacteria can seek shelter from stringent environmental stresses such as drying out, immune defense, protozoal predation, and antibiotic treatments. Research findings highlight three fundamental phases in the process of biofilm development: the first phase includes permanent adherence to a substrate; the second phase includes the microbial propagation and the secretion of the matrix; and the final phase includes the deterioration of the matrix and subsequent bacterial dispersion. This biorhythm permits bacteria to harness persistence to survive challenging environments and proliferate into new territory (O'Neill et al., 2008). Biofilm development is closely linked to quorum sensing (QS), a communication mechanism used by bacterial communities that activates when the population density

DOI: 10.1201/9781003519041-4

reaches a critical threshold. During the initial phase, when a bacterium is freely motile, bacteria are unable to acquire sufficient QS signal molecules. However, when neighboring bacteria accumulate and start to multiply, creating abundant micro-colonies, the bacterial population proliferates. This permits the activation of QS signals, enhancing a regulated development and growth of biofilm to further mature and, ultimately, disperse in a synchronized manner. Studies support this concept of QS, in which numerous bacteria depend on communication and synchronized biofilm dispersion. The dispersion stage is, however, the most crucial phase in determining bacterial survival because this phase allows bacteria to escape from a nutrient-depleted habitat and other waste-filled niches devoid of nutrients to establish on a new favorable site. Bacteria use techniques that involve stopping the synthesis of matrix components, breaking the noncovalent connections holding matrix materials together, or enzymatically breaking down an existing matrix to accomplish dispersion. Together, these processes ensure a smooth transition from a community of sessile biofilm cells to cells that move freely and disperse to infect. At minimal bacterial population densities, AI levels can be too weak to trigger a response. However, with the increase in population, AIs tend to begin accumulating up to a critical threshold level; they bind to, and ultimately activate, a transcriptional regulator. This regulator then triggers the expression of target genes, many of which encode virulence proteins that increase the pathogenicity or survival of bacteria. Thus, once a sufficient cell density, and, consequently, a sufficient concentration of AI, are reached, the system functions as a density-dependent switch, enabling a synchronized gene expression. QS can be disrupted in several ways, including prohibiting the synthesis of AIs, an essential signaling molecule for QS; neutralizing or enzymatically deteriorating AI and keeping their concentration at bay from reaching the optimal threshold concentration; introducing receptor antagonists or analog to compete AI bindings, which result in hampering signal–receptor interactions; suppressing transcription regulators or silencing target genes which may be generated during QS signals. These collective strategies interfere and obstruct communication amongst bacteria, which promptly inhibit coordination within the population, preventing virulence expression and biofilm formation (Weiland-Bräuer, 2021). Another strategy for disrupting microbial communication is to degrade or break down the enzymes of QS signaling molecules. For example, acyl-homoserine lactone (AHL), a QS autoinducer, is broken down by the chief enzymes AHL lactonase, which cleave the lactone ring structure, and AHL acylase, which hydrolyze the amide bond. Other enzymes, such as AHL oxidoreductases and AHL oxidases, also inactivate communication signals. These enzymatic processes collectively suppress AHLs, keeping them from achieving the optimal threshold concentration to initiate QS-synchronized pathways (Paluch et al., 2020). AHL lactonase enzyme, whose primary role is to disrupt the QS signaling pathway, offers other beneficial roles to bacterial survivability. Apart from interfering with QS signal molecules like AHLs, some lactonase enzymes, which are generated by *Bacillus* species, embody characteristics that detoxify and neutralize harmful mycotoxins (aflatoxin B1 and zearalenone). These additional functionalities show how certain bacterial enzymes are used to tackle the environmental stress and regulate synchronized bacterial communication (Tong et al., 2015).

4.3 QUORUM-SENSING SIGNAL–RESPONSE SYSTEMS IN GRAM-NEGATIVE BACTERIA

A regulated mechanism that enables bacteria to communicate with each other while collectively altering their behavior in response to the same or mixed species and the density of the adjacent microbial colonies is known as quorum sensing (QS). This mechanism collectively functions to produce, release, and detect extracellular signaling molecules such as AI (Figure 4.1). Autoinducers then accumulate in the environment as the bacterial population grows (Papenfort & Bassler, 2016). Most Gram-negative bacteria utilize many autoinducers and display some similar characteristics in quorum-sensing systems. For instance, acyl-homoserine lactones (AHLs) and compounds derived from S-adenosylmethionine (SAM) are used as autoinducers in the sensing system, and these

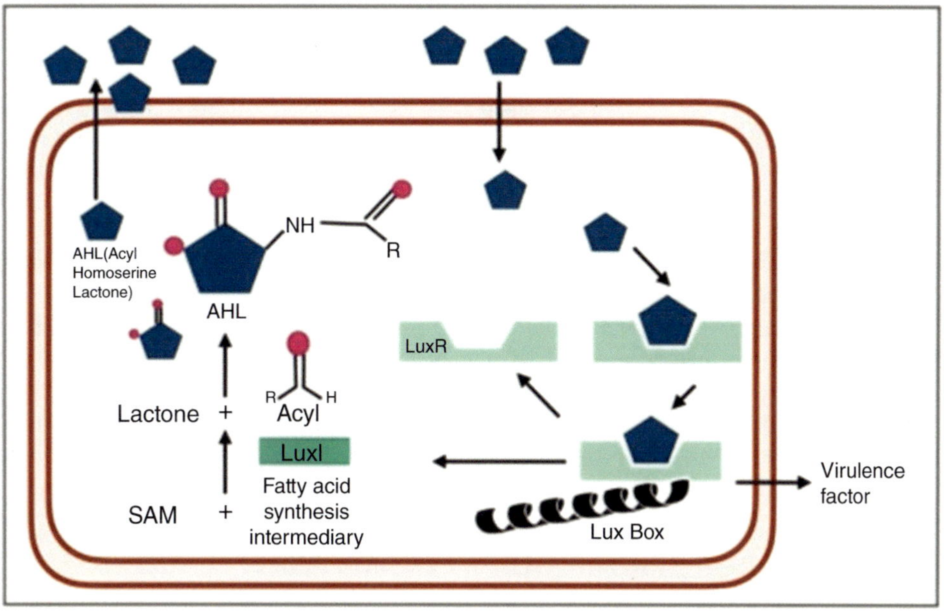

FIGURE 4.1 Quorum-sensing mechanism in Gram-negative bacteria.

autoinducers possess the capability to freely permeate the bacterial cell membrane. Autoinducers are bound by certain receptors present either in the cytoplasm or in some receptors that are found in the cytoplasm or inner membrane. QS is responsible for altering a large number of genes that support different biological operations. Autoinduction, a mechanism where AI-driven quorum sensing activation triggers autoinducer synthesis, creates a feed-forward loop that is speculated to support coordinated gene regulation within the population. Bacteria living in diverse communities are likely to come into contact with complicated combinations of autoinducers produced by their clonal offspring, near cousins, and non-kin neighbors, all of which could be bitter rivals. Therefore, the task of obtaining information from both related and unrelated sources poses a challenge for bacteria. AHLs are a particularly prevalent type of autoinducers found in Gram-negative bacteria. They consist of 4–18 carbon acyl chains, which may contain modifications and an underlying N-acylated homoserine-lactone ring. These AHLs are produced by LuxI-type synthases found in hundreds of bacterial species. AHLs are produced by LuxI enzymes, which acquire the lactone molecule from SAM (Galloway et al., 2011).

QS is carried out when bacteria recognize, process, and incorporate extracellular chemical information and further translate this information into alterations in gene expression. Bacteria accomplish these tasks through the employment of numerous autoinducers within bacterial consortia of diverse species (Elias & Banin, 2012). Additionally, the information could be affected by some external factors such as pH, temperature, osmolarity, internal factors (variations in transcript or protein numbers), or by competing bacteria that produce or consume autoinducers. Each one of these factors demands compensation. These problems are now resolved by using standard network design concepts found in quorum-sensing systems. *Pseudomonas* spp. and *Vibrio* spp. are used as examples to demonstrate these concepts.

P. aeruginosa makes use of a complex system of regulators and quorum-sensing receptors. LuxR-type receptors are the *P. aeruginosa* receptors, and, after binding of autoinducers within the cytoplasm, serve as DNA-binding transcriptional activators. *P. aeruginosa* has four recognized quorum-sensing pathways: the PqsR-controlled quinolone mechanism, the IQS system that operates in phosphate-limiting conditions, and two LuxR and LuxI-type systems known as LasR and LasI, as well as RhlR and RhlI. Autoinduction results from LasR's association with 3-oxo-C12-HSL, which activates a wide regulon of downstream genes, including the LasI synthase gene. LasR can activate rhlI and rhlR, which, in turn, encode the secondary quorum-sensing pathway (Lee & Zhang, 2015). Similarly, the genes pqsABCDH and pqsR encode the PQS system. RhlR carries similar functions to LasR and initiates the second autoinduction feed-forward loop through the activation of the RhlI receptor when bound to C4-HSL. The rhlRI, which links the three signal modules, is activated by the PqsR–PQS complex. RhlR is found to be an important quorum-sensing element in *P. aeruginosa*, playing the pivotal role of regulating the expression of virulence genes.

Quorum sensing influences the virulence component synthesis and biofilm formation (Rutherford & Bassler, 2012). Similarly, intracellular second-messenger signaling molecules, such as cyclic adenosine monophosphate (cAMP) and cyclic dimeric guanosine monophosphate (c-di-GMP), are required for biofilm formation as well as pathogenicity (Pesavento & Hengge, 2009). In *B. cenocepacia*, RpfR, a protein containing GGDEF and EAL domains, links with the DSF-family autoinducer cis-2-dodecenoic acid. When BDSF interacts with RpfR, the intracellular concentration of c-di-GMP decreases, thereby affecting the degree of virulence, biofilm formation, and swarming motility (Deng et al., 2012).

4.4 QUORUM-SENSING INHIBITION BY GRAM-POSITIVE BACTERIA

Quorum sensing (QS) is a bacterial cell-to-cell communication system that allows bacterial consortia to generate, recognize, and respond to extracellular signaling molecules known as autoinducers. Infections exhibited by Gram-positive bacteria pose a devastating health risk across the globe, and QS has been shown to influence the pathogenicity of these infections. Gram-positive and Gram-negative bacteria make use of distinct types of signaling molecules in their QS systems. Quorum sensing in Gram-positive bacteria often relies on small peptide-signaling molecules that undergo post-translational modifications. These peptide-signalling molecules interact with the sensory domain of a two-component histidine kinase signalling cascade. Two families of transcription factors with binding domains, RNPP and Rgg, have been found in Gram-positive bacteria. The RNPP protein family consists of Rap, NprR, PlcR, and PrgX. Those Gram-positive QS mechanisms where bacteria attach directly to their signalling peptides in the recipient cell are affiliated with this protein family. RNPP contributes to regulating several operations that contribute to bacterial virulence, such as sporulation, conjugation, biofilm formation, and pathogenic response.

Additionally, Rgg transcription factors, ubiquitous in many Gram-positive bacteria, function to regulate the expression of certain genes by binding to cytosolic pheromones. Rgg takes part in controlling diverse processes such as the production of glucosyltransferase in *Lactococcus lactis*, antibiotic synthesis in *Streptococcus mutans*, and secretion of proteins in *Streptococcus pyogenes*. Rgg also takes part in utilizing resources and aids in adapting to a new environment (Zheng et al., 2011). To modulate signalling swiftly, a variation in Gram-positive quorum-sensing systems requires different quorum-quenching strategies. In Gram-positive bacteria such as *S. aureus*, its pathogenicity depends upon the adhesion molecules and immune-modulating compounds, which are regulated by the quorum-sensing (QS) system. In *S. aureus*, its QS system is chiefly regulated by the agr locus, which is a gene regulator (Rutherford & Bassler, 2012). This system is composed of three major transcripts, called RNA, RNAII, and RNAIII, where RNAII encodes four essential genes, namely, agrB, agrD, agrC, and agrA (Figure 4.2). In *S. aureus*, the signaling cascade commences once the agrD gene starts encoding a 46-amino-acid peptide, subsequently modified by the AgrB protein to produce an autoinducing peptide (AIP). AgrA and AgrC operate as histidine kinases, hence creating a two-component network with an AIP-binding domain. Upon the stimulation of the two-component network, the RNAII operon is then activated and therefore serves as a transcription element for the transcript RNAIII itself (Kirchdoerfer et al., 2011). RNAIII promotes α-toxin production while suppressing the expression of coagulase, peptide A, fibronectin-binding proteins A and B, and other surface proteins. This regulatory cascade suppresses virulence from surface adhesion to the rise of

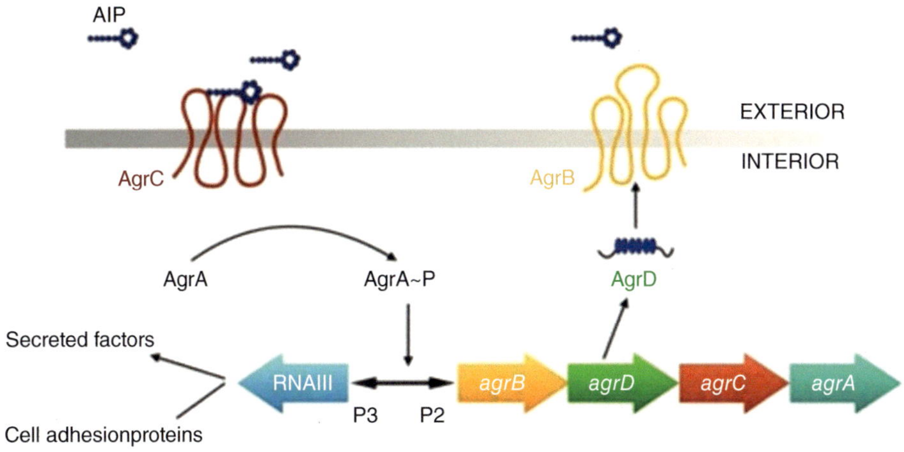

FIGURE 4.2 Quorum-sensing mechanism in Gram-positive bacteria.

toxin secretion. The agr system also controls biofilm formation in *S. aureus*, allowing *S. aureus* to transition from biofilm growth to virulence secretion as cell density increases.

Bacillus cereus, a Gram-positive pathogenic bacterium, colonizes the human intestine and causes illnesses such as diarrhea by producing virulence factors like hemolysin. The secretion of these pathogenic agents is regulated by the transcription factor PlcR in conjunction with an autoinducing peptide (AIP). The QS pathway initiates with the expression of the *PapR* gene, which encodes a 48-amino acid precursor peptide. This peptide, PapR, is exported outside the cell via a membrane transporter. Once extracellular, PapR undergoes proteolytic processing by protein B, encoded by the *NprB* gene, resulting in four peptide fragments of 5, 7, 8, and 11 amino acids. Among these, only the pentapeptide (5-amino acid) and heptapeptide (7-amino acid) fragments function as active AIPs, binding to and activating PlcR to trigger virulence gene expression. This phenomenon indicates that pathogenic traits are interlinked with the population density of bacteria (Banerjee & Ray, 2017).

S. pneumoniae, a Gram-positive pathogen, is the cause of several illnesses, including pneumonia, otitis media, and meningitis. It makes use of a 17-residue competence-stimulating peptide (CSP) to carry out quorum sensing. A pathogen's competence development occurs through two regulatory pathways: the initial phase, which is controlled by a transcription factor ComE; and the final phase, which is governed by another transcription factor ComX. These two pathways facilitate the expression of different gene sets, resulting in competence in bacteria over time. In *S. pneumoniae*, an operon known as ComCDE operates by inducing competence and biofilm formation. In *S. pneumoniae*, the ComCDE operon governs QS and virulence. The comC gene encodes precursor CSP, which is processed into its active form by ComA and ComB. The ABC transporter ComAB then exports mature CSP, triggering the QS cascade. Extracellular CSP binds to the transmembrane receptor ComD, inducing phosphorylation of the response regulator ComE. Activated ComE drives transcription of key operons (e.g., *comAB*, *comCDE*) and the *comX* gene, which encodes the sigma factor ComX and ComW proteins critical for bacterial competence and pathogenicity. This regulatory network synchronizes virulence gene expression with population density (Waters & Bassler, 2005).

4.5 NATURAL COMPOUNDS INHIBITING QUORUM SENSING

Natural compounds derived from diverse plant parts, such as roots, leaves, and fruits, possess therapeutic properties that can be structurally modified to enhance their medicinal efficacy. Studies have extensively investigated plant extracts for their intrinsic antioxidant, antibacterial, anti-inflammatory, and antiseptic activities, alongside their utility as precursors in synthesizing pharmaceutical agents.

Curcumin, a bioactive curcuminoid derived from *Curcuma longa* (a medicinal plant in the ginger family), is celebrated for its diverse pharmacological properties, including anti-Alzheimer's, anti-cancer, apoptotic, anti-inflammatory, and antioxidant effects. Recognized for its broad-spectrum antibacterial and antifungal activities, curcumin has long been utilized in traditional medicinal practices across Asia, particularly in India, China, and Southeast Asian countries. An emerging study highlights its role as a QS inhibitor due to its strong efficacy against pathogens, including *Candida albicans* and *Aeromonas hydrophila*. These potentialities have sparked research interest in curcumin's ability to disrupt biofilms caused by *Pseudomonas aeruginosa*, making it a potential candidate for the development of novel antimicrobial methods.

Another study also found that curcumin suppresses QS activities and biofilm formation in *P. aeruginosa* by disturbing iron homeostasis and oxidative stress responses. Their study showed that curcumin possesses the ability to alter the expression of proteins that are involved in reactive oxygen species (ROS) detoxification, iron uptake, the tricarboxylic acid (TCA) cycle, and the synthesis of pyoverdin and pyocyanin.

Cinnamaldehyde accounts for approximately 65% of cinnamon's composition and has potent antibacterial properties due to its α,β-unsaturated aldehyde group. This structural feature contributes its potential to disrupt microbial activity. Notably, cinnamaldehyde defies the traditional antibiotic resistance pathways, thereby increasing its potency as a therapeutic agent. Recent studies have extended their research beyond its antibacterial properties to further explore its potential function in suppressing microbial biofilm formation. Studies have proven that cinnamaldehyde can effectively disrupt biofilms by *Pseudomonas aeruginosa* via multiple approaches, demonstrating its effectiveness in inhibiting biofilm formation and their survival (Kosari et al., 2020). Paving the way for cinnamaldehyde as an effective alternative for combating persistent bacterial diseases. Another study demonstrated the synergistic action of cinnamaldehyde and conventional antibiotics, namely tobramycin and colistin. This collaboration showed promising results in disrupting *P. aeruginosa* biofilms by suppressing the QS mechanism. Through further research, cinnamaldehyde was found to possess the ability to downregulate critical genes (lasB, rhlA, and pqsA), suppressing QS and pairing with tobramycin promotes QS disruption. Also, when cinnamaldehyde is paired with an antibiotic (tobramycin or colistin), it facilitates effective eradication of biofilms and dispersal than single treatments alone (Topa et al., 2020).

Carvacrol, a phenolic monoterpenoid with potent antioxidant, anticancer, and antibacterial properties, is abundantly present in essential oils (EOs) of powerful herbs, namely oregano (*Origanum vulgare*), thyme (*Thymus vulgaris*), and pepperwort (*Lepidium flavum*) (Javed et al., 2021). Studies have shown commendable efficacy against *Pseudomonas aeruginosa* biofilms, suppressing biofilm formation by 74%–88% and lowering biofilm enzymatic function by up to 100% (Walczak et al., 2021). Biofilms from *Staphylococcus epidermidis* and *P. aeruginosa* were found disrupted by carvacrol through another study (Walsh et al., 2019). In *Mentha longifolia* EO, carvacrol is seen as a major component and demonstrates significant efficacy against *P. aeruginosa* alongside fungal biofilms on medical implants and devices, posing as a potent therapeutic strategy to combat device-associated infections (Pazarci et al., 2019). Another study explored the combined effects of proteolytic enzymes (pepsin and trypsin) with carvacrol to eliminate *Pseudomonas aeruginosa*-associated biofilms. These mammalian digestive enzymes break down proteins, specifically targeting those amino acid bonds in biofilm formations. When paired with carvacrol, the enzymes greatly enhanced biofilm reduction over carvacrol alone. The methodical steps involved initial treatment of biofilms with pepsin or trypsin, either individually or in tandem, followed by carvacrol treatment, displayed the most potent result, which additionally resulted in acquiring greater biomass and decreased survivability than the single enzyme treatment approach. Epifluorescence visualization confirmed this synergy: enzymes dispersed biofilm matrices, allowing carvacrol to exert a strong antimicrobial effect on exposed cells. This synergistic strategy indicates the promising potential of pairing enzymes with antimicrobial drugs to boost biofilm disruption efficacy against resistance in infections.

Eugenol or 2-methoxy-4-[2-propenyl] phenol is a phenolic aromatic molecule, procured from plant-derived essential oils (EOs), namely cloves, basil, nutmeg, bay leaves, and cinnamon (Hassan et al., 2018). Eugenol can penetrate cell membranes because of its lipophilic nature, which exerts influence on the membrane's fluidity and permeability and consequently disturbs its structural integrity. A study found that eugenol interferes with the QS system in a commensal bacterium, *Pseudomonas aeruginosa*, resulting in biofilm disruption. Another study showed eugenol's capability to lower the extracellular level of 3-oxo-C12-homoserine lactone, a QS signaling molecule, by altering the MexAB-OprM efflux pump, thereby disturbing the QS synergy. Additionally, eugenol was found capable of altering architectural structures of biofilm by declining the concentration of the EPS components, namely proteins and carbohydrates, and diminishing QS-associated proteins such as LasA and LasB, and another factor responsible for virulence, such as pyocyanin and rhamnolipid. These influences collectively alter the QS signaling pathway, degrade the EPS integrity, and restrict the production of virulence factors, thereby suppressing the overall growth of biofilm.

4.6 HORMONES AND QUORUM SENSING

A mechanism that regulates bacterial gene expression by using hormone-like autoinducers is known as quorum sensing. In many parts of the world, outbreaks of diseases such as severe diarrhoea and haemolytic uremic syndrome (HUS) are induced by pathogen enterohemorrhagic *Escherichia coli* (EHEC) serotype O157:H7 (Sperandio et al., 2002). Enterohemorrhagic *Escherichia coli* (EHEC) becomes pathogenic when it makes its way into the large intestines, causing damage to epithelial cells of the intestines, allowing attachment and effacing (AE) lesions. These lesions are identified by cytoskeleton rearrangement and microvilli damage, resulting in pedestal-like structures that anchor individual bacteria. The occurrence of attaching and effacing (AE) lesions in EHEC is controlled by the enterocyte effacement locus (LEE), which is not present in commensal *E. coli*. LEE encodes a type III secretion system, intimin adhesin, and its receptor Tir, organized into five operons (LEE1–5). The transcriptional activator Ler regulates LEE gene expression. EHEC also produces Shiga toxin (Stx), causing hemorrhagic colitis and HUS. Virulence genes, including LEE and Stx, are activated by QS, utilizing autoinducer-2 (AI-2) synthesized by the luxS gene, an enzyme, that detoxify S-adenosylmethionine (SAM) and further take part in the transition of ribose-homocysteine into homocysteine and 4,5-dihydroxy-2,3-pentanedione (precursor of AI-2) (Chen et al., 2002).

The regulation of various physiological processes, such as bioluminescence, genetic competence, antibiotic synthesis, sporulation, biofilm development, and the secretion of virulence factors, are governed by QS mechanisms (Wu & Luo, 2021). Bacteria living in the gut play a crucial role in tailoring the metabolic and immune function of the host by maintaining the varying composition of their population along the gastrointestinal tract. This mechanism supports bacterial production, detection, and response to signaling molecules, autoinducers, which, in turn, facilitate communication within the population. In the dynamic gut microbiome, the QS system aids the bacterial population in regulating their survival and spatial competence. Small bacterial masses having high cell density tend to exhibit a stronger QS response, increased autoinducer secretion, and a higher QS-mediated gene expression than those bigger bacterial aggregates which have a lower density (Gao et al., 2016).

In eukaryotic cell-to-cell interactions are performed via hormones, which are categorized into three major classes, namely, polypeptide hormones, steroid hormones, and amino acid-derived hormones, including catecholamines such as norepinephrine (NE) and epinephrine (Epi). Hormones such as NE have been shown to elevate bacterial growth (Lyte et al., 1996) and also actively get absorbed by bacteria. Research investigation shows tests with purified Epi and NE hormones at functional doses of 50 μM exerted activation of type III secretion in luxS mutants and induced LEE1 transcription. These results show the influence of host hormones in modulating bacterial signaling, thereby illustrating the significant role NE and Epi contribute to bacteria–host interactions, whilst altering their pathogenic behaviors. In *E. coli*, norepinephrine (NE) is shown to enhance the growth of bacteria by elevating their iron uptake level through enterobactin synthesis and iron uptake pathways (Burton et al., 2002). NE's role in bacterial pathogenesis goes beyond, as studies illustrate its involvement in activating virulence genes, including those responsible for encoding fimbriae (adhesion factors) and Shiga toxin. NE's dynamic influence is notable due to its dual functions of promoting bacterial growth and increasing pathogenicity, which implies not only its role in microbial behavior but also its capability to increase infection severity (Lyte et al., 1996).

The luxS system of QS is found in many bacterial species and may have evolved to assist in microflora-host signaling. Enterohemorrhagic *E. coli* (EHEC) uses this system to evade detection of its virulence. In the large intestine, where commensals like *E. coli*, *Enterococcus, Clostridium*, and Bacteroides that contain luxS are abundant (Chen et al., 2002), EHEC detects a luxS signaling molecule called AI-3. In addition to AI-3, epinephrine (Epi) obtained from the host also activates virulence genes responsible for encoding the type III secretion system and flagella. Due to their hydrophobic nature, both signals bind to the same outer membrane receptor, which then binds to various periplasmic sensor kinases. For example, QseC kinase is known to control flagella but not the LEE genes, suggesting that these two genes form separate, yet somewhat interlinked, controlled systems (Sperandio et al., 2002). These interactions allow for a combination of bacterial (AI-3) and host (Epi) signals, which leads EHEC to temporally control the expression of virulence characteristics like secretion systems and motility. This enables the optimum expression of genes during different stages of infection of specific locations in the gut. Understanding this communication between different biological kingdoms may shed light on microbial pathogenesis and host-microflora interplay.

4.6.1 Hormones and Quorum Sensing in Gram-Positive Bacteria

Signaling molecules play a major role in the pathogenesis of bacteria and the largely studied family of signaling molecule are called the autoinducer or quorum-sensing pheromones and researches have highlighted that stress hormones, specifically catecholamines (norepinephrine, epinephrine and dopamine), function effectively as a modulator affecting pathogenicity in bacteria (Sarkodie et al., 2019).

Catecholamines elicit biofilm development in a range of bacterial species. Initial studies used catecholamines like norepinephrine (NE) and dobutamine with *Staphylococcus epidermidis*. In the presence of human plasma as the nutrient source, catecholamines significantly increased biofilm growth on polystyrene and silicone surfaces in correspondence with increased exopolysaccharide synthesis. The findings speculate that catecholamine-induced biofilm proliferation contributes to colonization and infection in intravascular catheters, suggesting a potential etiological connection between catecholamines and bacterial pathogenesis associated with medical devices (Lyte et al., 1996).

Catecholamines exhibit a strain-specific capability to influence biofilms. For example, catecholamines have been found to enhance biofilm formation and adherence in both pathogenic (CFU series) and probiotic (HR series) strains of *Enterococcus faecalis* (Cambronel et al., 2020). Furthermore, reactions to Epi and dopamine in *Staphylococcus lugdunensis, S. epidermidis*, and *S. aureus* depend on the type of strain, resulting in biofilm development either increasing or decreasing depending on the strain. Other examples of catecholamines stimulating biofilm dispersion include Epi and NE in *Mannheimia haemolytica* (Pillai et al., 2018). Another study elicited that NE disrupts *Cutibacterium acnes* biofilms, particularly in the presence of either iron-bound transferrin or free iron. The role of catecholamines in mixed biofilm dynamics is significant for our understanding of bacterial competitiveness, which remains a relatively unstudied area. A separate study investigated the effects of norepinephrine (NE) on monoculture and co-culture biofilms of two commensal bacteria, *Staphylococcus epidermidis* and *S. aureus*, and results demonstrated that NE significantly modulated biofilm production for both of these species, with effects in aerobic and anaerobic conditions. In particular, *S. epidermidis* outcompeted *S. aureus* in dual-species biofilms, which was enhanced by NE, emphasizing its function in altering staphylococcal interactions. This study underscores the necessity to investigate catecholamines in complex microbial communities to provide a better understanding of ecological competitive mechanisms.

In aquaculture, infections generally occur in the context of stress, especially when catecholamines cause an increase in bacterial virulence. Catecholamines can raise virulence by suppressing the host's immune defenses against the pathogen and by increasing bacterial growth, adhesion, biofilm formation, and expression of virulence factors (e.g., adhesins) that enhance infection severity—all very practical examples of virulence developed by the pathogen (Sarkodie et al., 2019). Catecholamines are nitrogen-containing organic molecules derived from the amino acid L-tyrosine. Their synthesis begins with tyrosine hydroxylase, which catalyzes the hydroxylation of L-tyrosine to produce L-dopa (L-dihydroxyphenylalanine). L-dopa is then decarboxylated by Dopa-decarboxylase into dopamine. Subsequent enzymatic reactions convert dopamine: dopamine β-hydroxylase produces norepinephrine (NE), which is then methylated by phenylethanolamine-N-methyltransferase to yield epinephrine (Epi). Bacterial signaling primarily relies on two-component systems, where a histidine kinase sensor detects environmental cues, undergoes autophosphorylation, and transfers the phosphate group to an aspartate residue on a response regulator. This phosphorylation activates the regulator, typically enabling it to function as a transcription factor. A study demonstrated that the QseC sensor kinase acts as a bacterial receptor for host-derived epinephrine (Epi) and norepinephrine (NE). QseC, part of the QseBC two-component system, was initially linked to quorum sensing and regulates flagella biosynthesis and motility in enterohemorrhagic *E. coli* (EHEC) (Clarke et al., 2006). The QseC protein includes a periplasmic sensor domain, two transmembrane domains, and a kinase domain. The response regulator QseB features a receptor domain and a helix-turn-helix (HTH) domain. Upon activation, QseC transfers a phosphate group to QseB, which subsequently modulates its transcription and that of genes involved in flagella and motility (Clarke & Sperandio 2005).

More recently, these compounds have been exhibited in the physiological alteration of several Gram-positive bacteria, including *Enterococcus faecalis*. The stress-associated mammalian hormones, adrenaline and norepinephrine, activate a signaling cascade in *E. coli* and *S. enterica* by interacting with the QseC histidine sensor kinase protein.

Hormonal drugs modulate bacterial growth and biofilm formation, a correlation previously noted. A study showed that drug-induced shifts in bacterial growth typically align with biofilm changes. However, *Lactobacillus reuteri* displays reduced biofilm formation under sex steroids despite increased growth; this occurred potentially due to steroid-triggered biosurfactant production in *Lactobacillus* species, which inhibits biofilm (Clabaut et al., 2021). Progesterone may further influence bacterial phenotypes by altering quorum sensing, emphasizing hormones' role in microbial communication-dependent behaviors (Cadavid et al., 2018).

4.6.2 Hormones and Quorum Sensing in Gram-Negative Bacteria

In the QS system of Gram-negative bacteria, autoinducer (AI) molecules (Ziegert et al., 2024) N-acylhomoserine lactones (AHLs) are the predominant signaling molecules (Yajima, 2016). AHLs accumulate to coordinate gene expression once a population threshold is reached. This AI-driven process enables precise, density-dependent regulation of bacterial behaviors, serving as a highly efficient mechanism for activities like virulence or biofilm formation. Bacteria can actively sense their population density by measuring diffusible signaling molecule levels. When bacterial population density is low, the quorum-sensing (QS) pheromone concentration gets reduced, limiting bacterial activity. However, when population density grows and QS pheromone levels exceed a crucial threshold, bacteria then switch themselves to a coordinated group behaviour via a synchronized gene expression change. Quorum-sensing (QS) signaling molecules operate in a cell density-dependent approach, which requires an optimal concentration threshold to initiate coordinated group behaviours among the bacteria (Yajima, 2016).

Bacterial growth of different species shows different catecholamines' stimulating effects based on their type and concentration; for instance, hormones such as norepinephrine (NE), epinephrine (Epi), and dopamine (Dopa) show an ability to stimulate the proliferation of Gram-negative bacteria like *Escherichia coli, Yersinia enterocolitica*, and *Pseudomonas aeruginosa*. The virulence factor of bacteria is derived from the dynamic host–pathogen interactions assist to determine the outcome. (see Orji et al., 2018). Epi and NE both activate the QseC sensor kinase in E. coli K-12 MC1000, which causes the biofilm thickness to increase on polyvinyl chloride surfaces (Yang et al., 2014). Siderophores and ferric iron transport mechanisms have identified roles in catecholamine-mediated biofilm regulation (Feraco et al., 2016). In *P. aeruginosa* H103, flow-cell studies on the surfaces of glass under dynamic conditions revealed that biofilms showed little structural change when exposed to Epi. The Epi at 10 μM, however, significantly increased biofilm thickness and volume compared to untreated controls (Cambronel et al., 2019). These findings highlight the strain- and context-dependent roles of catecholamines in bacterial biofilm dynamics (Borrel et al., 2019). Epi and NE have also been shown to induce biofilm dispersal in *Mannheimia haemolytica* (Pillai et al., 2018). Catecholamine inotropes also improved the recovery of *Pseudomonas aeruginosa* after exposure to tobramycin (Freestone et al., 2012). The growth-stimulatory effect of catecholamines stems from their catechol group, which forms complexes with iron bound to host proteins such as beta globulin or lactoferrin (Lf). This weakens iron (Fe) binding, enabling bacteria to scavenge otherwise inaccessible host iron stores. Paradoxically, while catecholamines aid vertebrates in short-term survival during stress (e.g., "fight-or-flight" responses), their promotion of bacterial growth and biofilm formation sabotages long-term host survival. By avoiding the innate immune strategy of iron sequestration—a critical nutrient-limiting defense—catecholamines inadvertently enhance bacterial resilience, tipping the balance in favor of pathogens (Feraco et al., 2016). Stress hormones norepinephrine (NE) and dopamine (Dopa), elucidating the "fight-or-flight" response, stimulate growth in bacterial species such as *Escherichia coli*, *Yersinia enterocolitica*, *Listeria monocytogenes*, *Aeromonas hydrophila*, *Pseudomonas aeruginosa*, and *Vibrio parahaemolyticus* under serum-rich conditions (Sarkodie et al., 2019).

The hormones epinephrine and dynorphin have also been shown to stimulate QS in *Escherichia coli* and *P. aeruginosa*, respectively (Kim et al., 2020). A study revealed that *Pseudomonas aeruginosa* in the gut detects host physiological stress via dynorphin, an opioid-like compound released during severe stress. Upon exposure to dynorphin, *P. aeruginosa* activates quorum sensing—a bacterial communication system—to trigger toxin production and suppress beneficial probiotic bacteria critical for intestinal health. This stress-responsive virulence mechanism enables the pathogen to enhance its harmfulness by exploiting host stress signals, illustrating how bacterial pathogens dynamically adapt to host environments.

A study suggests that enterohemorrhagic *E. coli* (EHEC) employs QS to detect its intestinal environment and activate colonization genes. While EHEC utilizes the LuxS/autoinducer-2 (AI-2) system—a key player in interspecies bacterial communication, researchers found that even *luxS* mutants, though unable to produce AI-2, still triggered virulence gene expression via host-derived signals. This signal was identified as epinephrine, with bacterial responses blocked by adrenergic antagonists (alpha/beta blockers). Further experiments revealed that AI-2 (a LuxS-dependent furanosyl borate diester) was not involved; instead, EHEC produces a novel autoinducer, AI-3, whose synthesis also requires LuxS. This highlights cross-kingdom signaling between the bacterial LuxS/AI-3 QS system and host epinephrine pathways. The findings imply QS may serve as a shared "language" for communication between bacteria and eukaryotic hosts, bridging hormonal and microbial signaling mechanisms (Sperandio et al., 2002).

Human sex hormones, including estradiol and estriol, modulate QS in *Pseudomonas aeruginosa*. A study by Vidaillac and coworkers found how physiological concentrations of sex steroids influence membrane stress and virulence, contrasting earlier findings by Beury-Cirou et al. (2013), which demonstrated that supra-physiological doses of estrogens (estradiol, estrone, estriol) reduced QS signal molecules (C4-HSL, 3OC12-HSL), indicating suppression of QS pathways. On the other hand, Vidaillac et al. (2020) observed that nanomolar (physiological) levels of testosterone and estriol enhanced the virulence in *P. aeruginosa* PAO1 during infection models. These divergent results suggest a dose-dependent role of sex steroids, inhibiting QS at high concentrations but promoting virulence at physiological levels. This duality implies potential cross-talk or competition between bacterial QS systems and host endocrine signaling, highlighting how hormonal environments may shape microbial pathogenicity.

Klebsiella pneumoniae, an opportunistic pathogen, causes severe infections such as pneumonia, meningitis, bloodstream infections, and surgical site infections (McConville et al., 2017). Like many bacteria, *K. pneumoniae* utilizes QS to coordinate virulence factors, biofilm formation, and antibiotic resistance in both planktonic (free-floating) and biofilm-associated populations (Abraham, 2016). A study investigated whether certain drugs induce biofilm formation in *Klebsiella pneumoniae* ATCC 13884, potentially exacerbating bacterial resistance. In vitro assays revealed that acetaminophen, hydrochlorothiazide, and progesterone significantly enhanced biofilm production. Acetaminophen altered quorum sensing (QS) dynamics by boosting the synthesis of the signaling molecule N-hexanoyl-homoserine lactone (C6-HSL), increasing bacterial populations by 27%, biofilm formation by 49%, and fostering gentamicin-resistant biofilms while promoting urethral catheter colonization. Hydrochlorothiazide exhibited the strongest biofilm induction (51%), whereas progesterone primarily drove bacterial mass adherence without affecting population growth. These results illustrate how antibiotic drugs may inadvertently elevate resistance in bacteria by modulating biofilm-related pathways (Cadavid et al., 2018).

Enteric bacterial pathogens exploit host stress hormones, such as catecholamines, to enhance their virulence and survival. Hormones, which are secreted during host stress, modify host–pathogen dynamics by regulating bacterial pathogenicity, motility, and infection susceptibility (Freestone et al., 2012). For instance, pathogens like *Helicobacter pylori*, *Vibrio* spp., *Klebsiella pneumoniae*, *Pseudomonas aeruginosa*, *Escherichia coli*, and *Staphylococcus aureus* leverage catecholamines to boost growth, motility, and virulence gene expression (Moreira et al., 2016). In *P. aeruginosa* PA14, norepinephrine induces the Las QS system, thereby increasing virulence. This example, and others throughout, demonstrate how stress hormones can be used as conversion signals for infectious pathogens to target host physiology, illustrating their significance in infections.

A new study illustrates that microbes can interact with hormones produced by humans, suggesting human microbiomes' interaction with the host. These signaling molecules involved in their interaction pose multiple impacts on bacteria, influencing their growth, metabolism, and pathogenicity in a species-specific manner. A well-studied hormone, catecholamines, for instance, are stress hormones comprising epinephrine, norepinephrine, and dopamine. They serve as siderophores, activating virulence mechanisms and altering iron availability, which, in turn, hinders or boosts bacterial growth. For example, in Gram-negative bacteria, epinephrine and norepinephrine activate the QseBC quorum-sensing system, which increases their pathogenicity. Alongside catecholamines, other hormones also influence the microbiome's composition and function. This interplay emphasizes the vital importance of including hormonal impacts in studies on microbiome–host interactions and their consequences for human health (Luqman, 2023).

4.7 CONCLUSION

The bacterial quorum sensing dynamic with host-derived hormones shows an interlinking connection of cross-kingdom interplay that influences both the pathogenicity of microbes as well as the infection outcomes. Hormones play a crucial role in regulating bacterial growth as well as virulence. Several studies elicit stress hormones such as catecholamines (norepinephrine, epinephrine) and sex steroids (estradiol, testosterone), showcasing their capabilities to alter bacterial behavioral patterns which can potentially contribute to increased virulence, promoting biofilm formation as well as iron uptake while suppressing the host's immunological barrier system, placing the human immune system at risk. QS systems, such as LuxS/AI-3 in *E. coli* and LasR/RhlR in *P. aeruginosa*, function as molecular highways, allowing pathogens to exploit host stress signals to facilitate virulence expression. Contrariwise, hormonal and QS-disrupting approaches (enzymatic degradation of autoinducers, nanoparticle-mediated drug delivery) present promising strategies for addressing antimicrobial resistance (AMR) through a novel, multifaceted approach to target biofilm resilience and bacterial communication. The dose and context-dependent effects of hormones emphasize the complexities of host–microbe interactions, demanding a prospect for a more targeted approach to AMR mitigation in agriculture and healthcare.

REFERENCES

Abraham, W. R. (2016). Going beyond the control of quorum-sensing to combat biofilm infections. *Antibiotics*, *5*(1), 3. https://doi.org/10.3390/antibiotics5010003

Banerjee, G., & Ray, A. K. (2017). Quorum-sensing network-associated gene regulation in Gram-positive bacteria. *Acta Microbiologica et Immunologica Hungarica*, *64*(4), 439–453. https://doi.org/10.1556/030.64.2017.040

Beury-Cirou, A., Tannieres, M., Minard, C., Soulere, L., Rasamiravaka, T., Dodd, R. H., Queneau, Y., Dessaux, Y., Guillou, C., Vandeputte, O. M., & Faure, D. (2013). At a supra-physiological concentration, human sexual hormones act as quorum-sensing inhibitors. *PLoS ONE*, *8*(12), e83564. https://doi.org/10.1371/journal.pone.0083564

Borrel, V., Thomas, P., Catovic, C., Racine, P. J., Konto-Ghiorghi, Y., Lefeuvre, L., et al. (2019). Acne and stress: Impact of catecholamines on *Cutibacterium acnes*. *Frontiers in Medicine*, *6*, 155. https://doi.org/10.3389/fmed.2019.00155

Burton, C. L., Chhabra, S. R., Swift, S., Baldwin, T. J., Withers, H., Hill, S. J., & Williams, P. (2002). The growth response of *Escherichia coli* to neurotransmitters and related catecholamine drugs requires a functional enterobactin biosynthesis and uptake system. *Infection and Immunity*, *70*(11), 5913–5923. https://doi.org/10.1128/IAI.70.11.5913-5923.2002

Cadavid, E., Robledo, S. M., Quiñones, W., & Echeverri, F. (2018). Induction of biofilm formation in *Klebsiella pneumoniae* ATCC 13884 by several drugs: The possible role of quorum sensing modulation. *Antibiotics*, *7*(4), 103. https://doi.org/10.3390/antibiotics7040103

Cambronel, M., Nilly, F., Mesguida, O., Boukerb, A. M., Racine, P. J., Baccouri, O., et al. (2020). Influence of catecholamines (epinephrine/norepinephrine) on biofilm formation and adhesion in pathogenic and probiotic strains of *Enterococcus faecalis*. *Frontiers in Microbiology*, *11*, 1501. https://doi.org/10.3389/fmicb.2020.01501

Cambronel, M., Tortuel, D., Biaggini, K., Maillot, O., Taupin, L., Rehel, K., et al. (2019). Epinephrine affects motility, and increases adhesion, biofilm and virulence of *Pseudomonas aeruginosa* H103. *Scientific Reports*, *9*, 20203. https://doi.org/10.1038/s41598-019-56666-7

Chen, X., Schauder, S., Potier, N., Van Dorsselaer, A., Pelczer, I., Bassler, B. L., & Hughson, F. M. (2002). Structural identification of a bacterial quorum-sensing signal containing boron. *Nature*, *415*(6871), 545–549. https://doi.org/10.1038/415545a

Clabaut, M., Suet, A., Racine, P. J., Tahrioui, A., Verdon, J., Barreau, M., et al. (2021). Effect of 17β-estradiol on a human vaginal *Lactobacillus crispatus* strain. *Scientific Reports*, *11*, 7133. https://doi.org/10.1038/s41598-021-86628-x

Clarke, M. B., Hughes, D. T., Zhu, C., Boedeker, E. C., & Sperandio, V. (2006). The QseC sensor kinase: A bacterial adrenergic receptor. *Proceedings of the National Academy of Sciences of the United States of America*, *103*(27), 10420–10425. https://doi.org/10.1073/pnas.0604343103

Clarke, M. B., & Sperandio, V. (2005). Transcriptional regulation of *flhDC* by QseBC and sigma (FliA) in enterohaemorrhagic *Escherichia coli*. *Molecular Microbiology*, *57*(6), 1734–1749. https://doi.org/10.1111/j.1365-2958.2005.04792.x

Deng, Y., Schmid, N., Wang, C., Wang, J., Pessi, G., Wu, D., et al. (2012). *Cis*-2-dodecenoic acid receptor RpfR links quorum-sensing signal perception with regulation of virulence through cyclic dimeric guanosine monophosphate turnover. *Proceedings of the National Academy of Sciences*, *109*(38), 15479–15484. https://doi.org/10.1073/pnas.1205037109

Elias, S., & Banin, E. (2012). Multi-species biofilms: Living with friendly neighbors. *FEMS Microbiology Reviews*, *36*(5), 990–1004. https://doi.org/10.1111/j.1574-6976.2012.00325.x

Feraco, D., Blaha, M., Khan, S., Green, J. M., & Plotkin, B. J. (2016). Host environmental signals and effects on biofilm formation. *Microbial Pathogenesis*, *99*, 253–263. https://doi.org/10.1016/j.micpath.2016.08.015

Freestone, P. P., Hirst, R. A., Sandrini, S. M., Sharaff, F., Fry, H., Hyman, S., & O'Callaghan, C. (2012). *Pseudomonas aeruginosa*-catecholamine inotrope interactions: A contributory factor in the development of ventilator-associated pneumonia? *Chest*, *142*(5), 1200–1210. https://doi.org/10.1378/chest.11-2614

Galloway, W. R., Hodgkinson, J. T., Bowden, S. D., Welch, M., & Spring, D. R. (2011). Quorum sensing in Gram-negative bacteria: Small-molecule modulation of AHL and AI-2 quorum sensing pathways. *Chemical Reviews*, *111*(1), 28–67. https://doi.org/10.1021/cr100109t

Gao, M., Zheng, H., Ren, Y., Lou, R., Wu, F., Yu, W., Liu, X., & Ma, X. (2016). A crucial role for spatial distribution in bacterial quorum sensing. *Scientific Reports*, *6*, 34695. https://doi.org/10.1038/srep34695

Hassan, H. A., Geniady, M. M., Abdelwahab, S. F., Abd-Elghany, M. I., Sarhan, H. A., Abdelghany, A. A., et al. (2018). Topical eugenol successfully treats experimental *Candida albicans*-induced keratitis. *Ophthalmic Research*, *60*(2), 69–79. https://doi.org/10.1159/000488907

Javed, H., Meeran, M. F. N., Jha, N. K., & Ojha, S. (2021). Carvacrol, a plant metabolite targeting viral protease (Mpro) and ACE2 in host cells, can be a possible candidate for COVID-19. *Frontiers in Plant Science*, *11*, 601335. https://doi.org/10.3389/fpls.2020.601335

Kim, C. S., Gatsios, A., Cuesta, S., Lam, Y. C., Wei, Z., Chen, H., et al. (2020). Characterization of Autoinducer-3 structure and biosynthesis in *E. coli*. *ACS Central Science*, *6*(2), 197–206. https://doi.org/10.1021/acscentsci.9b01076

Kirchdoerfer, R. N., Garner, A. L., Flack, C. E., Mee, J. M., Horswill, A. R., Janda, K. D., Kaufmann, G. F., & Wilson, I. A. (2011). Structural basis for ligand recognition and discrimination of a quorum-quenching antibody. *Journal of Biological Chemistry*, *286*(19), 17351–17358. https://doi.org/10.1074/jbc.M111.231258

Kosari, F., Taheri, M., Moradi, A., Alni, R. H., & Alikhani, M. Y. (2020). Evaluation of cinnamon extract effects on *clbB* gene expression and biofilm formation in *Escherichia coli* strains isolated from colon cancer patients. *BMC Cancer*, *20*, 267. https://doi.org/10.1186/s12885-020-06736-1

Lee, J., & Zhang, L. (2015). The hierarchy quorum sensing network in *Pseudomonas aeruginosa*. *Protein & Cell*, *6*(1), 26–41. https://doi.org/10.1007/s13238-014-0100-x

Luqman, A. (2023). The orchestra of human bacteriome by hormones. *Microbial Pathogenesis*, *180*, 106125. https://doi.org/10.1016/j.micpath.2023.106125

Lyte, M., Frank, C. D., & Green, B. T. (1996). Production of an autoinducer of growth by norepinephrine cultured *Escherichia coli* O157:H7. *FEMS Microbiology Letters*, *139*(2–3), 155–159. https://doi.org/10.1111/j.1574-6968.1996.tb08196.x

McConville, T. H., Sullivan, S. B., Gomez-Simmonds, A., Whittier, S., & Uhlemann, A. C. (2017). Carbapenem-resistant Enterobacteriaceae colonization (CRE) and subsequent risk of infection and 90-day mortality in critically ill patients: An observational study. *PLoS ONE*, *12*, e0186195. https://doi.org/10.1371/journal.pone.0186195

Moreira, C. G., Russell, R., Mishra, A. A., Narayanan, S., Ritchie, J. M., Waldor, M. K., Curtis, M. M., Winter, S. E., Weinshenker, D., & Sperandio, V. (2016). Bacterial adrenergic sensors regulate virulence of enteric pathogens in the gut. *mBio*, *7*(3), e00826-16. https://doi.org/10.1128/mBio.00826-16

O'Neill, E., Pozzi, C., Houston, P., Humphreys, H., Robinson, D. A., Loughman, A., Foster, T. J., & O'Gara, J. P. (2008). A novel *Staphylococcus aureus* biofilm phenotype mediated by the fibronectin-binding proteins, FnBPA and FnBPB. *Journal of Bacteriology*, *190*(11), 3835–3850. https://doi.org/10.1128/JB.00167-08

Orji, F. A., Ugbogu, O. C., Ugbogu, E. A., Barbabosa-Pliego, A., Monroy, J. C., Elghandour, M. M. M. Y., & Salem, A. Z. M. (2018). Pathogenic flora composition and overview of the trends used for bacterial pathogenicity identifications. *Microbial Pathogenesis*, *121*, 139–146. https://doi.org/10.1016/j.micpath.2018.05.006

Paluch, E., Rewak-Soroczyńska, J., Jędrusik, I., Mazurkiewicz, E., & Jermakow, K. (2020). Prevention of biofilm formation by quorum quenching. *Applied Microbiology and Biotechnology*, *104*(5), 1871–1881. https://doi.org/10.1007/s00253-020-10349-w

Papenfort, K., & Bassler, B. L. (2016). Quorum sensing signal-response systems in Gram-negative bacteria. *Nature Reviews Microbiology*, *14*(9), 576–588. https://doi.org/10.1038/nrmicro.2016.89

Pazarci, O., Tutar, U., & Kilinc, S. (2019). Investigation of the antibiofilm effects of *Mentha longifolia* essential oil on titanium and stainless steel orthopedic implant surfaces. *Eurasian Journal of Medicine*, *51*(2), 128–132. https://doi.org/10.5152/eurasianjmed.2019.18432

Pesavento, C., & Hengge, R. (2009). Bacterial nucleotide-based second messengers. *Current Opinion in Microbiology*, *12*(2), 170–176. https://doi.org/10.1016/j.mib.2009.01.007

Pillai, D. K., Cha, E., & Mosier, D. (2018). Role of the stress-associated chemicals norepinephrine, epinephrine and substance P in dispersal of *Mannheimia haemolytica* from biofilms. *Veterinary Microbiology*, *215*, 11–17. https://doi.org/10.1016/j.vetmic.2017.11.025

Rutherford, S. T., & Bassler, B. L. (2012). Bacterial quorum sensing: Its role in virulence and possibilities for its control. *Cold Spring Harbor Perspectives in Medicine*, *2*(11), a012427. https://doi.org/10.1101/cshperspect.a012427

Sarkodie, E. K., Zhou, S., Baidoo, S. A., & Chu, W. (2019). Influences of stress hormones on microbial infections. *Microbial Pathogenesis*, *131*, 270–276. https://doi.org/10.1016/j.micpath.2019.04.013

Sperandio, V., Torres, A. G., & Kaper, J. B. (2002). Quorum sensing *Escherichia coli* regulators B and C (QseBC): A novel two-component regulatory system involved in the regulation of flagella and motility by quorum sensing in *E. coli*. *Molecular Microbiology*, *43*(3), 809–821. https://doi.org/10.1046/j.1365-2958.2002.02803.x

Tong, S. Y., Davis, J. S., Eichenberger, E., Holland, T. L., & Fowler, V. G. Jr. (2015). *Staphylococcus aureus* infections: Epidemiology, pathophysiology, clinical manifestations, and management. *Clinical Microbiology Reviews*, *28*(3), 603–661. https://doi.org/10.1128/CMR.00134-14

Topa, S. H., Palombo, E. A., Kingshott, P., & Blackall, L. L. (2020). Activity of cinnamaldehyde on quorum sensing and biofilm susceptibility to antibiotics in *Pseudomonas aeruginosa*. *Microorganisms*, *8*(3), 455. https://doi.org/10.3390/microorganisms8030455

Vidaillac, C., Yong, V. F. L., Aschtgen, M. S., Qu, J., Yang, S., Xu, G., Seng Z. J., Brown, A. C., Ali, M. K., Jaggi, T. K., & Sankaran, J. (2020). Sex steroids induce membrane stress responses and virulence properties in *Pseudomonas aeruginosa*. *mBio*, *11*(5), 10–1128. https://doi.org/10.1128/mBio.01774-20

Walczak, M., Michalska-Sionkowska, M., Olkiewicz, D., Tarnawska, P., & Warżyńska, O. (2021). Potential of carvacrol and thymol in reducing biofilm formation on technical surfaces. *Molecules*, *26*(9), 2723. https://doi.org/10.3390/molecules26092723

Walsh, D. J., Livinghouse, T., Goeres, D. M., Mettler, M., & Stewart, P. S. (2019). Antimicrobial activity of naturally occurring phenols and derivatives against biofilm and planktonic bacteria. *Frontiers in Chemistry*, *7*, 653. https://doi.org/10.3389/fchem.2019.00653

Waters, C. M., & Bassler, B. L. (2005). Quorum sensing: Cell-to-cell communication in bacteria. *Annual Review of Cell and Developmental Biology*, *21*, 319–346. https://doi.org/10.1146/annurev.cellbio.21.012704.131001

Weiland-Bräuer, N. (2021). Friends or foes: Microbial interactions in nature. *Biology*, *10*(6), 496. https://doi.org/10.3390/biology10060496

Wu, L., & Luo, Y. (2021). Bacterial quorum-sensing systems and their role in intestinal bacteria-host crosstalk. *Frontiers in Microbiology*, *12*, 611413. https://doi.org/10.3389/fmicb.2021.611413

Yajima, A. (2016). Recent advances in the chemistry and chemical biology of quorum-sensing pheromones and microbial hormones. In *Studies in Natural Products Chemistry* (Vol. 49, pp. 331–355). Elsevier. https://doi.org/10.1016/B978-0-444-63603-4.00010-3

Yang, K., Meng, J., Huang, Y. C., Ye, L. H., Li, G. J., Huang, J., et al. (2014). The role of the QseC quorum-sensing sensor kinase in epinephrine-enhanced motility and biofilm formation by *Escherichia coli*. *Cell Biochemistry and Biophysics*, *70*, 391–398. https://doi.org/10.1007/s12013-014-9924-5

Zheng, F., Ji, H., Cao, M., Wang, C., Feng, Y., Li, M., Pan, X., Wang, J., Qin, Y., Hu, F., & Tang, J. (2011). Contribution of the Rgg transcription regulator to metabolism and virulence of *Streptococcus suis* serotype 2. *Infection and Immunity*, *79*(3), 1319–1328. https://doi.org/10.1128/IAI.00193-10

Ziegert, Z., Dietz, M., Hill, M., et al. (2024). Targeting quorum sensing for manipulation of commensal microbiota. *BMC Biotechnology*, *24*, 106. https://doi.org/10.1186/s12896-024-00937-3

5 Regulatory Factors Behind Biofilm Formation

Niladri Sarkar, Sumit Nath, and Rina Rani Ray
Department of Biotechnology, Maulana Abul Kalam Azad University of Technology, Kolkata, India

5.1 INTRODUCTION

Biofilm is a three-dimensional complex structural stratum, generally produced by Gram-positive and Gram-negative bacteria. However, several fungi, archaea and algae might also form a differential biofilm layer. The formation of the biofilm consists of multiple stages including cell surface adherence, maturation, and dispersion. Bacteria are thought to be the primary producers of biofilm, although recent studies have shown fungi and algae can also form biofilm, but the structural and strategic characteristics are somewhat different in these cases. The initial attachment of the bacterial and algal biofilm are influenced by the appendages such as flagella, pili, fimbrial adhesins while electrostatic interaction, hydrophobic interaction plays in cell surface attachment in fungi (Liu et al., 2022). The different extracellular matrix is also observed in their biofilm formation. Numerous pathogenic species have been found to be capable of biofilm formation, including *Streptococcus pyogenes, Streptococcus mutans, Staphylococcus aureus, Staphylococcus epidermidis, Pseudomonas aeruginosa, Escherichia coli, Listeria monocytogenes*, and *Klebsiella pneumoniae*. Most of the Gram-negative bacteria, such as *Pseudomonas aeruginosa* and *Klebsiella pneumoniae*, synthesize Glycocalyx, which allows bacteria to bond with host cells, thereby creating micro colonies, which help in the formation of biofilms and shield the bacteria from the phagocyte system and antimicrobial substances. Another important pathogenic species is enteropathogenic *Escherichia coli* (EPEC), which is the primary cause of diarrhea in children and is linked to outbreaks of foodborne infections through the growth of biofilms. Additionally, *E. coli*-secreted biofilms are often characterized for catheter-induced chronic and persistent infections of the urinary tract (UTIs) (Mendhe et al., 2023). Simultaneously, Gram-positive bacteria exhibit another mechanism of biofilm formation, e.g. *Listeria monocytogenes*, which can survive for years in environments due to the formation of autoinducing peptides that helps them to produce biofilms.

The formation and characteristics of biofilms are driven by a complex teamwork of factors, including surface properties, hydrodynamic conditions, nutrient availability, quorum sensing, and species interactions and other environmental factors. Knowledge of these factors is pivotal for developing strategies to control and manipulate biofilm formation for beneficial or detrimental purposes. In this chapter we discuss the influential factors behind the formation and development of biofilm. It also unveils the different mechanisms of forming the biofilm of bacteria, fungi and algae.

5.2 STAGES OF BIOFILM FORMATION

Biofilm is a complex and dynamic structural layer of Gram-positive and Gram-negative bacteria. The developmental process of biofilm is often lengthy and relies on initial attachment, irreversible attachment, and maturation, including cell proliferation, differentiation and dispersion as shown in Figure 5.1. The formation of biofilm initiates with the preliminary attachment of planktonic microbial cells to the living or nonliving surfaces. Several factors, such as temperature, pH, surface

DOI: 10.1201/9781003519041-5

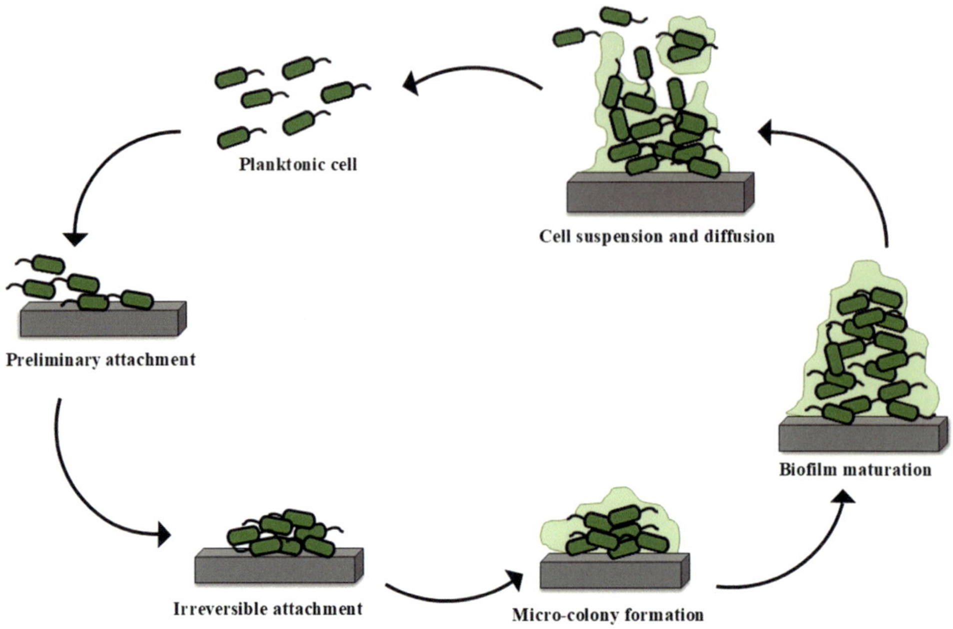

FIGURE 5.1 Schematic representation of the stages of biofilm formation.

condition, nutrient availability, and genetic influences, play a pivotal role in the initial cell–surface attachment. This initial attachment is often reversible, resulting in microorganisms detaching and exploring other potential colonization sites. Following the initial attachment, an irreversible attachment arrives while the bacterial cells are tightly adhered to the surface and metabolic activity is gradually increasing.

The engagement of several adhesins and the extracellular matrix contribute key roles to irreversible attachment. Quorum sensing (QS) is a cell-to-cell interaction process which exhibits microbial cells to sense their population and coordinate gene expression accordingly (Dutta et al., 2024). QS can trigger the expression of genes involved in EPS production, adhesin synthesis, and other biofilm-related processes. Maturation of a biofilm is a very programmed sequential process which include the irreversible attachment of cells, their growth and division, EPS accumulation, microcolony formation, coaggregation, rigid multi-layer formation, cell differentiation and gene transfer (Ghosh et al., 2022). The proximity of cells within the biofilm facilitates the transfer of genetic material, such as plasmids, between cells. This horizontal gene transfer can contribute to the spread of antibiotic resistance genes and other virulence factors within the biofilm. During the maturation process, bacteria have able to form a stable, complex structure, allowing them to counteract host defenses and antimicrobial agents, and the promotion of persistent or chronic infections. The matured sessile cells are then dispersed due to shortage of space, nutrients and looking for the new surfaces for the continual cycle.

5.3 FACTORS INFLUENCING BIOFILM FORMATION

Structured communities of microorganisms embedded in an EPS that they produce themselves are called biofilms. This matrix, which is mostly made up of proteins, lipids, polysaccharides, and nucleic acids, acts as a barrier to protect against environmental stressors such as nutrient limitation desiccation antimicrobial agents and host immune reactions. The intricate and dynamic process of biofilm formation includes sequential phases of initial irreversible attachment on a substratum, followed by microcolony formation through quorum sensing, maturation and finally cell dispersal. In order to control undesired biofilm formation in different settings it is essential to comprehend the factors that govern each stage of biofilm development.

5.3.1 Environmental Factors

Environmental factors are highly engaged in the formation and development of biofilm. Environmental parameters, such as temperature, pH, salinity, air flow, surface tension, surface quality, and nutrients, are considered to be the key influential factors in several metabolic activity and growth kinetics of microorganisms as shown in Figure 5.2.

5.3.1.1 Temperature

Temperature, a fundamental environmental parameter, has a major impact on microbial physiological and behavioral characteristics. Microorganisms express optimal growth rates within a specific temperature range. Temperature is an important factor in the regulation of enzyme activity, membrane fluidity, and gene expression. Consequently, it can improve the surface chemistry of microorganisms, the production of the extracellular matrix, and, ultimately, forming a mature biofilm (Dixit et al., 2023). However, the optimum temperature of the bacterial growth and metabolic activity depends on the species and the other environmental conditions. The diversity starts from psychrophiles, which can survive in low temperature (below 15°C) regions such as glaciers, icebergs, and mountains. This is followed by the mesophiles, bacteria which have an optimal growth temperature at 20–45°C. A wide range of pathogenic and non-pathogenic bacteria are considered to be mesophiles. In addition, another type of bacteria, thermophiles, can survive at high temperatures (45–122°C), which are found generally in hot springs, deep sea hydrothermal vents, and geothermal areas (Isnawati and Trimulyono, 2018).

Higher temperatures involve expressing surface hydrophobicity, facilitating initial adhesion via hydrophobic interactions, while lower temperatures may decrease cell motility and energy production. Specific adhesins, such as pili and fimbriae, often exhibit temperature-dependent expression, further influencing the adhesion process. The gene expression of the specialized protein involved in adaptation, stress responses, and biofilm formation has been widely affected by changes in

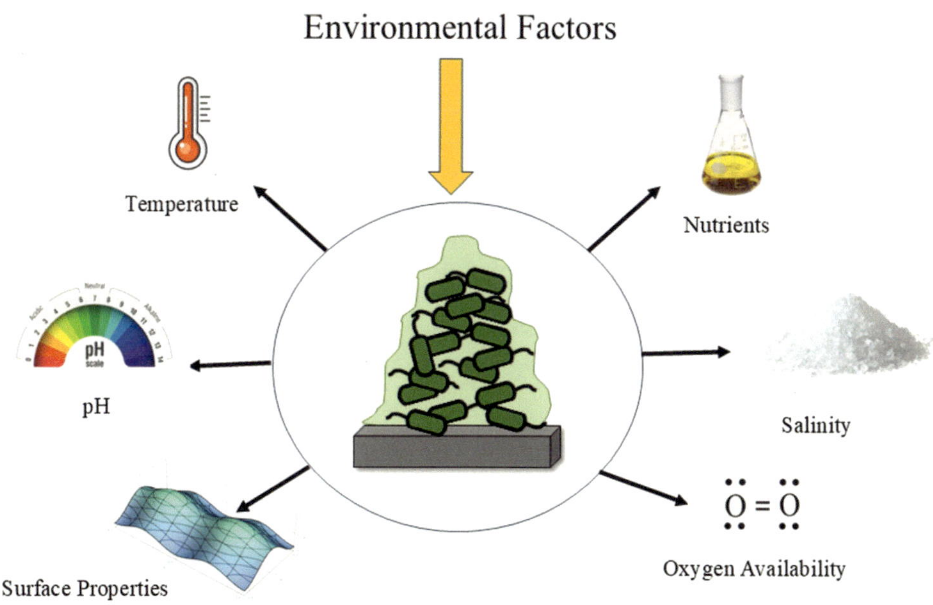

FIGURE 5.2 Environmental factors involved in formation of biofilm.

temperature. For example, heat shock and cold shock proteins provides protectivity to the biofilm, forming cells in response to the temperature stress by preventing protein denaturation and degradation. Heat shock proteins such as Hsp 60, Hsp 70 and small Hsp play a crucial role in thermophilic archaea for folding and refolding of the proteins, particularly when the proteins are damaged by high temperatures (Hu et al., 2022). Temperature is also considered as the important parameter in regulating quorum sensing and bacterial EPS production. The stability, diffusion, or degradation of QS autoinducer molecules (AI), are highly temperature-dependent which might interfere several processes such as motility, EPS production, virulence factors etc. The opportunistic bacteria such as *ESKAPE, Escherichia coli, Bacillus subtilis, Bacillus cereus* is showing their highest growth and ability to form mature multilayered biofilm at temperature ranges of 25–37°C. However, a study by Marín-Sanhueza et al. (2022) has shown that the temperature range of favorable biofilm formation for thermophilic bacilli is 50–65°C. A higher amount of biofilm was found to be produced by thermophilic *Bacillus haynesii* CamB6 (Marín-Sanhueza et al., 2022) at a temperature of 50°C.

5.3.1.2 pH

The most significant environmental condition after temperature is pH, or the acidic and alkaline nature of the medium or surfaces. Microorganisms typically express their highest growth activity within a specific pH range. pH stresses can hinder growth, or even promote cell death, by inhibiting enzyme activity and deranging the protein structure, and membrane integrity. Sub-optimal pH can also change the metabolic pathways which are directly associated with signaling molecules involved in QS and EPS production of biofilm development. pH has an impact on production of the types and relative abundance of various EPS components. Several bacteria are found to produce more acidic polysaccharides at lower pH, while others synthesize more neutral polysaccharides at higher pH. Recent studies have shown the dental biofilms (plaque) formation by *S. mutans*, leading to dental caries. This bacterium thrives in the acidic conditions produced by the fermentation of sugars. The acidic environment favors the dissolution of tooth enamel, promoting the development of cavities. Research indicates that *S. mutans*, when exposed to acidic conditions, expresses genes related to acid tolerance and EPS production to thrive in acidic medium. The EPS produced in acidic environments is often more tolerant, leading to the persistence of the biofilm. The EPS content generally encourages physical properties, such as viscosity and elasticity, which in turn lead to the overall stable construction of the biofilm. Extremes of pH can lead to the degradation or disassembly of EPS, thereby compromising the structural integrity of the biofilm. Acidic conditions can hydrolyze glycosidic bonds within polysaccharides, while alkaline conditions can cause the deprotonation of acidic polysaccharides, leading to their solubilization.

Initial attachment of the planktonic cells to the surfaces is a primary stage of biofilm formation which might be affected by the pH of the surfaces. Many bacteria depends on surface-exposed adhesins, such as fimbriae (pili) and non-fimbrial adhesins, to initiate attachment to surfaces. The structural and functional activity of these adhesins can be greatly affected by pH. Changes in pH can modulate the conformation and attraction of adhesins, enhancing their ability to bind with specific receptors. On the surface of *L. monocytogenes*, a foodborne pathogen, for example, they are potentially able to form biofilms on surfaces in food-processing instruments. It has been observed in several experimental studies that acidic pH can influence the adhesion and biofilm formation of *L. monocytogenes* on stainless steel and other surfaces. Generally, acidic conditions facilitate mostly the formation of hyphae in *C. albicans*, which are indispensable for the production of EPS, providing robustness and antifungal resistance (Tarifa et al., 2023).

5.3.1.3 Nutrient Availability

The formation of a biofilm is a dynamic process, involving complex processes such as biotic factors, including cell–cell signaling and genetic regulation, and abiotic factors, with nutrient availability. Nutrients provide the building blocks and energy required for several metabolic activities, such as cell growth, metabolism, and the production of the EPS matrix that is considered as the architecture of the biofilm. Nutrient availability plays a pivotal role in regulating the motility and surface attachment of the microorganism. Chemotaxis is the well-studied mechanism by which bacteria can move towards the chemical nutrient gradients, and it promotes better surface attachments of the motile cells. Conversely, nutrient limitation might reduce flagellar biosynthesis and motility, thereby impacting the rate of surface exploration.

Furthermost, different nutrient sources play separate roles in cell proliferation and biofilm maturation. In most of the cases, carbon sources are considered to be the primary energy sources, providing building blocks for cell growth and EPS synthesis. Specific carbon sources or growth conditions triggers the transcription of the genes associated with exopolysaccharide biosynthesis (Nie et al., 2024), regulating the quantity and composition of the EPS matrix. Similarly, nitrogen sources are essential for the synthesis of different proteins, nucleic acids and exopolysaccharides. For instance, under conditions of nitrogen starvation *Bacillus subtilis* facilitate the production of the exopolysaccharide EPS, leading to enhanced biofilm formation. Again, phosphorus is another key component of DNA, RNA, phospholipids, and ATP. Phosphorus limitation can restrict cell growth and EPS production, downregulating biofilm development. Some bacteria engage phosphate solubilization mechanisms to overcome phosphorus limitation. Other than the major elements, several trace elements e.g. iron, magnesium, manganese, cobalt, copper, zinc, are required for enzyme activation, and to catalyse the metabolism and EPS production. Several studies have shown that iron limitation may affect the growth of *Vibrio cholerae*, enhancing the production of the VPS (Vibrio polysaccharide), leading to enhanced biofilm formation (Pramanik and Vibhuti, 2022). Additionally, bivalent metallic cations, such as calcium (Ca2+) and magnesium (Mg2+), are able to link with negatively charged molecules of the EPS to develop the mechanical stability in *P. aeruginosa* and *Bacillus subtilis*.

5.3.1.4 Oxygen Availability

Oxygen (O2) is a rudimentary element for every life and their biochemical cycles. Depending on their use of oxygen, microorganisms are categorized into four classes, including obligate aerobic organisms, which can grow in the presence of oxygen, and obligate anaerobic microorganisms, which are not capable of surviving within the oxygen system. For instance, *Pseudomonas aeruginosa* is commonly found in soil and water, and also on plant surfaces, and supports aerobic respiration for its energy production. *P. aeruginosa* exhibited a well-designed electron transport chain with multiple oxidase enzymes which is helpful for adapting to various oxygen concentrations. The versatility of these bacteria assist them in colonizing diverse environments (Arai, 2011). Again, *Mycobacterium tuberculosis*, the causative agent of tuberculosis, necessitates oxygen for its survival and growth within host tissues. *Acinetobacter baumannii, Bacillus subtilis*, and *Nocardia asteroides* are among the other examples of obligate aerobic bacteria, where the bacteria of a few particular genera, namely, *Clostridium, Actinomyces, Bacteroides, Fusobacterium, Peptostreptococcus, Porphyromonas, Prevotella, Propionibacterium*, and *Veillonella*, are found to be obligate anaerobes. Further facultative anaerobes and aerotolerant microorganisms are those who can carry out their metabolic activity with or without oxygen or survive in the presence of oxygen but who do not require oxygen for respiration respectively. A large group of opportunistic bacteria such as *Staphylococcus aureus, Enterobacter spp. Enterococcus faecium, Klebsiella pneumoniae, Clostridium botulinum, Streptococcus spp., Escherichia coli* acts as facultative anaerobes, however *Lactobacillus rhamnosus, Parabacteroides distasonis*, and *Streptococcus pyogenes* are well-known aerotolerant bacteria (Santajit and Indrawattana, 2016). Also, some microorganisms can tolerate only a little amount of oxygen, called microaerophiles, such as *Helicobacter pylori*. Oxygen acts as an important electron recipient in aerobic respiration, enabling the formation of ATPs. Oxygen attainability directly impacts the metabolic pathways employed by microorganisms, significantly influencing their growth rate, nutrient acceptance, and production of EPS, which definitely facilitate microorganism to form biofilm.

5.3.1.5 Salinity

Salinity is another key environmental factor, particularly in marine, estuarine, and saline terrestrial environments. The osmotic stress formation due to enhanced salinity, resulting in the assemblage of congenial solutes within the cell to maintain turgor pressure. This osmotic stress can elicit the augmentation of genes involved in EPS biosynthesis. Furthermost, the saline profile may interfere in intracellular enzyme activity, and the overall metabolic pathways of microorganisms. Consequently, salinity plays a critical role in modeling biofilm formation dynamics, determining several stages of the process in complex and often species-specific ways. Research has shown that high salinity can modify the surface charge profile, either enhancing or inhibiting adhesion. The improved ionic strength then exhibits electrostatic repulsion between negatively charged bacterial cells and surfaces, increasing cell attachment. Certain salts can escalate the hydrophobicity of surfaces, facilitating the adhesion of hydrophobic microorganisms. Based on the increase in activity, two types of bacteria are found: halophiles, which are able to grow at different levels of salt concentration; and non-halophiles, which are unable to survive in such environments. *Erythrobacter flavus* is a halophilic bacterium which can survive in only 2–5% salinity, whereas moderate halophiles, such as *Vibrio costicola, Micrococcus halodenitrificans*, and *Bacillus halophilus*, show metabolic activity within 20–50% saline environments. Notably several bacteria, such as *Halobacterium salinarum, Haloferax mediterranei, and Haloarcula marismortui*, are considered to be extreme halophiles.

5.3.1.6 Hydrodynamic Conditions

Hydrodynamics, inclusive parameters such as flow rate, shear stress, turbulence, and surface roughness, exhibited a broader interference on the physical and chemical environment surrounding microorganisms. These factors can regulate the nutrient delivery, waste removal, QS of the bacteria, resulting in enhancing or depleting the mechanical stability of the biofilm structure.

Surface roughness, an important phenomenon for biofilm initiation, creates varying hydrodynamic environments at the microscale. Rough surfaces act as good shelter areas where microorganisms can adhere and commence biofilm formation, even under high shear stress conditions. The increased surface area of rough surfaces also enhances the availability of attachment sites. Conversely, highly irregular surfaces can promote turbulent flow patterns, potentially increasing detachment rates. A study of Choi et al. (2023) reveals that *Streptococcus gordonii, Fusobacterium nucleatum*, and *Porphyromonas gingivalis*, a causative agent of dental plaque, periodontitis, and gingivitis and shows the different levels of biofilm activity on different surfaces. For biofilm formation by bacteria, the rough titanium surfaces are more approachable than the smooth zirconium surfaces (Choi et al., 2023). The optimal surface roughness for biofilm formation is often species-specific and dependent on the prevailing hydrodynamic conditions.

5.3.2 Host Factors for Biofilm Formation

According to standard definitions, bacterial biofilms are permanent microbial communities that are encased in EPS. Microbial cells that are trapped in EPS and exhibit distinct phenotypes in gene transcription and growth rates are characterized by differences in their irreversible adherence to substrates surfaces or to one another. One microorganism, or a mix of bacteria fungi archaea protozoa and yeasts, make up a bacterial biofilm. Its channel-like structure controls the flow of nutrients, gases and antibiotics. Microbial consortia that originate from bacterial cultures of one or more species are known as bacterial biofilms. Due to interspecies interactions, multispecies biofilms exhibited the following novel characteristics in comparison to single-species biofilms: increased biofilm biomass, a greater number of community cells, increased metabolic activity of community constituents, increased antimicrobial tolerance and changes in spatial organization and architecture. Because of competition or collaboration, multispecies biofilms have unique characteristics that set them apart from floating. Numerous species of bacteria interact through mutual benefit, cooperation, competition, antagonism, utilization, and synergy. Among these, the formation of high-order spatial structures of multispecies biofilms and the regulation of bacterial microbial activities are significantly influenced by synergistic interactions. Multispecies biofilm formation is facilitated by bacterial cells' cooperation and competition with one another (Zhao et al., 2023).

5.3.2.1 Host Components as Building Blocks

It is traditionally understood that the term biofilm refers to a bacterial community trapped within the self-produced extracellular substances that are causative agents for 80% of bacterial infections. During the development of biofilm, bacteria are converted from planktonic growth form into an aggregated growth form specifically embedded within a biofilm matrix in which individual bacteria (reversible) binds to a substratum and, subsequently, become permanent after bacterial cells begin to develop EPS, such as protein, polysaccharide, and extracellular DNA (eDNA). EPS constitution becomes altered when susceptible to the range of micronutrients by generating particular enzymes by the bacterial cells. Biofilm facilitates the cells of bacteria to offer specific circumstances and generating virulence considerations due to the absence of the local oxygen and micronutrients, making them resistant to different antibacterial agents. Research has established that dormant bacterial cells within a biofilm exist and that they are highly active to develop tolerance to many different antibiotics. Further, the immune system is not fully capable of eliminating biofilm. Biofilm's matrix proteins safeguard it from human immune responses like opsonization and phagocytosis of bacterial biofilm (Mirzaei and Ranjbar, 2022). Interestingly, recent evidence discovered that bacteria are able to integrate host components, including DNA, collagen, fibronectin, hyaluronan, mucin, and filamentous polymers (F-actin), into their polysaccharide matrix to produce a dense biofilm. In *Pseudomonas* spp., rhamnolipid actively contributes in the formation of channel of the biofilm and proves to be harmful for the host. The neutrophils present in the zone of biofilm are able to produce waste derived from the host, such as actin filaments and DNA that are responsible the supply of more material for biofilm EPS formation and for inducing biofilm formation. It has been discovered that such host debris is present in sputum from cystic fibrosis (CF) patients, and the disturbance of these polymers can interrupt the formation of biofilm formation in *Pseudomonas aeruginosa.* When the immunity of the host does not eliminate the infection caused due to bacteria, host-derived materials from necrotic immune and non-immune cells may serve as the biological matrix to induce the formation of the biofilm of the cells of bacteria (Fux et al., 2005).

5.3.2.2 Inflammatory Response

A biofilm can form in the case in which the bacterial infection is not controlled early enough. An exopolysaccharide matrix is released during biofilm formation and can protect bacteria from antibiotics as well as the body's immune system. The amount of inflammatory mediators increases disproportionately in the presence of bacteria as well as endotoxins, driving the wound into a vicious cycle of inflammation. Inflammatory cytokines, such as TNF-α and IL-6, and Matrix Metalloproteinases (MMPs), are deposited due to biofilms' stimulation of pro-inflammatory macrophages and neutrophils and their response with the host's immune system. This triggers chronic inflammation. Polymicrobial biofilm, constituted by different microbial species, with their complex structure of the extracellular matrix, reduced permeability to antimicrobial agents, and rapid development of resistance mechanisms pose a potent challenge for effective biofilm eradication. Antibiotic resistance, which becomes a cause of serious concern to public health, is strongly promoted by the existence of diverse microorganisms within biofilms, especially in chronic wounds (Wu et al., 2019).

5.3.2.3 Host Immune Evasion

An instant reaction is created inside the body by the cells of the immune system, receptors, and a number of humoral systems such as mannose-binding lectins, and antibodies of the innate immune system upon the infections of the microbes. Microbial biofilms are said to evade the immune system of the body. For example, *S. aureus* can escape the immunological response of the host through macrophage dysfunction, the inhibition of phagocytosis mediated by leukocytes, and defenses mediated by lower antibodies due to the presence of EPS matrix. A 25% suppression of polymorphonuclear neutrophil (PMN's) burst due to oxidative stress response has been reported in *P. aeruginosa* sessile biofilm-bound cells compared to its planktonic equivalent. Alginate (a glycoprotein-like exopolysaccharide) and rhamnolipids (glycolipids) have been shown to prevent phagocytosis by leukocytes from attacking *P. aeruginosa* biofilm. Similarly, monocyte phagocytosis is inhibited in mature biofilms of *C. albicans*, thereby exerting an immunosuppressive action. Through the secretion of biofilm-enhancing compounds such as exopolysaccharides, proteins, and other peptides, *S. epidermidis* modulates the evasion of the immune system and avoids being killed. Alginate and rhamnolipids produced by *P. aeruginosa* aid in the necrosis of polymorphonuclear leukocyte and gradually leads to the evasion of the action of macrophages, respectively. *S. epidermidis* exerts immunity against phagocytosis by modulating the IgG deposition on the surface of bacteria and the activation of complement protein C3b, thereby escaping from the immune system and developing biofilm (Rather et al., 2021).

5.3.2.4 Specific Host–Pathogen Interactions

The immune response of the host is initiated by the recognition of different microorganism's virulence factors which are the essential characteristics to initiate the process of infection and interact with the host cells directly. In contrast, pathogenic microorganisms become more proficient in evading the immune response of the host in order to get through in a hostile environment and propagate to other tissues. Lipopolysaccharides (LPS) have been identified as pathogen-associated molecular patterns (PAMPs) and were identified as being among the strongest inducers of immune responses that drive the Toll-Like Receptor 4 (TLR4) signal pathway, and, occasionally, a prolonged reaction. LPS is structurally altered with effects on the lipid A and polysaccharide components stimulating the enhanced production of interleukin-6 (IL-6) and TNF (Tumor Necrosis Factor) in the biofilm stage compared to their counterpart of the planktonic cells and partly presenting to the elevation of the response of inflammation. Thus, the immune response is ineffective due to LPS immunogenicity and conformational differences, as assessed in the TLR5 knockout mice model, resulting in a critical lung infection and demonstrating the hypersusceptibility of the host's immune system towards the pathogen. But even if TLR5 exists, the immune system is not sufficient to manage the infection once biofilm formation occurs (Batoni et al., 2021). Two mechanisms are involved in the motility of *P. aeruginosa*, which includes type IV pili and flagella. As is known, type IV pili is related to motility and adhesion as well as the microcolony organization, which facilitates the biofilm formation. Conversely, the flagellum assists in the development and maturation stages, producing a vigorous biofilm, whereby it has been proven that mutants of type IV pili and flagellum in *P. aeruginosa* cannot establish an accurate structure, and they end up with a weak biofilm that is susceptible to antibiotics. Additionally, pyocyanin and pyoverdine are unique virulence factors within the *Pseudomonas* genus to be charged with providing colonies along with the production of fluorescent dye and siderophores and host proteins in distinct ways. In terms of primary infection and *Staphylococcus* planktonic cells, the immune system stimulates an enormous number of macrophages and neutrophils, which can largely neutralize the initial phase of staphylococcal infection. But once an irreversible adhesion and a primary biofilm are formed, *Staphylococcus* biofilms can increase the expression of many virulence factors that diminish the capacity of neutrophils and macrophages to eliminate staphylococcal host cells (Cangui-Panchi et al., 2023).

5.3.3 Intrinsic Factors for Biofilm Formation

Intrinsic factors play an important role in regulating the development and maturation of biofilms. Biofilm development is a complex procedure which depends on various intrinsic factors such as genetic factors, surface appendages, metabolic pathways, quorum-sensing systems, and motility mechanisms. Understanding these intrinsic factors is essential for developing effective strategies to control and prevent biofilm formation in various settings. Future research should focus on elucidating the intricate regulatory networks that govern biofilm formation and exploring novel approaches to target these intrinsic factors for biofilm disruption and prevention.

5.3.3.1 Role of Physiological Factors

5.3.3.1.1 Role of Fimbrial Structure

Bacterial surface appendages, especially fimbriae, are key players in the early stages of biofilm development, influencing initial attachment, cell–cell interactions, and subsequent maturation. Initially, fimbriae are categorized into four divisions, including type 1, containing adhesins and mannose-binding subunit (FimH) at the tip, which is found in most of the Gram-negative bacteria. The type 1 fimbriae facilitate the cells to bind to mannosylated glycoproteins present on surfaces. Secondly, the curly fimbriae, often considered as amyloid fibers that promote surface attachment and cell–cell aggregation. Enterobacteriaceae like *E. coli* and *Salmonella enterica* are shown to produce curli fimbriae in very regular cases. The other types of fimbriae are type P and type IV pili, which regulate the dynamic extension and retraction due to the creation of "twitching motility" that helps in surface exploration and biofilm formation. *V. cholerae* produces the toxin-coregulated pilus (TCP), which comes under type IV pilus, and which plays a role in the colonization of the human intestine and biofilm formation. TCP promotes microcolony formation and adherence to intestinal epithelial cells. Furthermore, TCP acts as a receptor for the CTXΦ bacteriophage, which carries the cholera toxin genes (Ligthart et al., 2020).

Several Gram-positive bacteria use surface appendages, such as fimbriae and flagella, to facilitate this initial interaction. These hair-like structures mediate attachment to a variety of surfaces, including living (host cells, other bacteria) and nonliving (medical devices, industrial equipment) materials. For example, *Streptococcus pneumoniae* utilizes pili to adhere to epithelial cells which become matured and colonized in the respiratory tract, leading to the formation of biofilm. The biofilm of these bacteria causes severe respiratory tract syndrome. The RrgA protein, a key pilin subunit, contributes to pilus assembly and bacterial virulence. Similarly, the Esp (Enterococcal surface protein) pilus of *Enterococcus faecalis* helps in surface attachment and biofilm formation on abiotic surfaces (Basset et al., 2011). In addition, *E. feacalis* also produce collagen-binding MSCRAMMs (Microbial Surface Components Recognizing Adhesive Matrix Molecules), including Ace adhesins protein, a vital component of the extracellular matrix which promotes biofilm formation on implanted devices. Although the primary function of the pili and fimbriae is to control the motility of the microorganism, it also promotes the biofilm formation indirectly by improving the cell surface adherence. Several studies have shown that *Bacillus subtilis*, flagella interfere with the initial attachment to a surface, followed by the synthesis of exopolysaccharides for biofilm maturation. The functionality of flagellar motor helps in sensing surface contact, triggering the development of the sessile cells. However, in some other Gram-positive bacteria, flagella are down-regulated once biofilm formation is underway, possibly due to the reduced need for motility within the established biofilm architecture.

5.3.3.1.2 Role of Afimbrial Structure

Bacterial adhesins, proteins that arbitrate the initial attachment of bacteria to biotic or abiotic surfaces, contribute a decisive role in the inception of biofilm formation. In addition to the fimbrial adhesins, afimbrial adhesins (AFAs) are also gaining acceptance for their huge contributions. AFAs are monomeric or oligomeric proteins that do not assemble into extended filaments. Like fimbrial adhesins, AFAs are also categorized into autotransporter or type V secretion system, two-partner system, chaperon-usher and Gram positive adhesins. Afimbrial mechanisms exhibit a wide range of properties where some of the afimbrial adhesins can bind preferentially to recognition sites on host cells, promoting initial attachment to the host surface (Patel et al., 2017). Ag43 in *Escherichia coli* interacts with the surface of other *E. coli* cells, promoting auto-aggregation and biofilm development.

In *S. aureus*, FnBPs associate to the EPS component fibronectin, assisting adhesion to implanted medical devices and other surfaces coated with fibronectin. Beside this, AFAs can bind with various cellular matrix components, such as collagen, laminin, and glycosaminoglycans, thereby fostering bacterial adherence to surfaces and contributing to biofilm framework.

Additionally, two-partner systems consist of two proteins in which a macro exoprotein (TpsA) has the domain of adhesion and a micro-outer membrane beta-barrel protein (TpsB) that facilitates secretion (Fan et al., 2012). The well-known example of this two-partner adhesin system is filamentous hemagglutinin (FHA) of *Bordetella pertussis*, the causative agent of whooping cough.

5.3.3.1.3 Role of Outer-Membrane Vesicles (OMVs)

Outer membrane vesicles (OMVs) are one of the physiological factors which contribute to the ability of bacteria to form their biofilm. Outer membrane vesicles (OMVs) are spherical structures released by Gram-negative bacteria, which help the bacteria to form their biofilm. These vesicles, which originate from the outer membrane, take active part in communication and interaction among bacteria, their surroundings, and host cells. OMVs have appeared to be determining mediators of bacterial reciprocal action and physiological processes which definitely interfere in various aspects of biofilm formation (Jan 2017). OMVs serve as transporters of essential extracellular matrix components (ECM), such as polysaccharides, proteins, and eDNA. For example, OMVs in *Pseudomonas aeruginosa* carry the Pel polysaccharide, which is considered to be a crucial component of the ECM, directly involved in developing biofilms, and in enhancing their structural integrity and stability (Van Loon et al., 2025).

In addition, hydrolytic enzymes present in the OMVs can be precise to nutrient-rich surfaces within the biofilm or the surrounding environment. The enzymes can disintegrate the complex macromolecules into smaller, more readily accessible micronutrients, which can then be introjected by the bacterial cells. The unique features of the OMVs rely on the production of siderophores under iron deficit medium. Again, OMVs can relocate the waste products away from the biofilm, preventing the aggregation of toxic metabolites that may sometimes inhibit growth and survival.

5.3.3.2 Role of Genetic Factors

The processes of biofilm formation and dispersal are tightly regulated by both environmental cues and genetic regulation (Ruhal and Kataria, 2021; Wolska et al., 2016). Bacterial biofilms are primarily regulated in three main ways, including quorum sensing (QS), bis-(3-5)-cyclic guanosine monophosphate (c-di-GMP) and small RNAs (sRNAs), at least in several Gram-negative species. Using tiny self-produced signal molecules called autoinducers, QS is a special language which is used for intercellular communication. When there are sufficient bacteria and the autoinducer concentration hits a certain value, the bacteria start to receive cellular signals for their critical mass and react either by the activation or the suppression of target genes. Bacterial genomes can contain roughly 10% genes that are regulated by QS (Wagner et al., 2003). Throughout the formation and spread of bacterial biofilms, QS controls are crucial. These systems are essential for the later development of the biofilm and are the primary regulators of biofilm dispersal, even though they are not involved in attachment or early biofilm formation.

5.3.3.2.1 Genes Required for Biofilm Formation

All of the genes under investigation had SNPs with the homB gene having the most (279 variants) and the cheY genes having the fewest (1 variant). SNPs from the adhesion genes, such as alpA, alpB, homD, cagE, and futB, were also represented. Apart from the genes that are involved in the regulation of cell shape and metabolism other genes, such as gluP, cgt, csd4, csd5, murF, and amiA, were also discovered to have significant associations. At A223V, G160S, and N156K, three SNPs of the alpB gene showed a significant correlation with the high-biofilm former. The high-biofilm former isolates showed higher alpB gene expression than the low-biofilm former isolates (Fauzia et al., 2023).

5.3.3.2.1.1 Molecular Mechanism Gene Regulation and Bacterial Biofilm Development The assembly and synthesis of the components of a biofilm matrix is a highly intricate task, the passage connecting its various stage involves the coordination, expression and simultaneous procurement of numerous genes through highly intricate genetic networks encompassing all levels of gene regulation. The transition of most bacteria from a motile planktonic to a sessile biofilm-forming mode of life is governed by post-transcriptional processes that are centered on the tightly controlled intracellular concentration of the chemical second messenger molecule cyclic-di-GMP. Other post-transcriptional regulatory mechanisms, including riboswitches RNases, small non-coding RNAs, RNA-binding proteins, and toxin-antitoxin systems, are known to regulate the formation and assembly of biofilms. Thus, post-transcriptional regulation is a potent molecular mechanism that bacteria use to quickly adapt to their changing environment and to emphasize gene expression in response to the developmental needs of the cell. The complex, multifactorial aspects of biofilm formation involve several mechanisms that either directly or indirectly influence the spatial and temporal coordination of the components essential for biofilm development (Martínez and Vadyvaloo, 2014).

5.3.3.2.1.2 Role of Transcriptional Regulator There are many transcriptional regulatory factors responsible for biofilm formation, but the most well-known are SpoVG, Spo0A, NosP and PNPase, respectively. Nitric oxide (NO) has been reported to cause the formation of biofilm in most bacteria through ligation to H-NOX (heme-NO/oxygen binding) domains. The majority of NO-responsive bacteria, however, do not have H-NOX domain-containing proteins. NosP or its co-cistronic kinase nahK (formerly hnoS)-deficient strains form biofilms that are immature in state, whereas hnoK and hnoX (NO/H-NOX-reactive kinase) mutants lead to wild-type biofilm structures. We show that NosP controls the activity of autophosphorylation of NahK as well as HnoK. HnoK and NahK have been reported to control three main types of response regulators (HnoB, HnoC, and HnoD) that are part of a NO-reactive multicomponent c-di-GMP signaling network. H-NOX and NosP resist each other through a mechanism of push–pull; NosP/NahK supports the development of biofilm by inhibiting the H-NOX/HnoK signaling pathway, leading to a decrease in the extension of biofilm development. NO integration leads to c-di-GMP reduction and biofilm formation, mainly by disinhibiting the HnoK activity (Nisbett et al., 2019). SpoVG, first found in *B. subtilis* as a sporulation factor, is an independent form of DNA-binding protein units that is readily available in the bacteria. It is widely preserved, particularly among Gram-positive bacterial strains like *B. subtilis*, *B. anthracis*, *B. cereus*, *Listeria monocytogenes*, *S. aureus*, and *Clostridium*. SpoVG may change the cell division asymmetrically, reducing hemolysin production, and controlling sporulating phenotypes in *B. subtilis*. The homolog of SpoVG in the non-sporulating organisms *L. monocytogenes* and *S. aureus* was demonstrated to be a virulence factor modulator and antibiotic resistance, which could assist the organisms in coordinating environmental growth and survival (Huang et al., 2021). Spo0A belongs to the response regulator that is induced by phosphorylation via a complex (multipart) phosphorelay. This phosphorelay is apparently necessary for biofilm formation by the induction of Spo0A. Mutated strains impaired for Spo0F, a response regulator protein whose phosphorylation plays a central role in integration of cellular and extracellular signals responsible for initiating sporulation, or for Spo0B, a phosphotransferase, were equally found defective in biofilm formation and the effects are similar to that of a mutant impaired for Spo0A, a master regulator for sporulation, a stage of dormancy in response to environmental stress. This phosphorelay is used in the integration of several signaling molecules for the beginning of sporulation. The primary signaling process that seems to trigger this phosphorelay for sporulation is starvation, but this phosphorelay can be controlled by signaling molecules derived from quorum sensing, damaged DNA, partitioning, and the replication of chromosomes and TCA cycle (Hamon and Lazazzera, 2001). There are also two main proteins identified to control biofilm development in different bacterial strains: the RNA-binding protein and polynucleotide phosphorylase (PNPase). PNPase are conserved 3′–5′ exo-ribonuclease, which are responsible for the catalysis and subsequent processing of RNA and are responsible for enhancing the virulence factors in various bacterial species. PNPase is an important enzyme in *L. monocytogenes* responsible for the processing and functioning of a CRISPR element. In Gram-negative bacterial strain *Escherichia coli* K-12, *Salmonella enterica* and *Serovar Typhimurium*, the inactivation of PNPase was demonstrated in the disruption of the formation of biofilm. By contrast, in *E. coli* C strain, the PNPase-deletion mutant exhibited enhanced biofilm formation (Quendera et al., 2023).

5.3.3.2.1.3 Role of Complex Regulatory Network

Biofilm formation is affected by various regulatory signals, which include small diffusible molecules, known as autoinducers for QS; bacterial second messengers like cyclic di guanylate (c-di-GMP), that plays a crucial role in regulating various cellular processes, like biofilm formation and interact with different regulatory components, such as two-component regulatory systems, alternative sigma factors, and small non-coding RNAs (ncRNAs). Generally, these regulatory pathways can interact with, or influence, one another. RNA regulators, such as ncRNA, can facilitate the transition from free-floating planktonic cells to a biofilm state. Typically, small ncRNAs vary in size from 50 to 400 base pairs, are not translated into proteins, and play multifunctional roles in metabolism, stress responses, pathogenesis, and biofilm development. These ncRNAs can activate or repress mRNAs or proteins. For mRNAs, ncRNAs exist as antisense RNA which pairs with target mRNAs through sequence complementarity, forming a duplex structure and modulating gene expression and responses to environmental changes. Pairing between ncRNAs can be classified as either cis or trans, depending on the proximity of the target RNA. Cis-acting non-coding RNA can be located directly across from their target mRNA, while trans ncRNAs are found in intergenic regions elsewhere in the genome, exhibiting limited similarity to the target mRNA. The Hfq RNA chaperone is required for the interaction between trans-acting ncRNAs and their target mRNAs. Base-pairing ncRNAs affect mRNA translation and stability by altering access to ribosome binding sites or facilitating degradation by ribonucleases, which impacts the expression of target genes. The second type of ncRNA can bind to RNA-binding proteins, inhibiting or counteracting their functions. These ncRNAs resemble sequences typically found in mRNAs, preventing their corresponding target mRNA from interacting with target proteins. The third group of ncRNAs regulates the retention of nucleic acids acquired horizontally, which can be mediated by bacteriophages or plasmids. Recent studies have demonstrated that regulatory RNAs from eukaryotic cells found in extracellular vesicles also influence bacterial biofilm formation and may serve as a control mechanism for bacterial infections in eukaryotic hosts. Identifying and characterizing small non-coding RNAs that regulate biofilm formation is crucial for developing new strategies and therapeutics to combat various biofilm-forming pathogens (Mitra & Mukhopadhyay, 2023).

5.3.3.3 Role of Antimicrobial Components

Antimicrobials, which include antibiotics, antivirals, antifungals, and antiparasitic, are chemical agents that can eliminate microbes, hinder their growth, and either avert or treat infections in humans, animals, and plants. Antimicrobial resistance denotes the ability of microorganisms to persist in the presence of an antimicrobial agent at higher concentrations for an extended duration, and it is measured as the minimum inhibitory concentration (MIC). Resistance to biofilms can manifest as either antibiotic resistance or antibiotic tolerance. Microorganisms achieve resistance against antimicrobials either by obtaining foreign genetic materials that encode resistance factors via horizontal gene transfer (HGT) within the extracellular polymeric substances (EPS) of a biofilm or through gene mutations. The mechanisms of antimicrobial resistance (AMR) include restricting access or reducing the permeability of antimicrobials, alterations in antibiotic targets through mutations, modifications to targets, and the enzymatic breakdown or chemical modification of antimicrobials. Antibiotic resistance (ABR) is a subset of AMR, as antibiotics are effective against bacteria, yet bacteria can evolve resistance to antibiotics.

Antimicrobial resistance can be classified as intrinsic (naturally occurring) due to genetic mutations, intrinsic genes that confer structural and functional features to the molecular target, acquired

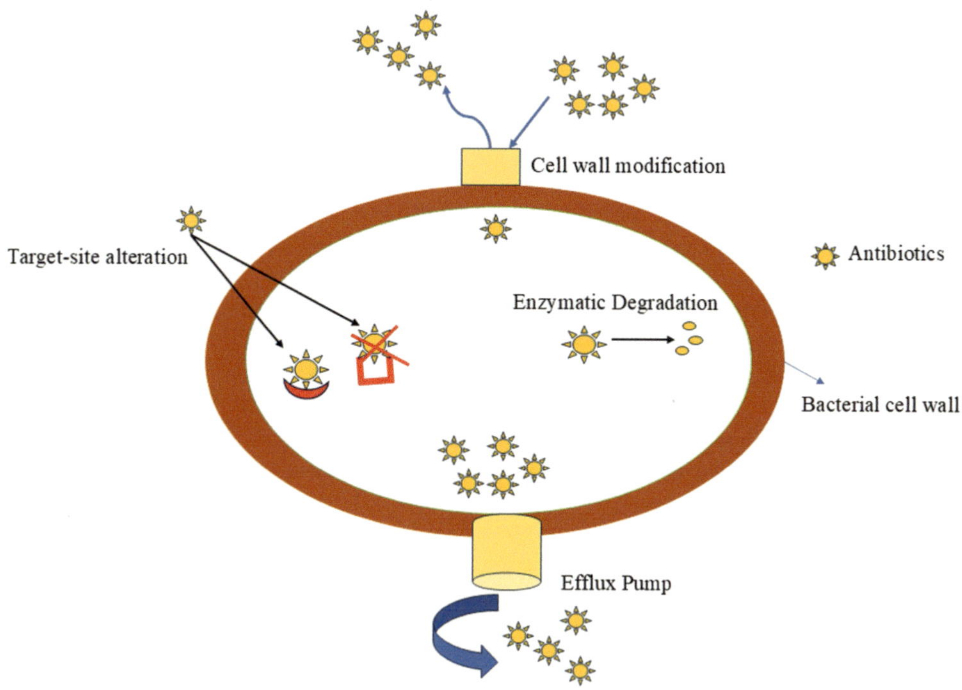

FIGURE 5.3 Diagrammatic representation of the process by which bacteria can adapt antibiotic resistance.

from external sources, or adaptive. Intrinsically, certain antibiotics, such as vancomycin and daptomycin, are potent against Gram-positive bacteria but ineffective against Gram-negative bacteria, owing to differences in cell wall structure, referred to as cell wall modification, as depicted in Figure 5.3. On the other hand, acquired resistance arises from genetic changes facilitated either by HGT or by mutations. Furthermore, bacteria can quickly adjust their resistance mechanisms through gene expression and protein production in response to stress or environmental changes, even when certain antibiotics are present. Genes like Thioredoxin A (Ttrx A), thioredoxin reductase (trxB), D-Ala-D-Ala carboxypeptidase, DacA, FabI, and SapC contribute to the intrinsic resistance against antibiotics such as fluoroquinolones, β-lactams, and aminoglycosides (Dutt et al., 2022).

5.4 ROLE OF DIFFERENT FACTORS ON BIOFILM FORMATION OF VARIOUS BACTERIA

5.4.1 In Gram-Positive Bacteria

Gram-positive bacteria, including extremely popular pathogens like *Staphylococcus aureus*, *Staphylococcus epidermidis, Streptococcus pneumoniae, Enterococcus faecalis*, and *Bacillus subtilis*, are considered to be the most opportunistic bacteria in severe chronic infections. Their ability to form sturdy biofilms depends on a sophisticated and regulated process, influenced by numerous features that govern initial attachment, cell–cell aggregation, matrix production, and eventual disbursement. The peptidoglycan layer in Gram-positive bacteria providing structural rigidity to the bacterial cell wall and contributes in various cellular processes. The peptidoglycan layer, along with teichoic and lipoteichoic acids, which are anionic polymers, plays a role in bacterial adhesion to surfaces. Teichoic acids are covalently bonded to the peptidoglycan layer, while lipoteichoic acids associate with the bacterial membrane. These polymer components bestowed negative charge to the bacterial cell surface, aided electrostatic interactions with positively charged surfaces and facilitated bacterial adhesion. Studies have shown that teichoic acid of *Bacillus subtilis* is orchestrated by operon tagA-O. Deletion of tagD leads to no synthesis and inhibits planktonic growth. However, restoring tagD expression using xylose inducible promoter-restored planktonic growth. The deletion of enzymes involved in the modification of cell wall teichoic acid leads to impaired biofilm development.

Gram-positive bacteria form more robust structures and the stability of the biofilm is often higher due to their thick peptidoglycan layer. Furthermost, the extracellular matrix produced by many Gram-positive bacteria plays an important role in bacterial aggregation and cell surface attachment. Extracellular matrix composed of proteins, polysaccharides, and eDNA.

The unique QS system of Gram-positive bacteria principally depends on producing autoinducing peptides which assist cell-to-cell communication and coordination. The well-known accessory gene regulatory (Agr) system of *Staphylococcus aureus* activates a signal transduction cascade, leading to transform gene expression and insisting to form the biofilm. Similarly, the feromone-sensitive system of *Enterococcus faecalis* produced autoinducing peptides and membrane-bound histidine kinase, typically works on the gene expression related to adhesion and virulence.

5.4.2 In Gram-Negative Bacteria

Initial adhesion in Gram-negative bacteria can be facilitated by c-di-GMP-dependent regulation, where, generally, elevated intracellular levels are required for the initial attachment. The QS system of the Gram-negative bacterium *P. aeruginosa* is characterized by the production of n-acylated homoserine lactones (AHLs), which act as signaling molecules and autoinducers (AI). This system suggests that AI is released when the population of bacterial cells reaches a certain threshold, allowing other bacterial cells to detect this AI as a signal that influences the LasI and RhlR systems, potentially leading to the production of rhamnolipids and activation of the PQS (Pseudomonas Quinolone signal). These components are essential for biofilm formation, specifically in the release of extracellular DNA (eDNA) and the production of exopolysaccharides Psl and Pel. In contrast, the cidA gene expression in *S. aureus*, which is not part of QS, has been shown to regulate the release of eDNA. Ultimately, QS plays an important role in mature biofilms due to the aggregation of the population, enabling them to determine the appropriate time to disperse. Various environmental factors surrounding the pathogens might induce the dispersal of biofilms that are thought to be formed by planktonic cells. However, studies with different pathogens reveal a significant difference between dispersed cells and the planktonic phenotype. Furthermore, the breakdown of the biofilm matrix can also lead to biofilm dispersal in both Gram-positive and Gram-negative bacteria. Enzymes such as proteases, nucleases, or phenol-soluble modulins that act as surfactants have been implicated in the dispersal of *S. aureus* Gram-positive bacterial biofilms, while rhamnolipids are responsible for those of *P. aeruginosa*. PslG is a glycoside hydrolase that belongs to the psl operon within *P. aeruginosa*, exhibiting a variety of regulatory factors involved in biofilm formation. Some researchers suggest that biofilm dispersal may serve as a potential therapeutic option for chronic infections related to biofilms. In *V. cholerae*, several sensory mechanisms regulate the timing of dispersion. The interplay between a generalized stress response via rpoS and QS supports the decision-making process for dispersal. Secondary messengers, like nucleotides and c-di-GMP, are recognized for their roles in Gram-negative bacteria, as demonstrated by *P. aeruginosa*. Molecular analysis of *S. aureus* showed that the activation of QS facilitated the detachment of biofilms. Thus, commonalities in the dispersal of biofilms in both bacteria include the inhibition of QS and the influence of matrix-degrading enzymes, such as glycoside hydrolase and DNase. Nevertheless, recent studies indicate that the dispersed cells exhibit a virulent and antibiotic-resistant phenotype. In one such study, the dispersed phenotype of *Streptococcus pneumoniae* was shown to be more virulent than another phenotype, while the *in-vitro*-derived dispersed phenotype of *P. aeruginosa* displayed resistance to antibiotics (Singh et al., 2017).

5.4.3 In Plant-Associated Bacteria

Plant-associated bacteria dwell in various niches, including the rhizosphere (the soil surrounding plant roots), the phyllosphere (aerial plant surfaces), and the plant endosphere (internal tissues). Plant-associated bacteria exhibit heterogeneous effects on plant health, including beneficial interactions such as growth development and disease restraint to deleterious effects such as pathogenesis and disease. One critical strategy employed by many plant-associated bacteria for survival and communication with their host immune system is the formation of biofilms. *Agrobacterium tumefaciens* is a well-known plant pathogen that causes crown gall disease. *A. tumefaciens* forms biofilms on wounded plant tissues, facilitating its attachment, colonization, and the transfer of its T-DNA into plant cells. The T-DNA encodes genes that cause the formation of tumor-like galls on the plant, which provide a niche for the bacteria. Biofilm formation is essential for the virulence of *A. tumefaciens*. Several studies have shown *Xanthomonas campestris pv. campestris*, the causative agent of black rot in crucifers, forms biofilms both in the phyllosphere and within the vascular system of the plant.

Additionally, biofilms may also benefit the plant such as *Pseudomonas fluorescens*, a plant growth-promoting rhizobacterium (PGPR) that accumulates plant roots and can inhibit the plant diseases. *P. fluorescens* forms robust biofilms on root surfaces, eliciting its potentiality to fight against other microorganisms and to deliver beneficial compounds to the plant. The EPS in *P. fluorescens* biofilms play an important role in root surface hydration and nutrient availability, indirectly promoting plant growth. Additionally, *P. fluorescens* strains produce antibiotics within the biofilm, suppressing other pathogens. Similarly, nitrogen-fixing bacteria *Azospirillum brasilense* forms their biofilm in the root nodules under the soil, which also indirectly facilitates plants by fixing nitrogen in the soil.

5.4.4 In Fungi

Previously, it had been thought that only bacteria are capable of forming biofilm. Recent studies, however, have revealed that not only bacteria, but also several fungi, are potentially forming biofilm in their lifespan for resisting antimicrobial sensitivity and survival. Fungal biofilms preferentially differ from their bacterial counterparts in terms of cellular composition, morphology, and ECM structure. *Candida albicans* is the most opportunistic fungal pathogen in humans that can form biofilms on various surfaces, including medical devices, oral mucosa, and implanted catheters. *Candida albicans* biofilms show a differential characteristic framework, comprising yeast cells, pseudohyphae, and true hyphae embedded in an ECM primarily composed of β-1,3-glucan. Other Candida species, such as *Candida glabrata, Candida parapsilosis*, and *Candida auris*, are also capable of forming biofilms. Other than candida, *Aspergillus fumigatus*, a common causative agent of aspergillosis, particularly in immunocompromised individuals, can form biofilms in the lungs of patients with cystic fibrosis and chronic obstructive pulmonary disease (COPD).

5.4.5 In Algae

Algae, particularly microscopic diatoms to macroscopic seaweeds, can also form biofilms on various surfaces, including rocks, underwater structures, and even other organisms. Exploration of this new field of algal biofilms unveil multifaceted insights into aquatic ecosystems, influencing nutrient cycling, primary production, and overall biodiversity. Diatoms are a leading group of single-celled algae that are potentially able to form prolific biofilm in both freshwater and marine environments. Several species, such as *Navicula, Nitzschia*, and *Amphora*, readily form biofilms on surfaces, and play a significant role in primary production in these ecosystems. *Chlamydomonas, Chlorella*, and *Ulva*-like green algae and dinoflagellates such as *Prorocentrum* also form biofilm and are gradually attracting interest in the biotechnological field for their beneficial properties. In addition, cyanobacteria, e.g. *Microcystis*, are the primary influencer for algal bloom production, and can also form biofilms.

5.5 CONCLUSION

Biofilm formation is a complex and multifaceted process influenced by a wide range of factors. Surface characteristics, nutrient availability, hydrodynamics, temperature, pH, microbial interactions, and signaling molecules all play crucial roles in determining the rate, extent, and characteristics of biofilm formation. Future research should focus on elucidating the complex interactions of several factors with microorganisms. This knowledge will be critical to addressing the challenges posed by biofilms in various fields, including medicine, industry, and environmental science. The development of targeted approaches that disrupt specific stages of biofilm formation, such as adhesion, EPS production, or quorum sensing, holds great promise for future biofilm management.

REFERENCES

Arai, H. (2011). Regulation and function of versatile aerobic and anaerobic respiratory metabolism in pseudomonas aeruginosa. *Frontiers in Microbiology*, *2*. https://doi.org/10.3389/fmicb.2011.00103

Basset, A., Turner, K. H., Boush, E., Sayeed, S., Dove, S. L., & Malley, R. (2011). Expression of the type 1 pneumococcal pilus is bistable and negatively regulated by the structural component RrgA. *Infection and Immunity*, *79*(8), 2974–2983. https://doi.org/10.1128/IAI.05117-11

Batoni, G., Martinez-Pomares, L., & Esin, S. (2021). Editorial: Immune response to biofilms. *Frontiers in Immunology*, *12*, 696356. https://doi.org/10.3389/fimmu.2021.696356

Cangui-Panchi, S. P., Ñacato-Toapanta, A. L., Enríquez-Martínez, L. J., Salinas-Delgado, G. A., Reyes, J., Garzon-Chavez, D., & Machado, A. (2023). Battle royale: Immune response on biofilms – host–pathogen interactions. *Current Research in Immunology*, *4*, 100057. https://doi.org/10.1016/j.crimmu.2023.100057

Choi, S., Jo, Y.-H., Luke Yeo, I.-S., Yoon, H.-I., Lee, J.-H., & Han, J.-S. (2023). The effect of surface material, roughness and wettability on the adhesion and proliferation of Streptococcus gordonii, Fusobacterium nucleatum and Porphyromonas gingivalis. *Journal of Dental Sciences*, *18*(2), 517–525. https://doi.org/10.1016/j.jds.2022.09.010

Dixit, S., Varshney, S., Gupta, D., & Sharma, S. (2023). Factors affecting biofilm formation by bacteria on fabrics. *International Microbiology*, *27*(4), 1111–1123. https://doi.org/10.1007/s10123-023-00460-z

Dutt, Y., Dhiman, R., Singh, T., Vibhuti, A., Gupta, A., Pandey, R. P., … & Priyadarshini, A. (2022). The association between biofilm formation and antimicrobial resistance with possible ingenious bio-remedial approaches. *Antibiotics*, *11*(7), 930.

Dutta, B., Mandal, D., Lahiri, D., Nag, M., & Ray, R. R. (2024). Elucidation of antibiofilm and quorum-quenching activities of leucocin against E. coli 25922 biofilm. *International Journal of Peptide Research and Therapeutics*, *30*(6), 61.

Fan, E., Fiedler, S., Jacob-Dubuisson, F., & Müller, M. (2012). Two-partner secretion of gram-negative bacteria. *Journal of Biological Chemistry*, *287*(4), 2591–2599. https://doi.org/10.1074/jbc.M111.293068

Fauzia, K. A., Aftab, H., Miftahussurur, M., Waskito, L. A., Tuan, V. P., Alfaray, R. I., Matsumoto, T., Yurugi, M., Subsomwong, P., Kabamba, E. T., Akada, J., & Yamaoka, Y. (2023). Genetic determinants of Biofilm formation of Helicobacter pylori using whole-genome sequencing. *BMC Microbiology*, *23*(1), 159. https://doi.org/10.1186/s12866-023-02889-8

Fux, C. A., Costerton, J. W., Stewart, P. S., & Stoodley, P. (2005). Survival strategies of infectious biofilms. *Trends in Microbiology*, *13*(1), 34–40. https://doi.org/10.1016/j.tim.2004.11.010

Ghosh, S., Saha, I., Dey, A., Lahiri, D., Nag, M., Sarkar, T., Pati, S., Rebezov, M., Shariati, M.A., Thiruvengadam, M., & Ray, R. R. (2022). Natural compounds underpinning the genetic regulation of biofilm formation: An overview. *South African Journal of Botany*, *151*, 92–106. https://doi.org/10.1016/j.sajb.2021.11.039

Hamon, M. A., & Lazazzera, B. A. (2001). The sporulation transcription factor Spo0A is required for biofilm development in *Bacillus subtilis*. *Molecular Microbiology*, *42*(5), 1199–1209. https://doi.org/10.1046/j.1365-2958.2001.02709.x

Hu, C., Yang, J., Qi, Z., Wu, H., Wang, B., Zou, F., Mei, H., Liu, J., Wang, W., & Liu, Q. (2022). Heat shock proteins: Biological functions, pathological roles, and therapeutic opportunities. *MedComm*, *3*(3), e161. https://doi.org/10.1002/mco2.161

Huang, Q., Zhang, Z., Liu, Q., Liu, F., Liu, Y., Zhang, J., & Wang, G. (2021). SpoVG is an important regulator of sporulation and affects biofilm formation by regulating Spo0A transcription in Bacillus cereus 0–9. *BMC Microbiology*, *21*(1), 172. https://doi.org/10.1186/s12866-021-02239-6

Isnawati, I., & Trimulyono, G. (2018). Temperature range and degree of acidity growth of isolate of indigenous bacteria on fermented feed "fermege". *Journal of Physics: Conference Series*, *953*, 012209. https://doi.org/10.1088/1742-6596/953/1/012209

Jan, A. T. (2017). Outer membrane vesicles (OMVs) of gram-negative bacteria: a perspective update. *Frontiers in microbiology*, *8*, 1053

Ligthart, K., Belzer, C., De Vos, W. M., & Tytgat, H. L. P. (2020). Bridging bacteria and the gut: Functional aspects of type iv pili. *Trends in Microbiology*, *28*(5), 340–348. https://doi.org/10.1016/j.tim.2020.02.003

Liu, S., Le Mauff, F., Sheppard, D. C., & Zhang, S. (2022). Filamentous fungal biofilms: Conserved and unique aspects of extracellular matrix composition, mechanisms of drug resistance and regulatory networks in Aspergillus fumigatus. *Npj Biofilms and Microbiomes*, *8*(1), 83. https://doi.org/10.1038/s41522-022-00347-3

Marín-Sanhueza, C., Echeverría-Vega, A., Gómez, A., Cabrera-Barjas, G., Romero, R., & Banerjee, A. (2022). Stress dependent biofilm formation and bioactive melanin pigment production by a thermophilic bacillus species from Chilean hot spring. *Polymers*, *14*(4), 680. https://doi.org/10.3390/polym14040680

Martínez, L. C., & Vadyvaloo, V. (2014). Mechanisms of post-transcriptional gene regulation in bacterial biofilms. *Frontiers in Cellular and Infection Microbiology*, *4*. https://doi.org/10.3389/fcimb.2014.00038

Mendhe, S., Badge, A., Ugemuge, S., & Chandi, D. (2023). Impact of biofilms on chronic infections and medical challenges. *Cureus*. https://doi.org/10.7759/cureus.48204

Mirzaei, R., & Ranjbar, R. (2022). Hijacking host components for bacterial biofilm formation: An advanced mechanism. *International Immunopharmacology*, *103*, 108471. https://doi.org/10.1016/j.intimp.2021.108471

Mitra, A., & Mukhopadhyay, S. (2023). Regulation of biofilm formation by non-coding RNA in prokaryotes. *Current Research in Pharmacology and Drug Discovery*, *4*, 100151. https://doi.org/10.1016/j.crphar.2022.100151

Nie, C., Xiong, Z., Zhang, H., Hou, C., Xie, F., Song, X., Xia, Y., & Ai, L. (2024). Effect of carbon sources on characterization of exopolysaccharides from Streptococcus thermophilus S-3. *Food Bioscience*, *61*, 104676. https://doi.org/10.1016/j.fbio.2024.104676

Nisbett, L.-M., Binnenkade, L., Bacon, B., Hossain, S., Kotloski, N. J., Brutinel, E. D., Hartmann, R., Drescher, K., Arora, D. P., Muralidharan, S., Thormann, K. M., Gralnick, J. A., & Boon, E. M. (2019). Nosp signaling modulates the NO/H-NOX-mediated multicomponent c-Di-GMP network and biofilm formation in *shewanella oneidensis*. *Biochemistry*, *58*(48), 4827–4841. https://doi.org/10.1021/acs.biochem.9b00706

Patel, S., Mathivanan, N., & Goyal, A. (2017). Bacterial adhesins, the pathogenic weapons to trick host defense arsenal. *Biomedicine & Pharmacotherapy*, *93*, 763–771. https://doi.org/10.1016/j.biopha.2017.06.102

Pramanik, A., & Vibhuti, R. K. (2022). Molecular mechanism of iron transport systems in vibrio. *Journal of Pure and Applied Microbiology*, *16*(1), 116–129. https://doi.org/10.22207/JPAM.16.1.77

Quendera, A. P., Pinto, S. N., Pobre, V., Antunes, W., Bonifácio, V. D. B., Arraiano, C. M., & Andrade, J. M. (2023). The ribonuclease PNPase is a key regulator of biofilm formation in Listeria monocytogenes and affects invasion of host cells. *NPJ Biofilms and Microbiomes*, *9*(1), 34. https://doi.org/10.1038/s41522-023-00397-1

Rather, M. A., Gupta, K., & Mandal, M. (2021). Microbial biofilm: Formation, architecture, antibiotic resistance, and control strategies. *Brazilian Journal of Microbiology*, *52*(4), 1701–1718. https://doi.org/10.1007/s42770-021-00624-x

Ruhal, R., & Kataria, R. (2021). Biofilm patterns in Gram-positive and Gram-negative bacteria. *Microbiological Research*, *251*, 126829. https://doi.org/10.1016/j.micres.2021.126829

Santajit, S., & Indrawattana, N. (2016). Mechanisms of antimicrobial resistance in ESKAPE pathogens. *BioMed Research International*, *2016*, 1–8. https://doi.org/10.1155/2016/2475067

Singh, P. K., Bartalomej, S., Hartmann, R., Jeckel, H., Vidakovic, L., Nadell, C. D., & Drescher, K. (2017). Vibrio cholerae combines individual and collective sensing to trigger biofilm dispersal. *Current Biology*, *27*(21), 3359–3366. https://doi.org/10.1016/j.cub.2017.09.041

Tarifa, M. C., Agustín, M. D. R., & Brugnoni, L. I. (2023). Biological control of foodborne pathogens by lactic acid bacteria: A focus on juice processing industries. *Revista Argentina de Microbiología*, *55*(4), 378–386. https://doi.org/10.1016/j.ram.2023.04.001

Van Loon, J. C., Le Mauff, F., Vargas, M. A., Gilbert, S., Pfoh, R., Morrison, Z. A., Razvi, E., Nitz, M., Sheppard, D. C., & Howell, P. L. (2025). Structural and functional analysis of *Pseudomonas aeruginosa* (*PelA*) provides insight into the modification of the Pel exopolysaccharide. *Journal of Biological Chemistry*, *301*(5), 108432. https://doi.org/10.1016/j.jbc.2025.108432

Wagner, V. E., Bushnell, D., Passador, L., Brooks, A. I., & Iglewski, B. H. (2003). Microarray analysis of Pseudomonas aeruginosa quorum-sensing regulons: effects of growth phase and environment. *Journal of Bacteriology*, *185*(7), 2080–2095. https://doi.org/10.1128/JB.185.7.2080-2095.2003

Wolska, K. I., Grudniak, A. M., Rudnicka, Z., & Markowska, K. (2016). Genetic control of bacterial biofilms. *Journal of Applied Genetics*, *57*(2), 225–238. https://doi.org/10.1007/s13353-015-0309-2

Wu, Y.-K., Cheng, N.-C., & Cheng, C.-M. (2019). Biofilms in chronic wounds: Pathogenesis and diagnosis. *Trends in Biotechnology*, *37*(5), 505–517. https://doi.org/10.1016/j.tibtech.2018.10.011

Zhao, A., Sun, J., & Liu, Y. (2023). Understanding bacterial biofilms: From definition to treatment strategies. *Frontiers in Cellular and Infection Microbiology*, *13*, 1137947. https://doi.org/10.3389/fcimb.2023.1137947

6 Peptide Hormones and Biofilm

Bandita Dutta, Debarati Chatterjee, and Rina Rani Ray
Department of Biotechnology, Maulana Abul Kalam Azad University of Technology, Kolkata, India

6.1 INTRODUCTION

Bacteria tend to form a cluster of communities and to make attachments irreversibly to a solid surface, leading to the formation of an infection. This type of sessile cell community formation enclosed in a slimy matrix of extracellular polymeric substances (EPS) is known as a biofilm. These specific sessile microcolonies eventually lead to an adaptive resistance to antibiotics and antibacterial drugs. The persistent bacterial cells develop antibiotic and multidrug resistance through the unregulated use of antibiotics and the genetic evolution of the species. The EPS matrix prevents the diffusion of antibiotic compounds into the biofilm. This prevents the conventional approach to medicines. Biofilms are formed on a wide range of non-sterile surfaces, from living tissues to medicinal devices and even on hot springs, deep-sea vents, rocks, soils, etc., eventually it leads to a threat in our society. Bacterial antibiotic resistance can occur by two mechanical pathways: intrinsic or extrinsic mechanisms (Blair et al., 2015). There are two principal criteria that support the formation of biofilm. Firstly, the contextual criterion constituent of a mono- or polymicrobial community at the level of a solid surface, including in some medicinal items such as prostheses, artificial valves, and catheters, acquires a three-dimensional structure (the sessile bacteria). Then comes the morphological criterion associated with the formation of the concerned microorganisms in an extracellular matrix consisting of complex polymers.

Quorum sensing (QS) is another important aspect of biofilm formation, whereby bacterial cell-to-cell communication is accomplished. This involves the detection, production, and generation of responses by extracellular inducer molecules which are known as autoinducers (AIs). As the bacterial cell density increases, more AIs accumulate in the environment, triggering alterations in gene expression in accordance with the level of population density. Both Gram-negative and Gram-positive bacteria use quorum sensing to control the diverse array of physiological activities. They achieve this through QS systems, which are communication systems based on chemical signals.

In Gram-positive bacteria, these AIs are frequently so-called quorum-sensing peptides (QSP). These serve as signalling molecules, regulating bacterial biofilm formation, the expression of virulence factors, and gene expression (Verbeke et al., 2017). In more complex animals, however, some peptides released by several endocrine glands or specialized cells act as chemical signals to trigger specific responses. In other words, it can be stated that peptide hormones are involved in communication amongst multicellular organisms, yet bacteria use peptides as a mode of communication amongst their populations through quorum sensing. Therefore, host peptide hormones are likely to play a significant role in the formation of microbial biofilms.

The present chapter attempts to identify the specific role, if any, played by the peptide hormone on the biofilm formation by various bacteria.

DOI: 10.1201/9781003519041-6

6.2 PEPTIDE HORMONES

Peptide hormones are important regulators of metabolism and energy balance. They have been linked to the regulation of energy consumption, hunger, reproduction, and the gastrointestinal and cardiovascular systems. There is also mounting evidence that peptide hormones and their receptors interact with both brown and white adipose tissue to help regulate energy balance. Peptides are compounds containing two or more amino acids bonded together by amide bonds (also called peptide bonds). They may contain as few as two or as many as thousands of amino acids. Amino acid-based hormones are either amino acid derivatives or peptides.

Peptide hormones are long-chain peptides composed of between 3 and 200 amino acids linked by peptide bonds. They are categorized into glycoproteins, short polypeptides, and small protein hormones. The first of these categories includes thyroid-stimulating hormone (TSH), luteinizing hormone (LH), and follicle-stimulating hormone (FSH) whereas the second category includes adrenocorticotropic hormone (ACTH), antidiuretic hormone (ADH), and oxytocin (OXT). Examples of protein hormones include human growth hormone (HGH), prolactin (PRL), and insulin. These hormones are mostly classified under large precursor protein moieties that can undergo or perform cleavage in enzymatic reactions.

Peptide hormones, which are signalling molecules made of amino acids, typically engage with G protein-coupled receptors (GPCRs) located on the surfaces of cells. Binding to these receptors by peptide hormones triggers a cascade of intracellular events, ultimately leading to a variety of physiological responses. Numerous peptide hormones, such as glucagon, calcitonin, and oxytocin, are recognized for their action via GPCRs. Recently, there has been increasing interest in how microbial growth and biofilm formation can influence the behaviours and pathogenicity of peptide hormones.

Biofilm are the complex microbial communities protected by EPS matrix, that making bacterial cells antibiotic-resistant and protecting them from environmental stress. It has also been suggested recently that peptide hormones can influence biofilm formation and microbial resistance. Some recent studies have suggested that the interplay or cascade of benign reactions occurs between bacterial biofilms and the peptide hormones (Gannesen et al., 2024). It was mentioned that the C-type natriuretic peptides on *Staphylococcus epidermidis* biofilms emphasize how protein expression varies in anaerobic environments. On the other hand, Patel et al. (2025) demonstrated and proposed a potential link between the bacterial cellular membrane and peptide hormones, which can serve as a good, potent antimicrobial therapy. This complete analysis was performed on microbes such as *E. coli*. Another study, by Otzen et al. (2025), suggests a broad link between the implications of peptide hormones in health and disease formation. It has been suggested that bacterial biofilms, which affect amyloid aggregation, are linked to neuropediatric disorders.

6.3 EFFECT OF HORMONES FROM THE PITUITARY GLAND AND BIOFILM

6.3.1 Effect of Anterior Pituitary Hormones

Pituitary hormones, including growth hormone, prolactin, and ACTH, primarily regulate important physiological functions such as growth, metabolism, and responses to stress, but they do not directly impact biofilm development or dispersal. As stated earlier, biofilms are structured microbial communities embedded within a self-produced extracellular matrix, usually regulated by bacterial quorum sensing and the availability of nutrients, pH, and surface chemistry. These microbial activities are controlled by local environmental signals instead of endocrine regulation from the brain. Thus, the role of the pituitary gland is largely distinct from microbial biofilm dynamics, which are more dependent on microbial communications and environmental conditions than on systemic hormonal regulation.

6.3.2 Effect of Posterior Pituitary Hormones

Oxytocin and vasopressin, two of the neuropeptides secreted by the hypothalamus but released by the posterior pituitary, are increasingly known for their functions beyond the realm of classical physiology, such as possible interactions with microbial biofilm. Oxytocin, for instance, has been seen to inhibit the formation of biofilm by some types of bacteria. This inhibition can be implemented through down-regulating the genes responsible for biofilm construction, thereby reducing the ability of bacteria to make stable, defensive communities. In addition, oxytocin might reinforce the efficiency of certain antibiotics to form a synergistic effect that enhances antimicrobial activities. This information implies that oxytocin has the potential to act as an effective adjunct for the treatment of biofilm-mediated infections, which are typically refractory to conventional antibiotic treatment. Oxytocin was found to act on the gut–brain axis and also influence the microbiome content, and, in doing so, modulate host immunity, metabolic status, and health.

Arginine vasopressin (AVP), however, has more subtle and less well-defined interactions with biofilm processes. AVP does not seem to directly stimulate biofilm production by bacteria such as *Pseudomonas aeruginosa*, but could affect the host's physiological reaction to such infections. Biofilms themselves can alter host sensory neuron activity, and thus inflammation, immune responses, and pain perception. Because AVP is a central regulator of stress and inflammatory responses, its function in host response to biofilm-related infections might be indirect but consequential. AVP also affects gastrointestinal function and immune regulation, which could influence gut microbiota composition. Microbiota changes, frequently organized in biofilm-like communities, can feed back into systemic health and disease susceptibility. Generally speaking, though oxytocin and vasopressin do not primarily act as antimicrobial molecules, their control over microbial function, host physiology, and microbiome relationships depicts a more complex role in controlling biofilm-based health effects.

6.3.3 The Impact of Alpha-MSH and Hydrophobic Peptides

The occurrence of infections brought on by different Gram-negative organisms, particularly *Escherichia coli*, is a persistent problem in both healthcare and community settings because of their propensity to build biofilms and their growing resistance to existing medicines. Although the majority of treatment regimens aim to eradicate *E. Coli* cells, the pathogenicity factors known as endotoxins (lipopolysaccharides), which are linked to the onset of sepsis and the primary cause of death in critical care units worldwide, are frequently disregarded. This study used microbiological, biophysical, and microscopic experiments to examine peptides' effectiveness based on alpha-melanocyte-stimulating hormones (Singh and Mukhopadhyay, 2014), namely Ana-9 and Ana-10, against *E. coli*. Confocal microscopy confirmed that Ana-9 and Ana-10 both showed increased activity against planktonic *E. coli* cells and maintained their effectiveness against biofilm. From a molecular standpoint, spectroscopic investigations showed that the peptides' structural changes resulted from their interaction with LPS (Lipopolysaccharide) and insertion into the hydrophobic environment of LPS. The peptide can penetrate the inner membrane, depolarize it, and eventually suppress the bacterial cells within the biofilm due to the electrostatic interaction between the peptide and LPS, which causes the outer membrane to become disorganized (Patel et al., 2025).

6.4 EFFECT OF HORMONES FROM THE THYROID GLAND AND BIOFILM

6.4.1 Impact of Thyroxine on Biofilm Formation

Thyroxine is an important thyroid hormone involved in the regulation of bacterial growth and biofilm formation. It was evident that thyroxine can modify the structure, stability, and formation of biofilms through influencing bacterial growth and surface colonization.

6.4.1.1 Impact on Bacterial Growth

The effect of thyroxine on biofilm formation is directly related to its effect on bacterial growth. Where bacterial growth is stimulated, as in Gram-negative species, thyroxine will stimulate increased biofilm formation. In bacteria where growth is inhibited, like Gram-positive strains, biofilm formation is frequently decreased. It was observed that the growth of Gram-negative bacteria, including *Escherichia coli*, is stimulated, while simultaneously the growth of Gram-positive bacteria like *Bifidobacterium longum* is inhibited. This differential effect highlights thyroxine's ability to selectively modulate bacterial populations. Therefore, it is apparent that thyroxine's modulating effect on biofilm formation is an extension of its effect on bacterial activity and viability.

6.4.1.2 Impact on Adherence to Surfaces

In addition to growth, thyroxine dramatically influences bacterial adherence to surfaces, an obligatory first step in the development of biofilms. It is notable that this impact does not seem to be entirely mediated via modifications in the hydrophobicity of bacteria. Additional mechanisms, potentially involving changes in surface proteins or signalling mechanisms, could be responsible for thyroxine's influence on bacterial adhesion. Thyroxine has been found to have distinct anti-biofilm activity against *Lactobacillus reuteri*, indicating a multifaceted, species-specific interaction. These differences point towards directed research for gaining an absolute insight into the role of the hormone in varied microbial communities.

6.4.1.3 Impact on Gut Microbiota

The effect of thyroxine extends to the gut microbiota, affecting bacterial growth kinetics, biofilm development, and surface adhesion in vitro. These observations imply that variations in thyroxine levels, such as those encountered in thyroid diseases, could profoundly alter the composition and functional properties of the gut microbiome. Consequently, thyroxine could indirectly influence host health through the modulation of biofilm-related bacterial populations within the gastrointestinal tract.

6.4.2 Effect of Triiodothyronine (T3) on Biofilm

Triiodothyronine (T3), one of the most significant thyroid hormones, not only controls human metabolism but also affects bacterial behavior, including biofilm formation. Recent research suggests that T3 may have either an enhancing or an inhibitory effect on biofilm formation depending on the bacterial strain, environmental conditions, and hormone concentration. In Gram-negative bacteria such as *Escherichia coli*, T3 has been reported to enhance biofilm formation. Such an effect can be modulated by the regulation of gene expression related to adhesion, motility, and extracellular matrix synthesis in a manner that allows bacteria can form and sustain biofilms better.

In contrast, in some Gram-positive bacteria such as *Bifidobacterium longum*, T3 has been found to inhibit biofilm production. The mechanisms behind this might be the disruption of bacterial communication networks, including quorum sensing, or the inhibition of the synthesis of essential biofilm components such as exopolysaccharides. This inhibitory action indicates that T3 may selectively affect microbial populations, perhaps favoring the growth of beneficial bacteria by inhibiting pathogenic bacteria (Nakamura et al., 2022).

T3's effect on biofilm formation is also regulated by the activities of other hormones, including norepinephrine or cortisol, which may synergize or antagonize T3's action.

Furthermore, T3 dosage is also a critical aspect, as low concentrations might induce biofilm development, while higher concentrations might suppress biofilm formation, depending on the involved bacterial strain.

These results highlight the intricate interplay between host hormonal signals and microbial communities. Insights into how T3 and other host factors regulate biofilm dynamics may reveal novel avenues for the control of bacterial infections, especially those involving biofilm-mediated antibiotic resistance.

6.4.3 Effects of Calcitonin on Biofilm

Calcitonin, a calcium homeostasis hormone, has been increasingly recognized to be involved in modulating the development of bacterial biofilms. Biofilms, a consortium of complex microorganisms embedded in an extracellular matrix that they have synthesized, play key roles in bacterial persistence, antibiotic resistance, and pathogenicity. Calcitonin's interaction with biofilm formation appears to be species-dependent and multifaceted.

One such aspect of this interaction involves the regulation by calcitonin of calcium concentration. Calcium ions are known to encourage biofilm development through the stabilization of the extracellular matrix, increasing cell adhesion, and promoting interspecies communication among bacterial cells. Higher calcium concentrations thus stabilize biofilm structure and improve resistance against environmental stresses. By modulating systemic calcium levels, calcitonin may indirectly influence these processes either by favoring or by impeding biofilm formation based on bacterial species and environment.

Notably, certain research indicates that calcitonin per se, independent of its calcium-regulating function, might also act directly on bacterial cells or modulate biofilm kinetics. For example, some studies have noted that calcitonin suppresses biofilm development or modifies biofilm structure in certain bacterial species. This might be mediated by mechanisms like the inhibition of bacterial signal transduction, the modulation of gene expression for biofilm production, or interference with extracellular matrix elements.

But the impact of calcitonin is not consistent for all bacteria. For some, it seems to increase biofilm resilience, perhaps through indirectly stabilizing biofilm organization through calcium-mediated processes. In others, it appears to decrease biofilm stability, implying an intricate and situation-dependent function.

In general, the interaction between calcitonin, calcium homeostasis, and bacterial biofilm development points to a fascinating nexus between host hormone function and microbial physiology. A more profound understanding of this interaction may have significant implications for the control of biofilm-related infections, especially in therapeutic contexts in which biofilm resistance presents a considerable problem.

6.4.3.1 The Effects of Calcitonin on *Staphylococcus Epidermidis*

Staphylococci are capable of detecting Substance P (SP) in the skin; however, this neuropeptide is usually released from nerve terminals along with another molecule, Calcitonin Gene-Related Peptide (CGRP). In several studies, it was discovered that CGRP significantly activated the virulence of *Staphylococcus epidermidis* at extremely low concentrations ($<10^{-12}$ M), whereas *Staphylococcus aureus* was insensitive to CGRP. Interestingly, CGRP-treated *S. epidermidis* induced the release of interleukin-8 (IL-8) from keratinocytes, an action associated with augmentation of the secretion of the antimicrobial peptide cathelicidin LL-37. While *S. epidermidis* was not found to alter the secretion of virulence factors, substantial modifications in its surface characteristics were noted. Upon exposure to CGRP, *S. epidermidis* showed an enhancement in adhesion to keratinocytes but reduced internalization and biofilm development, modifications that were associated with augmented surface hydrophobicity. The DnaK chaperone was found to be the CGRP-binding protein in *S. epidermidis*. In addition, the action of CGRP was blocked by gadolinium chloride ($GdCl_3$), an MscL mechanosensitive channel inhibitor. $GdCl_3$ also inhibited the membrane translocation of EfTu, the Substance P sensor (N'Diaye et al., 2016a).

6.4.3.2 Effects of Calcitonin on Other Bacteria

Research observed that calcium structurally stabilizes the biofilm matrix in *Bacillus subtilis* to prevent biofilm dispersion. Under the controlled conditions, the dispersion of the biofilm, followed by the escape of the cells from deteriorating biofilms, was observed in a modified LBGM medium. Dispersion of biofilm was driven synergistically by decreased expression of the *epsA* operon, which encodes exopolysaccharide synthetases, and by sporulation through activation of the *spoVF* operon. Constitutive expression of *epsA* in a sporulation-deficient mutant effectively blocked biofilm dispersion. Notably, the addition of calcium halted biofilm dispersion without affecting the transcription of matrix or sporulation genes, indicating a post-transcriptional stabilization of the biofilm. It was also demonstrated that calcium enhanced the physical robustness of the biofilm matrix, likely by neutralizing negative charges or forming ion bridges between matrix polymers. Removal of calcium weakened biofilm architecture without inhibiting biofilm formation itself. The study proposes that during sporulation under calcium-limited conditions, calcium is sequestered from the matrix into spores, weakening the biofilm and promoting dispersion. These findings reveal calcium to be a key structural component that enhances biofilm resilience and suggest that calcium manipulation could serve as a novel strategy for biofilm control (Nishikawa and Kobayashi 2021).

6.5 EFFECT OF PARATHYROID GLAND HORMONES ON BIOFILM

Parathyroid hormone (PTH), an important calcium homeostasis regulator in the blood, has been seen to have indirect effects on microorganism growth and development within biofilms, especially in dental and periodontal environments. Although the main function of PTH is to regulate calcium and phosphate balance, its effect on host tissues and immune responses sets the stage for influencing microbial communities.

In periodontal health, PTH has proved to have beneficial effects. It has been reported to enhance bone regeneration and attenuate inflammation in experimental models of periodontitis, a chronic infection strongly linked with biofilm formation. Through the promotion of bone healing and the modulation of the inflammatory response, PTH has the potential to restore periodontal structures lost due to biofilm-associated pathogens, indicative of potential therapeutic merit in the management of periodontitis and other biofilm-related oral infections.

In the case of dental caries, studies have investigated the connection between PTH gene polymorphisms and susceptibility to tooth decay. While the results have been equivocal, the studies suggest the potential for a genetic connection between PTH activity and the susceptibility to caries, an etiology inherently rooted in bacterial biofilm on the surfaces of teeth.

Outside of the oral cavity, PTH also interacts with gut microbiota. Changes in PTH levels can affect the balance and composition of gut microbial communities, potentially altering the ratio of beneficial to pathogenic bacteria. Because the gut microbiota has an important role in systemic health, these changes might have downstream effects on biofilm formation and microbial behaviour elsewhere in the body.Overall, though the major function of PTH is systemic regulation of calcium, its indirect role in microbial ecosystems, especially within biofilm-based communities such as the oral and gut, suggests its wider implications for health and disease. Much remains to be elucidated on the mechanisms and possible clinical uses for these interactions.

6.6 EFFECTS OF HORMONES FROM THE GASTROINTESTINAL TRACT ON BIOFILM

6.6.1 Effect of Gastrin

Gastrin plays a pivotal role in controlling biofilm formation within the gastrointestinal system, especially in stimulating the development of some bacteria like *Helicobacter pylori*. It was observed that gastrin stimulates *H. pylori* growth through decreasing the lag period, enhancing logarithmic phase growth, and doubling the final bacterial count. The effects of gastrin are moderated by competition from other peptides like cholecystokinin (CCK) and penta-gastrin that can suppress *H. pylori* growth. Other hormones, like ghrelin, have unique and sometimes contrasting effects on bacterial growth and biofilm production from gastrin (Zavros et al., 2005). Gastrin also plays a critical role in ensuring the integrity of the gastric mucosa. Modulation of mucosal composition directly affects biofilm formation; e.g., fucosylated mucus has been demonstrated to inhibit biofilms inhabiting intestinal crypts, while non-fucosylated mucosa is not as resistant. Moreover, host and bacterially derived proteases also affect biofilm development and pathogenicity by degrading structural proteins necessary for maintaining biofilm integrity. The effect of Gastrin on bacterial growth and mucosal structure regulates biofilm properties like size, architecture, and enzymatic content. Pathological conditions like gastronomes, in diseases like Zollinger–Ellison Syndrome, which lead to the overproduction of gastrin, cause ulcers, and change the gut microbiota to favour biofilm-associated alterations. Chronic atrophic gastritis, linked with increased gastrin levels, also predisposes to gastric cancers and alters microbial populations.

6.6.2 Effect of Ghrelin

Ghrelin, a 28-amino acid peptide that is expressed mainly in the stomach and pancreas, occurs in two forms: acylated ghrelin (AG) and deacylated ghrelin (DAG). Though both forms occur in plasma and saliva at approximately 100–200 pg/ml, AG alone interacts with the growth hormone secretagogue receptor (GHS-R1a), promoting the release of growth hormone and modulating functions, including appetite regulation, memory, glucose homeostasis, thymopoiesis, and neuronal survival. Even though AG is well documented to have anti-inflammatory action on gingival epithelial cells, direct antimicrobial actions of AG and DAG on microbes were not investigated thoroughly.

Given that the chemical properties of ghrelin—namely, a positive net charge and putative amphipathic helix structure—have similarities to established antimicrobial peptides such as vasoactive intestinal peptide (VIP), orexin B (ORXB), and LL-37, it was hypothesized by the researchers that ghrelin could have antimicrobial action as well. To do this, they compared the bactericidal activities of AG and DAG against Gram-negative (*Escherichia coli* and *Pseudomonas aeruginosa*) and Gram-positive (*Staphylococcus aureus* and *Enterococcus faecalis*) bacteria. They also examined the impact of AG and DAG on the negative surface charge of *E. coli* cells. This research sought to extend the knowledge of ghrelin's possible antimicrobial activity beyond its known functions in host biological processes (Min et al., 2012).

6.7 EFFECTS OF HORMONES FROM THE PANCREAS ON BIOFILM

6.7.1 Effect of Glucagon

A few studies demonstrate that glucose concentration directly affects biofilm formation by *Staphylococcus epidermidis* and *Staphylococcus aureus* in vitro, showing a distinct threshold response. Since it is known that hyperglycemia is associated with enhanced postoperative and nosocomial infections, the authors examined how glucose levels within the clinically relevant range of 20–300 mg/dL affect biofilm production, an important determinant of implant-related infections.

Biofilm formation was assessed by culturing *S. epidermidis* and *S. aureus* in Lennox broth supplemented with incremental glucose concentrations. The biofilms were stained using crystal violet, and the biomass was determined by absorbance measurements after 24 hours and 48 hours of incubation, respectively. Statistical analysis indicated that the biofilm mass rose in a threshold-dependent manner with increasing glucose concentrations. *S. epidermidis* showed two thresholds, between 0–20 mg/dL and 160–200 mg/dL, leading to 36% and 26% augmentation in biofilm formation, respectively. For *S. aureus*, a major 196% increase was found at 200–240 mg/dL.

Notably, *S. epidermidis* consistently produced more biofilm than *S. aureus* at every glucose concentration, and species-specific differences were established. These observations show that hyperglycemia increases bacterial pathogenicity both by weakening the host immune system and by directly stimulating biofilm production, hence enhancing the risk of infection.

Although the study strongly establishes a glucose-dependent threshold for biofilm formation, it recognizes limitations. The in vitro model is not as complex as in vivo systems, with host proteins, changing pH, and glucose levels, etc. Only two species of bacteria were tested, and biofilms were monocultures and not polymicrobial communities found in clinical infections. Despite these limitations, the evidence strongly suggests that abnormal glucose levels greatly enhance pathogenic biofilm formation (Waldrop et al., 2014).

Antimicrobial peptides (AMPs) have become a promising substitute for antibiotics over the past two decades. The widely held belief is that these peptides' antibacterial properties are forcing bacteria to develop resistance mechanisms. Thus, among the several methods created to spread or eliminate biofilm, AMP, either alone or in combination with antibiotics, has been suggested as a potential anti-biofilm weapon (Reffuveille et al., 2014). GLP-1, a crucial signalling hormone that increases insulin secretion and satiety while lowering glucagon secretion and stomach emptying, is already employed as treatment target for type 2 diabetes (T2D). According to preliminary research, when giving rats a high-fat diet, the prebiotic oligofructose raised their GLP-1 levels as determined by the portal vein and prevented them from gaining weight (Cani et al., 2005; LeValley et al., 2020).

It is found that GLP-1 is an incretin hormone that plays a vital role in maintaining glucose homeostasis and regulating appetite. The results show that the common gut commensal E. *faecalis* can alter host metabolism by breaking down GLP-1 with its protease GelE. The regulation of glucose and the emergence of metabolic disorders, including type 2 diabetes and obesity, may be significantly influenced by these interactions. Novel therapeutic strategies that aim to modulate microbial variables to preserve metabolic health can be derived from knowledge of these host-microbiome interactions (Louis et al., 2023).

Some major pieces of information were received from a study about how the Osteocrin (OSTN), a natriuretic peptide receptor agonist, can break up biofilms made by *Pseudomonas aeruginosa*. This bacterium is responsible for chronic infections due to its ability to form biofilms. With doses comparable to hANP (Human Atrial Natriuretic Peptide), OSTN was observed to disperse established biofilms of the *P. aeruginosa* PA14 strain. Nevertheless, OSTN's dispersal effect was marginally weaker than hANP's (–61% versus –73%). The therapeutic applications of OSTN in clinical settings require more exploration, especially in light of its effectiveness against different bacterial strains and its function in biofilm-mediated infections.

6.7.2 Effect of Insulin

Insulin has an important role to play in the regulation of *Escherichia coli* biofilm. The research revealed that insulin acts as a QS molecule, regulating biofilm production based on environmental conditions and the availability of nutrients. With minimal nutrient media (such as Yeast Nitrogen Base with 1% peptone, YNBP), insulin alone greatly increased biofilm production compared to complex nutrient-rich media (Mueller Hinton, MH). This indicates that limiting nutrients increases the insulin-mediated biofilm response.

Temperature conditions and aeration also affect the effects of insulin. At 23°C, insulin increased levels of biofilm in both static and shaking cultures, particularly where minimal media were employed. However, when maintained at 37°C (the temperature of the human body), biofilm production was typically greater in static culture and was increased further by the addition of insulin. Notably, the inclusion of sugars varied the insulin impact. Upon the addition of glucose, insulin synergistically enhanced biofilm formation over either glucose or insulin alone. This was most significant under minimal media at 23°C, and levels of biofilm were increased up to 220-fold relative to elevated temperatures. Galactose plus insulin elicited the highest levels of biofilm formation, significantly more than with glucose or lactose, particularly at 37°C in complex media. This can be attributed to the role of galactose-specific lectins in increasing biofilm-associated aggregation.

On the other hand, lactose inhibited biofilm formation when used in combination with insulin, yielding much lower biofilm quantities than glucose or galactose.

In general, action of insulin is bifunctional: without sugars, it prefers planktonic (free-floating) forms, whereas in the presence of glucose or galactose, it supports adhesion and biofilm formation. These observations indicate that in diabetic conditions, where insulin and glucose levels are high, the potential for biofilm-related infections (such as urinary tract infections) could be higher.

Furthermore, modulation of sugar and insulin levels may be investigated as a method for regulating biofilm formation, particularly for infection control of antimicrobial-resistant *E. coli* strains (Patel et al., 2021).

6.7.2.1 Factors Affecting Biofilm Formation by Insulin

6.7.2.1.1 Intracellular c-di-GMP Levels

Insulin therapy dramatically elevates intracellular cyclic di-GMP (c-di-GMP) contents of *Pseudomonas aeruginosa*. Cyclic di-GMP is an important second messenger that controls the formation of biofilms by increasing bacterial surface adhesion, cell clumping, and the release of extracellular polymeric substances. Experiments in diabetic mouse models revealed that wounds infected with *P. aeruginosa* and subjected to insulin treatment exhibited increased c-di-GMP content, leading to stronger biofilm development than in untreated controls. In vitro, insulin stimulated c-di-GMP production directly, indicating a direct correlation between exposure to insulin and enhanced biofilm biomass.

6.7.2.1.2 Host Immune Modulation

Insulin not only influences bacterial behaviour but also modulates the host immune response. In diabetic wound models, insulin treatment biased immune polarization towards a Th2-type response, as evidenced by augmented IL-4 expression and the earlier peak of IFN-γ expression. This bias decreased effective bacterial clearance and extended the inflammatory phase of wound healing. Therefore, this immune modulation indirectly contributed to prolonged biofilm development, establishing a chronic infection environment that further delayed wound closure.

6.7.2.1.3 Nutrient Availability

Under diabetic conditions, insulin therapy changes nutrient distribution and availability in wound environments. The increased availability of nutrients during insulin therapy supports bacterial growth and biofilm development. The interplay between high glucose levels and insulin-mediated metabolic alterations creates an optimal niche for *P. aeruginosa* to form dense, resistant biofilms, hindering wound healing and enhancing antibiotic resistance (Wei et al., 2019).

6.7.2.1.4 Role of Insulin/Glucose Ratio in the Regulation of Pathogen Biofilm Formation

Insulin can function as an adjuvant in the regulation of biofilm levels. The modulation of insulin-regulated biofilm formation by environmental conditions has not been previously described. The present study measured the effects of various environmental growth conditions and nutrients on insulin-modulated levels of biofilm production (Patel et al., 2021). The polypeptide QS molecule insulin has a wide host specificity and is produced by bacteria, fungi, vertebrates, and protists.

Similar to human insulin, *E. coli*-derived insulin causes reactions in mammalian cells and tissues. Human insulin influences the development of *E. coli* biofilms and shows global rheostat-like control over all phases, including adhesion, motility, cell-population expansion, and the creation and maturation of biofilms. The modifications that have been shown to lead to increased virulence are related to the changes in cell wall design. The alterations seem to be related to glucose absorption and insulin binding to the cell, which is consistent with findings of insulin's interaction with mammalian cells (Klosowska and Plotkin, 2006). Hydrophobicity and capsule formation are examples of cell surface modification that took place in tandem with changes in biofilm levels. According to these results, insulin combined with glucose, which is found in the urine of people with type 2 diabetes, can activate the production of biofilms and raise the risk of urinary tract infections (UTIs) (Ostrowska et al., 2013). The standard practice used to keep human blood glucose levels under control for people who are unable to metabolize glucose on their own is the use of basal insulin, which is administered with every meal. The sole insulin approved for use as a basal therapy in patients with type 1 and type 2 diabetes is insulin detemir, along with a different kind of long-acting (basal) insulin called insulin glargine. According to recent research, insulin, a hormone essential for glucose metabolism, may also influence bacterial pathogenicity, especially in *E. coli*-mediated infections. There are several factors, including vascular problems, hyperglycemia, and compromised immune responses, that make diabetics more vulnerable to infections. UTIs caused by *E. coli* are especially prevalent among these diseases.

Based on experimental observations, insulin can exacerbate infection in diabetic individuals by serving as a stimulatory factor in the pathophysiology of *E. coli* (Wei et al., 2019). Furthermore, chronic wound infections in diabetics are prevalent, with *P. aeruginosa* being a common pathogen. These bacteria develop biofilms that hinder the healing process and lead to recurring infections. Research indicates that insulin may unintentionally promote the development of *P. aeruginosa* biofilms by enhancing intracellular c-di-GMP levels, even though insulin is necessary to regulate blood glucose levels in diabetes patients. This effect may contribute to delays in wound healing and exacerbate chronic infections. Clinicians managing diabetic wounds should be aware of this potential consequence and consider strategies to mitigate biofilm-related issues. Understanding this connection is crucial for developing comprehensive treatment programs that address both infection management and metabolic control in diabetic patients (Watters et al., 2014).

From the aforementioned research, we learn that patients with diabetes are more susceptible to chronic wound infections, often exacerbated by biofilm-forming bacteria such as *P. aeruginosa*. While insulin is crucial for controlling blood glucose levels in diabetic individuals, it may inadvertently affect the immune system in a way that encourages the growth of *P. aeruginosa* biofilms. This effect can increase antibiotic tolerance, leading to more chronic wound infections and treatment challenges. Despite its importance for glycaemic control, insulin treatment may have unexpected consequences on the host's immune system that promote the development of *P. aeruginosa* biofilm in diabetic wounds (Tang et al., 2023). In addition, scientists have established a novel treatment strategy to address the challenges associated with diabetic periodontitis, an inflammatory and chronic bacterial biofilm condition worsened by hyperglycaemia.

6.8 EFFECT OF OTHER PEPTIDE HORMONES ON BIOFILM

6.8.1 Impact of *Somatostatin* on *H. Pylori*

Helicobacter pylori is the only source of gastric ulcers and peptic ulcers in the human stomach. It colonizes the stomach and intestinal mucosa, releasing peptide hormones such as somatostatin and gastrin. Gastrin peptides exist in two active forms: gastrin-17 and gastrin-34. A study shows that the somatostatin effects are comparatively greater than gastrin against *H. pylori* growth. On the other hand, it also concluded that somatostatin did not cause any growth inhibition in *E. Coli* (Yamashita et al., 1998). The effects of somatostatin on human cells are mostly mediated by adenylate cyclase and rarely by guanylate cyclase, which affect intracellular cAMP and cGMP concentrations. Two cell-permeable analogues of cAMP and cGMP were investigated for their influence on *H. pylori*. *H. pylori* showed increased intra-bacterial concentrations of cAMP and cGMP when treated with somatostatin. Despite these inconsistencies, it appears that somatostatin targets a particular sensor in *H. pylori* (Yamashita et al., 1998).

Researchers identified binding studies using radiolabelled somatostatin-14 (SST-14) in order to characterize the somatostatin sensor in *H. pylori* (Lesouhaitier et al., 2019), This study discusses how somatostatin interacts with *H. pylori*, influencing its physiology and biofilm formation. It also highlights the role of SST-14 in modulating bacterial behaviour.

6.8.2 Impact of Natriuretic Peptides and *P. Aeruginosa* Hormone Presentation

Natriuretic peptides are principally renowned due to their cardiovascular and osmoregulation activities. The members of this peptide family are C-type natriuretic peptide (CNP), atrial natriuretic peptide (ANP), and brain natriuretic peptide (BNP) (Louis et al., 2022)

BNP, like other natriuretic peptides, can impact biofilm development and dispersion, especially in *Pseudomonas aeruginosa*. BNP may suppress biofilm development and disperse existing biofilms, potentially via activating the AmiC and AmiR pathways, which regulates biofilm formation in *P. aeruginosa*. BNP can also operate as an adjuvant, boosting the anti-biofilm activity of some antibiotics (Rosay et al., 2015).

It is found that human BNP (hBNP) has antimicrobial action against both Gram-positive and Gram-negative bacteria among the natriuretic peptides studied; hANP and urodilatin are also active, although with a lesser activity; hCNP does not exhibit any direct effect against bacteria. For this study to work properly, it is also required to evaluate the real potential of bacterial physiology. After the work, it was concluded that there has been no impact of hBNP and hCNP at 1 and 0.1 μM on *P. aeruginosa* growth. In addition, the study also concluded that HCNP was the most active peptide as it was still able, at 10 nM, to halve the biofilm development by *P. aeruginosa* and could inhibit biofilm formation in a dose-dependent way (Desriac et al., 2018).

6.8.3 Impact of Substance P on Bacteria

A cutaneous strain of *Bacillus cereus* was employed as the paradigm for the initial investigation of the impact of Substance P (SP) on bacteria. It was shown that SP had comparable effects on the production of biofilms in *S. aureus* and *Staphylococcus epidermidis*, with an increase in thickness following exposure to SP (1 μM) (N'Diaye et al., 2016b). Inflammation, stress, or even a psychological breakdown, can raise the concentration of SP in the skin, which is commonly linked to skin problems (Severini et al., 2002). This suggests that the effect of SP on bacterial biofilm should be assessed in connection with these skin conditions (Totté et al., 2016) and this idea served as the foundation for the creation of novel cosmetics and medications with therapeutic uses in dermatology.

6.9 MECHANISM OF ACTION OF PEPTIDE HORMONES CONCERNING BIOFILMS

Although many of the peptide hormones act in a different manner toward the bacterial biofilm, few of them promote biofilm formation, while others often function like antibacterial peptides (Table 6.1).

TABLE 6.1
A Brief Discussion on Peptide Hormones from the Hypothalamus, Along with Their Functions and their Effects on Biofilm Growth

Name of Hormone	Source	Major Function	Effect on Biofilm
Growth hormone releasing hormone	Hypothalamus	Stimulation of Somatotroph cells of Pituitary	No direct effect
Thyrotropin-releasing hormone	Hypothalamus	Stimulation of Thyrotroph cells of Pituitary	No direct effect
Corticotropin-releasing hormone	Hypothalamus	Stimulation of Corticotroph cells of Pituitary	No direct effect
Gonadotropin-releasing hormone	Hypothalamus	Stimulation of Gonadotroph cells of Pituitary	No direct effect
Somatostatin	Hypothalamus, pancreas, gut	Inhibits the release of growth hormone, insulin, glucagon, and other hormones.	Inhibiting biofilm formation and cell proliferation although, its direct impact on biofilm formation is not fully established.
Growth hormone	Pituitary (Anterior)	Promotes normal physical growth in children, Helps in protein synthesis, skeletal growth and impact on metabolic functions.	In some cases, somatotrophin may promote biofilm development, potentially enhancing bacterial virulence and resilience. Conversely, it can also inhibit biofilm formation, particularly in certain bacterial strains. The effects of somatotrophin on biofilms are complex and species-specific, often influenced by factors like the presence of other bacteria in mixed-species communities.
Thyroid-stimulating hormone	Pituitary (Anterior)	Stimulates thyroid gland to release T3 and T4.	TSH itself doesn't directly interact with biofilms.
Adrenocorticotropic hormone	Pituitary (Anterior)	Stimulates adrenal cortex to release different steroid hormones.	No direct action.
Follicle-stimulating hormone	Pituitary (Anterior)	Stimulates growth of ovarian follicles, promoting estrogen production and preparing the eggs for ovulation. In males, FSH stimulates the Sertoli cells in the testes to produce sperm.	Follicle-stimulating hormone (FSH) primarily affects biofilm formation by influencing the development of ovarian follicles and the production of estradiol.
Luteinizing hormone	Pituitary (Anterior)	stimulates ovulation and the production of progesterone, while in males, it stimulates testosterone production.	Luteinizing hormone (LH) does not directly affect biofilm formation.
Melanocyte stimulating hormone	Pituitary	responsible of the melanin synthesis, being the main actor of skin pigmentation.	Alpha-melanocyte stimulating hormone (α-MSH) and its derivatives have been shown to have anti-biofilm activity, meaning they can inhibit the formation and/or disrupt existing bacterial biofilms.

Prolactin	Pituitary (Anterior)	stimulates milk production in the mammary glands during pregnancy and after childbirth.	Prolactin's effect on biofilms is multifaceted and not fully understood, but research suggests it can influence both the formation and antibacterial properties of these bacterial communities. While some studies indicate prolactin may stimulate biofilm formation, others suggest it can also contribute to antimicrobial defense mechanisms.
Vasopressin	Pituitary (Posterior)	regulate body fluid balance, particularly water retention.	vasopressin can indirectly impact biofilm development through its effects on the immune system, stress responses, and gut microbiota.
Oxytocin	Pituitary (Posterior)	Plays a key role in childbirth and lactation.	Oxytocin can have a complex effect on bacterial biofilm formation, potentially both reducing and increasing it.
Insulin	Pancreas	Reduces blood glucose level.	Insulin can significantly enhance biofilm formation in certain bacterial species, particularly in the presence of specific nutrients and environmental conditions.
Glucagon	Pancreas	Increases blood glucose level.	Glucagon, a hormone that raises blood sugar, does not directly affect biofilm formation itself, but it indirectly influences it by influencing blood glucose levels.
Human chorionic gonadotropin	Placenta	Maintains pregnancy	does not directly influence biofilm formation.
Calcitonin	Thyroid	Lowers blood calcium level	can have varying effects on biofilm formation depending on the bacteria and the specific conditions. In some cases, calcium can promote biofilm formation, while in others, it can inhibit it. Calcitonin itself, or its related peptides like Calcitonin Gene-Related Peptide (CGRP), can also influence biofilm formation by modulating bacterial signaling pathways.
Parathyroid hormone	Parathyroid	Increases blood calcium level	PTH does not directly regulate biofilm formation, it can influence biofilm formation indirectly by impacting bone remodelling and the gut microbiome, which can then influence the composition and activity of biofilms.
Gastrin	Gut	Stimulate the secretion of gastric acid (HCl) in the stomach.	Gastrin does not appear to have a direct effect on biofilm formation, at least not in the same way as other hormones.
Secretin	Small intestine	Regulates digestive processes	Secretin, a gastrointestinal hormone, does not directly influence biofilm formation.
Cholecystokinin	Small intestine	Stimulates the release of bile and digestive enzymes.	CCK's effects on gut motility and immune responses can indirectly impact biofilm formation and bacterial activity.
Ghrelin	Stomach	Stimulates appetite and hunger.	May influence biofilm formation and dispersal
Leptin	White adipose tissue	Regulates body weight and energy balance	Indirect effect.
Asprosin	White adipose tissue	Regulates insulin sensitivity and thereby glucose homeostasis.	Is not directly known to affect biofilm formation or growth.
Somatomedin	Liver	Stimulates cell growth and division.	Somatomedins, a group of insulin-like growth factors, can influence biofilm formation by affecting bacterial growth and cell proliferation.XXXXXSome studies indicate that hormonal drugs, including those related to somatomedins, can affect the growth and biofilm formation of gut bacteria.
Atrial natriuretic peptide	Heart, cardiac muscle	Regulates blood volume and blood pressure by promoting sodium and water excretion	It can prevent biofilm formation and disperse established biofilms, even at low concentrations. ANP's effect appears to be mediated by the P. aeruginosa sensor AmiC and the AmiR regulator, and it can also enhance the effectiveness of certain antibiotics.

(*Continued*)

TABLE 6.1 (Continued)
A Brief Discussion on Peptide Hormones from the Hypothalamus, Along with Their Functions and their Effects on Biofilm Growt

Name of Hormone	Source	Major Function	Effect on Biofilm
Pancreatic polypeptide	Pancreas	Regulates pancreatic secretion and appetite	Has shown various effects on biofilm. While direct inhibitory effects of PP on biofilm formation in vitro are not widely reported, PP's role in modulating pancreatic secretion and potentially reducing bacterial viability within biofilms is suggested by some research.
Angiotensin	—	Regulates blood pressure	A decrease in Angiotensin II, has been linked to increased biofilm formation in some studies. XXXXXChanges in gut bacteria and their metabolites, as influenced by angiotensin II, may have implications for overall health, including the immune response and its ability to fight biofilm infections.
Erythropoietin (EPO)	Kidneys	Stimulates the bone marrow to produce more red blood cells	In some cases, EPO has been found to enhance the resolution of bacterial infections, particularly those involving E. coli and *Staphylococcus aureus*. This is likely due to EPO's ability to increase macrophage clearance of bacteria, potentially through the upregulation of CD36, a receptor involved in phagocytosis.XXXXXSome studies suggest that EPO may influence biofilm formation, potentially by modulating inflammation or cell signalling pathways. For example, in brain injury models, EPO has been observed to reduce neuronal apoptosis and cell biofilm damage, and it can also activate microglia, which play a role in tissue repair and inflammation.
Amylin	Pancreas	Regulates blood glucose levels and satiety.	Amylin, a hormone secreted with insulin, has been shown to affect biofilm formation, primarily by influencing amyloid protein aggregation, which is a key component of the biofilm matrix.
Brain natriuretic peptide	Cardiac ventricles	Regulates blood volume and blood pressure by promoting water and salt excretion through the kidneys,	Brain natriuretic peptide (BNP), like other natriuretic peptides, can influence biofilm formation and dispersion, particularly in *Pseudomonas aeruginosa*. BNP can inhibit biofilm formation and disperse established biofilms, possibly by activating the AmiC and AmiR pathway, a system involved in regulating biofilm formation in *P. aeruginosa*. BNP can also act as an adjuvant, enhancing the anti-biofilm effect of certain antibiotics.
Placental lactogen (PL)	Placenta	Prepares the mother's body for breastfeeding and ensures adequate fetal nutrition	PL has been shown to significantly inhibit biofilm formation by various microorganisms.XXXXXPL can decrease the hydrophobicity of both Gram-positive and Gram-negative bacteria, potentially impacting their ability to adhere to surfaces and form biofilms.XXXXXPL can alter the structure and morphology of biofilms, including affecting the nutrient channels within them.XXXXXPL can reduce the amount of eDNA in the biofilm matrix, which is a key component of the biofilm structure. XXXXXPL has demonstrated effectiveness in an in vitro wound biofilm model, suggesting its potential for therapeutic applications in infected wounds
Thrombopoietin (TPO)	Liver, kidney	Regulates platelet production	Thrombopoietin (TPO), a growth factor that primarily regulates platelet production, can have an inhibitory effect on bacterial biofilms, particularly those formed by *Staphylococcus aureus* and *Staphylococcus epidermidis*. Specifically, the thrombopoietin receptor agonist eltrombopag (EP) has shown promise in inhibiting biofilm formation and even eradicating mature biofilms.

Lipotropin	Pituitary	To mobilize lipids from adipocytes for use as an energy source	Lipotropins, specifically beta-lipotropin (β-LPH), can affect biofilm formation, particularly in Gram-negative bacteria. β-LPH can stimulate aldosterone production and also influence the growth and biofilm formation of bacteria like *B. fragilis* and *E. coli*. In contrast, some lipopeptides, like those produced by *Paenibacillus thiaminolyticus*, can inhibit biofilm formation in various bacterial species.
Adiponectin	Adipocytes	Regulates energy metabolism	Adiponectin's anti-inflammatory and anti-atherogenic properties may help to mitigate the inflammatory response and reduce the risk of complications associated with biofilm-related infections. While more research is needed to fully understand the effects of adiponectin on biofilm formation, its potential to modulate inflammatory processes and prevent tissue damage suggests that it could be a valuable tool in the fight against biofilm-related diseases.
Relaxin	Ovary, placenta	Prepares the body for childbirth	Relaxin has been shown to inhibit biofilm formation and reduce biofilm mass, likely by affecting extracellular matrix production and potentially by interfering with profibrotic factors. Specifically, relaxin has been observed to reduce collagen secretion and deposition, which are key components of the biofilm matrix. Additionally, relaxin may influence quorum sensing, a process that regulates biofilm formation and development.
Orexin/ hypocretins	Hypothalamus	Regulates sleep-wake cycles, arousal, and appetite	Research suggests that orexin/hypocretin signaling can influence biofilm formation in certain contexts, potentially through their effects on neuronal activity and immune responses.
Endothelin	Endothelial cells	Regulates blood pressure and vascular tone	Endothelin-1 (ET-1) primarily affects biofilm formation by influencing the inflammatory response and vascular function, potentially contributing to increased biofilm formation and/or resistance to treatment. ET-1 can trigger inflammation, vasoconstriction, and fibrosis, which could impact biofilm development in various contexts.
Hepcidin	Liver	Regulates iron homeostasis in the body	Hepcidin, specifically hepcidin 20, exhibits anti-biofilm activity, particularly in acidic environments, by reducing bacterial survival within biofilm structures. It disrupts the biofilm's extracellular matrix, altering the architecture of biofilms and making them more susceptible to antibiotics like vancomycin. Hepcidin's ability to bind to DNA may also play a role in destabilizing biofilm structure and reducing matrix accumulation
Encephalin	Central Nervous system, Adrenal medulla	Neurotransmitters that reduce pain signals	Do not directly inhibit biofilm formation or have a known effect on biofilm structure.
Anti-Mullerian hormone	Sertoli cells in the testes	Sexual differentiation during foetal development	Does not directly influence biofilm formation

Following naturally occurring antimicrobial and host defence peptides, the design of small synthetic peptides with broad-spectrum antibiofilm activity has provided a glimmer of hope in the fight against pan-antibiotic-resistant bacteria, the conventional superbugs (Louis et al., 2023; Pletzer et al., 2016). Some peptide hormones have been identified for their ability to modify the biofilm formation of both Gram-negative and Gram-positive bacteria. Many peptide hormones are related to biofilm formation, including Dynorphins, Somatostatin, and Natriuretic Peptides for Gram-negative bacteria, and Substance P, Calcitonin Gene-Related Peptide, and Natriuretic Peptides for Gram-positive bacteria. The impact on biofilm development might vary significantly depending on the bacteria and the exact circumstances. Calcium may either promote or prevent biofilm growth. Calcitonin and its associated peptides, such as Calcitonin Gene-associated Peptide (CGRP), can also impact biofilm development by regulating bacterial-signalling pathways (Roy et al., 2018).

6.9.1 Peptide Hormones Affecting Bacterial Biofilm Formation and Virulence

Previous studies have demonstrated that biofilm-associated wounds are typically sluggish to heal compared to non-biofilm-challenged normal wounds in diabetic mouse models. Recent studies showed that insulin treatment affects cyclic di-GMP signalling in *P. aeruginosa*-infected chronic wounds in db/db diabetic mice. Insulin treatment enhances *Pseudomonas aeruginosa* biofilm formation by increasing intracellular cyclic di-GMP levels, leading to chronic wound infection and delayed wound healing (Wei et al., 2019).

6.9.2 Effect of Human Insulin on the Formation of Catheter-Associated *E. Coli* Biofilms

Researchers have successfully developed a catheter material that can either prevent biofilm formation or assist individuals experiencing insulin loss in their urine in the selection of the optimal catheter material (Plotkin et al., 2014). The removal of excess insulin through urine occurs with or without the presence of glucose, depending on the individual's condition. During the pharmacokinetic peak immediately following injection, elevated insulin levels can arise in cases of insulin-dependent diabetes mellitus, peripheral insulin resistance, or dysregulated insulin synthesis as a response to high blood glucose levels, all of which lead to insulin excretion. Though glucose has been linked with enhanced bacterial colonization and biofilm development, the effect of urinary insulin on catheter-associated infection has not been sufficiently investigated (Muller et al., 2005).

The substances used as quorum signals can originate from within a species, across different species, or between various biological kingdoms. Specifically, in *E. coli*, human-recombinant insulin (insulin-r) functions as an inter-kingdom quorum signal. Previous research demonstrates that insulin-quorum signalling is critical for regulating the chemotactic responses, growth rates, and biofilm production of *E. coli*. Notably, insulin-r significantly enhances biofilm production in the presence of glucose (Klosowska and Plotkin, 2006). In summary, type 2 diabetes mellitus is a widespread condition marked by peripheral insulin resistance. Individuals with this condition exhibit hyperinsulinemia due to an increased secretion of insulin. This excess insulin is eliminated through urine, with or without glucose, depending on the state of the disease and glycaemic management. Furthermore, individuals with type 2 diabetes experience a higher prevalence of UTIs compared to those with type 1 diabetes mellitus or individuals with normal metabolic function.

Hence, different peptide hormones show different impacts on biofilm formation; some are inhibitory while some have a stimulatory effect, others are indifferent in nature (Figure 6.1). The overall impact is essentially dependant on the nature of the bacteria.

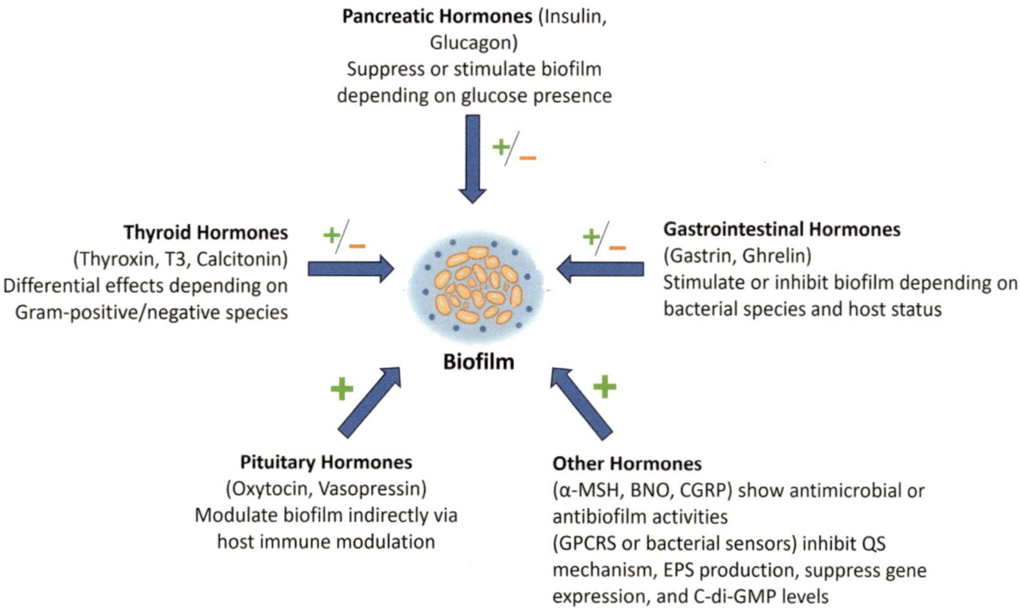

FIGURE 6.1 Different Hormones and Their Activities on Biofilm Inhibition. ('+' sign Denotes Biofilm Inhibition, and '-' Sign Denotes Biofilm Inhibition).

6.10 THERAPEUTIC POTENTIAL OF PEPTIDE HORMONES CONCERNING BIOFILM

Many recent studies focus on the key role of host-derived peptide hormones that principally focus on bacterial biofilm formation, its virulence, its eradication, and, most notably, on the new opportunities for therapeutic achievements. Various mechanisms that can influence bacterial biofilms are mostly known to us to date.

6.10.1 Through Modified AMPS

AMPs like defensins and cathelicidins can easily combat any type of microbial infection, but many barriers prevent their therapeutic application, such as enzymatic degradation and rapid clearance from the human body. Recently, researchers have modified AMPs further in order to overcome proteolytic degradation and to use them as a therapeutic (Patel et al., 2025). Created synthetic peptides derived from α-melanocyte-stimulating hormones (α-MSH) that have improved cationic and hydrophobic qualities, increasing their ability to penetrate bacterial biofilms. These peptides were interesting candidates for therapeutic development since they demonstrated broad-spectrum antibacterial action in addition to disrupting *E. coli* biofilms.

6.10.2 Hormone-Based Biofilm Disruptors

Potential biofilm disruptors include several peptide hormones, including glucagon, insulin, and natriuretic peptides. By destabilizing biofilm structures and disrupting bacterial-signalling pathways, these hormones can increase the susceptibility of bacteria to antimicrobial drugs. Bacterial metabolism and biofilm regulation are significantly influenced by insulin. *Pseudomonas aeruginosa* is one of the bacterial species with insulin-sensing systems that affect its capacity to build biofilms (Singh et al., 2023). It has been shown that QS system of *P. aeruginosa* is disrupted by insulin, which lowers biofilm density and increases antibiotic penetration. Insulin significantly enhances the removal of pathogens from lung infections caused by cystic fibrosis when combined with traditional antibiotics like tobramycin. Moreover, insulin has been shown to prevent sucrose-dependent biofilm development in *Streptococcus mutans*, the main pathogen responsible for dental plaque, by altering the bacterial glucose metabolism (Wang et al., 2022). Recent studies have indicated that CNP, a peptide hormone, plays a role in the control of microbial biofilms in addition to its primary involvement in cardiovascular and renal function. According to publications, the ability of *Staphylococcus epidermidis* to produce biofilms in anaerobic environments was diminished due to altered protein expression brought on by CNP exposure. CNP analogues can be explored as biofilm-targeting drugs, especially infections linked with indwelling medical devices (Gannesen et al., 2024).

6.11 PROTECTION OF MEDICAL DEVICES WITH PEPTIDE HORMONES

Biofouling mainly focuses on medicinal devices, like catheters, prosthetic joints, and ventilators, which are highly prone to multi-species bacterial colonization, leading to severe biofilm-related infections in patients. If such devices can be coated with peptide hormones by some innovative method, there are chances to prevent biofilm formation on their surfaces. One of the leading ways is antimicrobial peptide coatings, the LL-37-Infused Catheters. In this process, hydrogels are being infused with LL-37 into urinary catheters, which can reduce infection by *E. coli* and *K. pneumoniae* formation at a higher rate of 80% (Fontanot et al., 2024). Another process that is being identified is the use of titanium implants coated with AMPs and various other growth factors that are showing severely enhanced resistance to *S. aureus* biofilms. This is facilitating the orthopaedic surgeries to a greater extent than in the past.

6.12 CONCLUSION

This chapter presents a comprehensive overview of the dynamic interaction between peptide hormones and bacterial biofilms, highlighting an area of microbiological and biomedical science in rapid development. Biofilms, as resilient microbial communities encased in a defensive extracellular matrix, present enormous challenges in the clinical environment through their resistance to antibiotics and host immune systems. The function of quorum sensing and other environmental, structural, and molecular variables in the establishment and maintenance of biofilms is critical to deciphering their complex biology. Peptide hormones, long known to play regulatory roles in metabolic and physiological processes, have more recently been found to be significant determinants of microbial behaviour. Evidence suggests that they can dramatically alter biofilm development, stimulating or suppressing it as a function of hormone type, concentration, and microbial species involved. Insulin, somatostatin, natriuretic peptides, and glucagon, among other hormones, not only influence bacterial signalling cascades and growth but also provide new therapeutic options for biofilm-mediated infections.

The chapter also draws attention to the possibilities of exploiting hormone-based mechanisms and antimicrobial peptides in creating novel treatments and biofilm-resistant medical devices. From employing altered antimicrobial peptides to hormone-coated implants, these approaches suggest great promise in terms of eradicating recalcitrant biofilm-associated infections, especially in high-risk groups like diabetic patients. Overall, the complex interaction between peptide hormones and biofilms provides new avenues for interdisciplinary research that combines endocrinology, microbiology, and biomedical engineering. Through an understanding of these interactions, we can move closer to more targeted, efficient, and sustainable antimicrobial treatments in the era of increasing antibiotic resistance.

REFERENCES

Blair JM, Webber MA, Baylay AJ, Ogbolu DO, Piddock LJ. Molecular mechanisms of antibiotic resistance. *Nature Reviews Microbiology*. 2015 January;13(1):42–51.

Cani PD, Neyrinck AM, Maton N, Delzenne NM. Oligofructose promotes satiety in rats fed a high-fat diet: Involvement of glucagon-like peptide-1. *Obesity Research*. 2005 June;13(6):1000–1007.

Desriac F, Clamens T, Rosay T, Rodrigues S, Tahrioui A, Enault J, Roquigny L, Racine PJ, Taupin L, Bazire A, Dufour A. Different dose-dependent modes of action of C-type natriuretic peptide on *Pseudomonas aeruginosa* biofilm formation. *Pathogens*. 2018 April 24;7(2):47.

Fontanot A, Ellinger I, Unger WW, Hays JP. A comprehensive review of recent research into the effects of antimicrobial peptides on biofilms—January 2020 to September 2023. *Antibiotics*. 2024 April 9;13(4):343.

Gannesen AV, Ziganshin RH, Ovcharova MA, Mosolova AM, Loginova NA, Diuvenji EV, Nevolina ED, Mart'yanov SV, Plakunov VK. Changes in the protein profiles of *staphylococcus epidermidis* planktonic cultures and biofilms under anoxic conditions in the presence of the CNP hormone. *Microbiology*. 2024 December;93(6):799–811.

Klosowska K, Plotkin B. Human insulin modulation of *Escherichia coli* adherence and chemotaxis. *American Journal of Infectious Diseases*. 2006;2(4):197–200.

Lesouhaitier O, Clamens T, Rosay T, Desriac F, Louis M, Rodrigues S, Gannesen A, Plakunov VK, Bouffartigues E, Tahrioui A, Bazire A. Host peptidic hormones affecting bacterial biofilm formation and virulence. *Journal of Innate Immunity*. 2019 November 5;11(3):227–241.

LeValley SL, Tomaro-Duchesneau C, Britton RA. Degradation of the incretin hormone glucagon-like peptide-1 (GLP-1) by *Enterococcus faecalis* metalloprotease GelE. *MSphere*. 2020 February 26;5(1):10–128.

Louis M, Clamens T, Tahrioui A, Desriac F, Rodrigues S, Rosay T, Harmer N, Diaz S, Barreau M, Racine PJ, Kipnis E, Grandjean T, Vieillard J, Bouffartigues E, Cornelis P, Chevalier S, Feuilloley MGJ, Lesouhaitier O. *Pseudomonas aeruginosa* biofilm dispersion by the human atrial natriuretic peptide. *Advanced Science (Weinh)*. 2022 March;9(7):e2103262. doi: 10.1002/advs.202103262. Epub 2022 January 14. PMID: 35032112; PMCID: PMC8895129

Louis M, Tahrioui A, Tremlett CJ, Clamens T, Leprince J, Lefranc B, Kipnis E, Grandjean T, Bouffartigues E, Barreau M, Defontaine F. The natriuretic peptide receptor agonist osteocrin disperses *Pseudomonas aeruginosa* biofilm. *Biofilm*. 2023 December 1;5:100131.

Min C, Ohta K, Kajiya M, Zhu T, Sharma K, Shin J, Mawardi H, Howait M, Hirschfeld J, Bahammam L, Ichimonji I, Ganta S, Amiji M, Kawai T. The antimicrobial activity of the appetite peptide hormone ghrelin. *Peptides*. 2012;36(2):151–156. doi: 10.1016/j.peptides.2012.05.006

Muller LM, Gorter KJ, Hak E, Goudzwaard WL, Schellevis FG, Hoepelman AI, Rutten GE. Increased risk of common infections in patients with type 1 and type 2 diabetes mellitus. *Clinical Infectious Diseases*. 2005 August 1;41(3):281–288.

N'Diaye A, Mijouin L, Hillion M, Diaz S, Konto-Ghiorghi Y, Percoco G, Chevalier S, Lefeuvre L, Harmer NJ, Lesouhaitier O, Feuilloley MG. Effect of Substance P in *Staphylococcus aureus* and *Staphylococcus epidermidis* virulence: Implication for skin homeostasis. *Frontiers in Microbiology*. 2016b April 15;7:506.

N'Diaye AR, Leclerc C, Kentache T, Hardouin J, Poc CD, Konto-Ghiorghi Y, Chevalier S, Lesouhaitier O, Feuilloley MG. Skin-bacteria communication: Involvement of the neurohormone calcitonin gene related peptide (CGRP) in the regulation of *Staphylococcus epidermidis* virulence. *Scientific Reports*. 2016a October 14;6:35379. doi: 10.1038/srep35379. PMID: 27739485; PMCID: PMC5064375.

Nakamura Y, Suzuki S, Murakami S, Nishimoto Y, Higashi K, Watarai N, Umetsu J, Ishii C, Ito Y, Mori Y, Kohno M, Yamada T, Fukuda S. Integrated gut microbiome and metabolome analyses identified fecal biomarkers for bowel movement regulation by *Bifidobacterium longum* BB536 supplementation: A RCT. *Computational and Structural Biotechnology Journal*. 2022 October 25;20:5847–5858. doi: 10.1016/j.csbj.2022.10.026. PMID: 36382178; PMCID: PMC9636538.

Nishikawa M, Kobayashi K. Calcium prevents biofilm dispersion in *bacillus subtilis*. *Journal of Bacteriology*. 2021;203(14):e0011421. doi: 10.1128/JB.00114-21

Ostrowska K, Strzelczyk A, Różalski A, Stączek P. Bacterial biofilm as a cause of urinary tract infection–pathogens, methods of prevention and eradication. *Postepy higieny i medycyny doswiadczalnej (Online)*. 2013 October 25;67:1027–1033.

Otzen DE, Peña-Díaz S, Widmann J, Daugberg AO, Zhang Z, Jiang Y, Mittal C, Dueholm MK, Louros N, Wang H, Javed I. Interactions between pathological and functional amyloid: A match made in Heaven or Hell? *Molecular Aspects of Medicine*. 2025 June 1;103:101351.

Patel N, Curtis JC, Plotkin BJ. Insulin regulation of *Escherichia coli* abiotic biofilm formation: Effect of nutrients and growth conditions. *Antibiotics (Basel, Switzerland).* 2021;10(11):1349. doi: 10.3390/antibiotics10111349

Patel P, Abdullah SJ, Tiwari K, Bhattacharjya S, Mukhopadhyay K. Antimicrobial and antibiofilm potential of α-MSH derived cationic and hydrophobic peptides against *Escherichia coli*: Mechanistic insight through peptide–lipopolysaccharide interactions. *Biochimica et Biophysica Acta (BBA) –Biomembranes*. 2025 January 1;1867(1):184398.

Pletzer D, Coleman SR, Hancock RE. Anti-biofilm peptides as a new weapon in antimicrobial warfare. *Current Opinion in Microbiology*. 2016 October 1;33:35–40.

Plotkin BJ, Wu Z, Ward K, Nadella S, Green JM, Rumnani B. Effect of human insulin on the formation of catheter-associated *E. coli* biofilms. *Open Journal of Urology*. 2014 May 28;5(5).

Reffuveille F, De La Fuente-Ñũnez C, Mansour S, Hancock RE. A broad-spectrum antibiofilm peptide enhances antibiotic action against bacterial biofilms. *Antimicrobial Agents and Chemotherapy*. 2014 September;58(9):5363–5371.

Rosay T, Bazire A, Diaz S, Clamens T, Blier AS, Mijouin L, Hoffmann B, Sergent JA, Bouffartigues E, Boireau W, Vieillard J, Hulen C, Dufour A, Harmer NJ, Feuilloley MG, Lesouhaitier O. *Pseudomonas aeruginosa* expresses a functional human natriuretic peptide receptor ortholog: Involvement in biofilm formation. *mBio*. 2015 August 25;6(4):e01033-15. doi: 10.1128/mBio.01033-15. PMID: 26307165; PMCID: PMC4550695

Roy R, Tiwari M, Donelli G, Tiwari V. Strategies for combating bacterial biofilms: A focus on anti-biofilm agents and their mechanisms of action. *Virulence*. 2018 January 1;9(1):522–554. doi: 10.1080/21505594.2017.1313372. PMID: 28362216; PMCID: PMC5955472

Severini C, Improta G, Falconieri-Erspamer G, Salvadori S, Erspamer V. The tachykinin peptide family. *Pharmacological Reviews*. 2002 June 1;54(2):285–322.

Singh M, Mukhopadhyay K. Alpha-melanocyte stimulating hormone: An emerging anti-inflammatory antimicrobial peptide. *BioMed Research International*. 2014;2014:874610. doi: 10.1155/2014/874610. Epub 2014 July 23. PMID: 25140322; PMCID: PMC4130143

Singh R, Smith D, Lee W. Insulin-mediated biofilm inhibition in *Pseudomonas aeruginosa*: A novel anti-infective strategy. *Journal of Bacteriology*. 2023;205(5):e00456-23.

Tang R, Ren Y, Zhang Y, Yin M, Ren X, Zhu Q, Gao C, Zhang W, Liu G, Liu B. Glucose-driven transformable complex eliminates biofilm and alleviates inflamm-aging for diabetic periodontitis therapy. *Materials Today Bio*. 2023 June 1;20:100678.

Totté JE, Van der Feltz WT, Bode LG, van Belkum A, Van Zuuren EJ, Pasmans SG. A systematic review and meta-analysis on *Staphylococcus aureus* carriage in psoriasis, acne and rosacea. *European Journal of Clinical Microbiology & Infectious Diseases*. 2016 July;35:1069–1077.

Verbeke F, De Craemer S, Debunne N, Janssens Y, Wynendaele E, Van de Wiele C, De Spiegeleer B. Peptides as quorum sensing molecules: measurement techniques and obtained levels in vitro and in vivo. *Frontiers in Neuroscience*. 2017;11:183

Waldrop R, McLaren A, Calara F, McLemore R. Biofilm growth has a threshold response to glucose in vitro. *Clinical Orthopaedics and Related Research*. 2014;472(11):3305–3310. doi: 10.1007/s11999-014-3538-5

Wang T, Ishikawa T, Sasaki M, Chiba T. Oral and gut microbial dysbiosis and non-alcoholic fatty liver disease: The central role of *Porphyromonas gingivalis*. *Frontiers in Medicine*. 2022 March 2;9:822190.

Watters C, Everett JA, Haley C, Clinton A, Rumbaugh KP. Insulin treatment modulates the host immune system to enhance *Pseudomonas aeruginosa* wound biofilms. *Infection and Immunity*. 2014 January;82(1):92–100.

Wei Q, Zhang Z, Luo J, Kong J, Ding Y, Chen Y, Wang K. Insulin treatment enhances *pseudomonas aeruginosa* biofilm formation by increasing intracellular cyclic di-GMP levels, leading to chronic wound infection and delayed wound healing. *American Journal of Translational Research*. 2019 June 15;11(6):3261.

Yamashita K, Kaneko H, Yamamoto S, Konagaya T, Kusugami K, Mitsuma T. Inhibitory effect of somatostatin on Helicobacter pylori proliferation in vitro. *Gastroenterology*. 1998 November 1;115(5):1123–1130.

Zavros Y, Eaton KA, Kang W, Rathinavelu S, Katukuri V, Kao JY, Samuelson LC, Merchant JL. Chronic gastritis in the hypochlorhydric gastrin-deficient mouse progresses to adenocarcinoma. *Oncogene*. 2005;24:2354–2366. doi: 10.1038/sj.onc.1208407

7 Stress Hormones and Biofilms

Sreejita Ghosh, Sumit Nath, and Rina Rani Ray
Department of Biotechnology, Maulana Abul Kalam Azad University of Technology, Kolkata, India

7.1 INTRODUCTION

Biofilms are composed of a complex community of microbes that are embedded in a self-produced extrapolymeric matrix, which poses particular challenges, both clinically and environmentally. Biofilms can be deposited on any niches within the host body, including mucous membranes and other tissues in the orofacial complex, as well as on dental and medical devices. Biofilms are responsible for causing chronic and acute infections, which are difficult to treat, because of their antibiotic resistance and their capacity to evade the host immune response (Lories et al., 2020). Stress hormones, including epinephrine, norepinephrine, and cortisol, play a major role in the biofilm formation and are a crucial evolving field of research because of their substantial impacts on microbial behaviour as well as pathogenic attributes within the host environment (Parijs and Steenackers, 2018).

Increased levels of stress hormones, typically allied with physiological stress, affect microbial signalling pathways and can modify virulence factor expression in a few pathogens that can then, in turn, alter the dynamics of biofilm. This interaction between host neuroendocrine signalling and the establishment of microbial biofilms is a very complex form of interaction between the host and its microorganisms, which can readily lead to infection, especially in immunocompromised hosts or chronic stress states. Norepinephrine has been shown to interact with bacterial adrenergic receptors to enhance bacterial adherence, motility, and biofilm formation in various systems, including the gastrointestinal, respiratory, and urinary systems (Leinweber et al., 2018). The problem further leads to stress hormones that enhance the strength and antibiotic resistance of biofilms. Biofilm infections can activate the reactions caused by stress hormones, which are usually characterized by the increased production of exopolysaccharide (EPS), hence making it difficult to strengthen the biofilm structure and complicating therapeutic interventions. Therefore, understanding the role of stress hormones in biofilm formation can become important in developing targeted therapies, especially in relation to infections linked to chronic stress or infections that develop in high-stress environments, such as intensive care units.

This chapter reviews the biochemical pathways of the effects of stress hormones on biofilm development. Its findings have implications for human health and potential therapeutic approaches toward the treatments of infections caused by biofilms. The host–pathogen interactions show how the stress-induced changes within a human body create an environment even more favourably poised for microbial colonization and the persistence of biofilms. This knowledge is much needed to improve clinical approaches aimed at effectively addressing biofilm-associated infections and to enhance patient outcomes in conditions adversely affected by stress.

7.2 STRESS CONDITIONS

The human organism mounts an all-around stress response, to be caused by internal and external stressors-for example, by traffic congestion just as much as by fear of failure. This physiological mechanism for the response to dangers, often called "fight or flight," has its roots in evolutionary

DOI: 10.1201/9781003519041-7

strategies for survival. The mechanism is facilitated by the secretion of hormones that activate the body for quick responses to dangers: adrenaline and cortisol, most notably. There were immediate effects such as fast heart rate and breathing, increased muscle tension, and open airway passages to increase oxygenation.

The current physiological response has its basis in the hypothalamic–pituitary–adrenal (HPA) axis (Alborzkouh et al., 2015). If a stressor is present, the hypothalamus will release a corticotropin-releasing hormone, which stimulates the pituitary gland to produce an adrenocorticotropic hormone. This hormone stimulates the adrenal glands to produce cortisol, which keeps the organism in a heightened state of alertness. When the stressor has been dealt with, the level of cortisol falls; then the parasympathetic nervous system is triggered and this calms the body by restoring balance.

However, chronic or chronic stress—a characteristic of modern ways of life—-can result in over-expression of this system, endangering health in both its physical and its mental aspects. It can even disrupt sleep patterns and affect other cognitive functions, such as memory or problem-solving (Rogerson et al., 2024).

Another critical point is that the stress response does not operate in isolation, but also interconnects with other physiological networks. Chronic activation of the HPA axis has been considered to change the gut microbiota, which, in turn, will affect overall health. Moreover, biochemical interaction between stress hormones and biofilm formation by pathogens underlines how complex this response might be (Howarth et al., 2020). Stress may interfere with changes in the hormone balance that inadvertently create favourable micro-environments for some microbes and thus contributes to infections as well as complicates recovery.

A better understanding of the interplay between stress response and its consequent effects on other physiological systems forms the basis through which approaches aimed at relieving its prolonged effects can start. Techniques for relaxation, mindfulness-based practices, and changes in lifestyles can contribute significantly toward chronic stress reduction while promoting general well-being.

7.3 STRESS HORMONES

Under the effects of stress, the organism undergoes adaptive alteration in hormone levels, especially in regard to corticotropin-releasing hormone (CRH), cortisol, catecholamines, and thyroid hormones. These hormonal changes enable the "fight-or-flight" response to be turned on through which organisms are able to respond to direct stimulus. In mammals, both cortisol and cortisone are the glucocorticoids primarily involved, whereas, in birds and rodents, corticosterone plays a similar role. These hormones are recognized extensively as stress markers and are widely used in clinical and research settings to establish physiological responses (Ranabir and Reetu, 2011).

Exposure to chronic stress, however, causes long-term hormonal disorders that affect endocrine health. Over-activation by elevated levels of stress hormones may lead to metabolic, immune, and reproductive dysfunction, all of which are factors for the diseases mentioned: hyperthyroidism, adrenal fatigue, and insulin resistance. Stress may also contribute to the severity and progression of existing endocrine disorders, making treatments for these conditions all the more difficult to handle.

In addition to cortisol, the role of adrenaline and noradrenaline—two other key catecholamines—is important in enabling abrupt physiological changes, like increased heart rate and alertness, associated with episodes of acute stress. Chronic activation of this system can cause hypertension and cardiovascular stress (Botía et al., 2023).

There is an interactive link between stress and thyroid hormones: stress may adversely affect thyroid function, yet atypical thyroid activity can fuel the response to stress. Such complexity underscores the need for de-stressing through lifestyle and therapeutic interventions to safeguard endocrine as well as general health.

7.4 STRESS HORMONES: SOURCE AND CHEMISTRY

Stress hormones, generally defined as the stress proteins/universal stress proteins/heat shock proteins, are secreted to help organisms survive under stressful conditions and directly related to stress tolerance. The stress hormones are released through a hormone-signalling pathway by specialized glands, such as adrenaline hormone which released by the adrenaline glands present on the upper surface of the kidneys, and thyroxine hormone which is produced by thyroid glands. In addition to controlling the physiology of the human body, these biochemicals are also involved in controlling the pathogenesis, biofilm formation and inefficiency of the antibiotics of microorganisms (Rather et al., 2021). The description of the stress hormones is described in Table 7.1.

When stressful situations are encountered, the human body responds by releasing stress hormones such as cortisol, epinephrine (adrenaline), and norepinephrine. These hormones are part of the "fight-or-flight" response, preparing us to handle immediate threats by raising heart rate, blood pressure, and glucose levels. However, while this response is helpful in acute situations, it can have unintended effects on how our body handles infections.

Over recent years, research indicated that stress hormones significantly affect microbial behaviour, such as their capacity to infect (virulence). For example, cortisol is one of the major stress hormones that can directly stimulate the virulence and growth of some bacteria. In presence of

TABLE 7.1
Table Describing the Functions of Different Stress Hormones

Hormones	Description	References
Cortisol	Secreted when pituitary–adrenal axis is activated due to stress and is a part of neuroendocrine signalling system. Activation of this signalling system results in hypothalamic secretion of corticotrophin-releasing factor (CRF). The pituitary is further stimulated by CRF for the secretion of adrenocorticotropin (ACTH), 8-lipotropin and 3-endorphin respectively.	Herman et al. (2003)
Catecholamines	Released when a stimulating signal is generated along the pituitary–adrenal axis leading to increased cardiac output, skeletal muscle blood flow, sodium retention, reduced intestinal motility, cutaneous vasoconstriction, increased glucose, bronchiolar dilatation and behavioural activation.	Goldstein (1987)
Vasopressin	Acute stress causes fast vasopressin release from the paraventricular nucleus of the hypothalamus together with corticotrophin releasing hormone CRH. Vasopressin is responsible for the secretion of ACTH from the pituitary gland by acting on the V1b receptor. During chronic stress condition expression of hypothalamic vasopressin takes place.	Aguilera et al. (2008)
Gonadotropins	Stress causes a suppression of circulating gonadotropins and gonadal steroid hormones, therefore upsetting the usual menstrual cycle. Persistent exposure to stress condition leads to disablement of the reproductive system. Gonadotropins are responsible for the down regulation of pituitary signalling.	Cameron (1997)
Thyroid Hormones	The thyroid gland's function gets down-regulated under stressful conditions; the secretion of thyroid-stimulating hormone are inhibited in stress condition due to the action of glucocorticoids present in the central nervous system.	Helmreich et al. (2005)
Prolactin	Prolactin release during stress may be mediated by vasopressin and the peptide histidine isoleucine. It is unclear, therefore, what the teleological significance of the prolactin level change is. It might have an impact on homeostasis or the immune system.	Itoh et al. (1983)

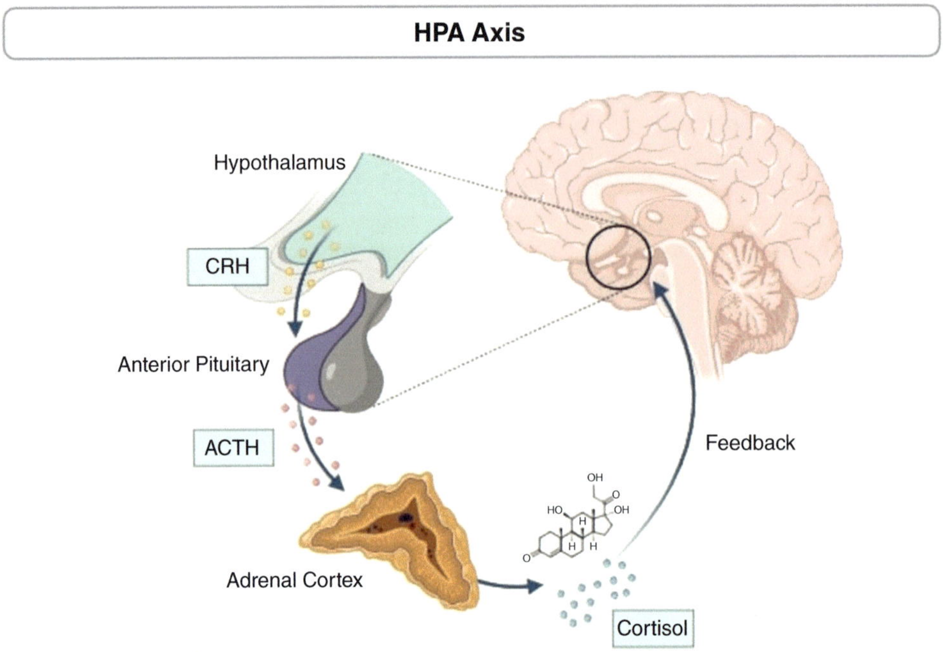

FIGURE 7.1 Source and process of formation of stress hormone cortisol through HPA axis.

magnified levels of cortisol, few pathogens can alter their gene expression, resulting in improved endurance and persistence inside the host (Sarkodie et al., 2019). This indicates that during chronic stress, the host body might become more susceptible to microbial infections because of the preventive effects of cortisol on immune responses and its capacity to encourage bacterial growth (Figure 7.1).

Fascinatingly, stress hormones not only stimulate microbial growth but also promote the formation of biofilms. Biofilms are shielding communities of bacteria, which are infamously hard to eliminate with the available antibiotics. A study on a fish pathogen called *Flavobacterium columnare* demonstrated that cortisol could modulate the formation of biofilms inversely, depending on the strain of bacteria (Declercq et al., 2019). Few bacterial isolates depicted enhanced formation of biofilms in presence of cortisol, whereas others displayed diminished production of biofilms. This is an indication that stress responses can generate an environment that either facilitates or restricts development of biofilm, depending on the explicit features of the pathogens.

In a nutshell, the communication between bacterial infections and stress hormones focuses on an interesting and multifaceted association, in which the psychological state of the host can indirectly stimulate microbial pathogenicity. This underlines the significance of handling chronic stress, not just for mental health but also for upholding vigorous immune response and averting the development of persistent infections.

7.5 INFLUENCES OF STRESS HORMONES ON MICROBIAL INFECTIONS

Glucocorticoids, catecholamines, and neuroendocrine factors are released into the host's bloodstream during stressful conditions. The direct impact of the stress-related hormones on the immune system and the intestinal barrier functions had long been the only reason why the infection progresses due to the stress condition. The progression of an infection and/or susceptibility to a microbe can be impacted by the shift from T helper 1-mediated cellular immunity to T helper 2-mediated humoral immunity that chronic stress is known to induce. Bacteria can, however, respond directly to host stress signals. Catecholamines can change the development, motility, biofilm formation, and/or pathogenicity of pathogens and commensal bacteria, influencing the outcome of infections in many hosts (Verbrugghe et al., 2012).

Hyper-cortisolemia is a disorder characterized by a sustained excess of cortisol blood levels, which can arise as a result of the hypothalamic–pituitary–adrenal axis dysregulation or exogenous drugs. Hyper-cortisolemia causes immunocompromise, making the patient susceptible to a variety of bacterial, viral, fungal, and parasitic illnesses. To promote appropriate care of hyper-cortisolemia, the main clinician must understand the many causes and illnesses associated with cortisol excess. In the hyper-cortisolemic patient, normal cortisol levels must be restored to lower the risk of infection or to improve the treatment and cure of an existing infection (Fareau and Vassilopoulou-Sellin, 2007). According to Cortês et al. (2021), catecholamines increased *Streptococcus mutans* biofilm formation in orthodontic brackets ($p < 0.05$), but did not affect acid production (pH reduction) or the expression of biofilm-related genes (gtfB, gtfC, gbpA, gbpB, gbpC, gbpD, and brpA), aciduric (relA), or acidogenic properties (ldh). Uropathogenic *Escherichia coli* (UPEC) causes pyelonephritis and ascending urinary tract infections (UTI), both of which can cause significant kidney damage. Collecting duct cells, which control hormonally regulated ion transport and water absorption via vasopressin, are the preferred intrarenal site of bacterial adhesion and the commencement of an inflammatory response. The effect of the powerful V2 receptor (V2R) agonist deamino-8-D-arginine vasopressin (dDAVP) on the activation of the innate immune response was investigated utilising established and primary cultured collecting duct cells, as well as an ascending UTI experimental paradigm. dDAVP inhibits Toll-like receptor 4-mediated nuclear factor κB activation and chemokine production in a V2R-specific way. dDAVP suppressed cystic fibrosis by activating protein phosphatase 2A and requiring an undamaged Cl– channel in the transmembrane conductance regulator. In mice injected with UPECs, in vivo infusion of dDAVP resulted in a significant reduction in proinflammatory mediators and neutrophil recruitment, as well as a substantial increase in the bacterial load in the kidney. In contrast, administering the V2R antagonist SR121463B to UPEC-infected mice increased both the local innate response and antibacterial host defence (Chassin et al., 2007). Gonadotropin was able to reduce the infection as well as the biofilm formation when combined with antibiotics to treat urinal tract infection in female candidates (Güner and Kisadere, 2022). L-thyroxine primarily suppressed Sp growth, certain derivatives have shown efficacy against other Gram-positive bacteria such as *Staphylococcus aureus* and *Enterococcus faecalis* while being inert against Gram-negative pathogens. Human l-thyroxine binds to *Sp*-flavodoxin, forming a 1:1 complex (Galano-Frutos et al., 2024). The results of the current investigation indicate that PIP, a protein derivative of prolactin, strongly inhibited a wide range of bacterial strains, including standard and environmental multiple drug-resistant (MDR) strains. The increased bactericidal potential of common antibiotics such as ampicillin in synergistic investigations of PIP makes it an important investigatory molecule for the development of combinatorial antibacterial treatments to deal with the rising burden of antimicrobial resistance (Yousuf et al., 2022).

7.6 GUT MICROBIOME VARIATION AND THE HYPOTHALAMIC–PITUITARY–ADRENAL (HPA) AXIS

The microbiome–gut–brain (MGB) axis is an important pathway through which stress dysregulation occurs, and disruptions in the development of front limbic regions, which are particularly sensitive to the hypothalamic–pituitary–adrenal (HPA) axis signalling, may play a critical role in the emergence of adolescent-onset depression. Disparities in the gut microbiome may disturb the HPA axis, leading to inflammation and elevated permeability of the gut. This allows endotoxins to gain entry inside the bloodstream, leading to neuroinflammation, thereby perturbing neurotransmitter biosynthesis, involving serotonin. Since most of the serotonin is produced by stomach, hindrance in this path may result in depression symptoms (Rosin et al., 2021).

The HPA axis plays a key role in the host immune response against stress, along with infections. When pathogens including viruses or bacteria infiltrate the host, the HPA axis gets activated, resulting in a flow of glucocorticoid hormones such as cortisol. This response is mandatory for endurance, since glucocorticoids aid in regulating inflammation and maintaining homeostasis at times of infections. Research indicated that disorder in this system, via adrenalectomy (adrenal gland removal) or hindering glucocorticoid receptors, worsens infections, thereby magnifying rates of mortality (Webster and Sternberg, 2004). Hence, the HPA axis is considered an important part of the host defensive arsenal.

Recently, it has been revealed that an astounding link is present amongst the gut microbiome as well as the HPA axis. The microbiome, the enormous community of microbes present within the gastrointestinal (GI) tract, seems to have the capacity of modulating the principal stress response system, especially at times of stressful or challenging situations (Figure 7.2).

Communication between the brain and the gut, very often named the gut–brain axis, includes numerous signalling molecules, involving neurotransmitters such as histamine, acetylcholine, and catecholamines, and also corticotropin-releasing factors. Interestingly, some of these neurochemicals can even originate directly from the microbiota itself (Kim and Yosipovitch. 2020).

Evidence from animal studies, particularly in rodents, indicates that early life exposure to a diverse microbiome can shape the normal development of the HPA axis. Germ-free animals—those raised without any exposure to microbes—often show exaggerated stress responses, which normalize when certain beneficial bacteria are introduced. While this connection is well-established in animal models, human studies are still evolving. However, there are strong indications that imbalances in gut microbial composition could influence psychological health, including cognitive development and susceptibility to mood disorders, which may be indirectly tied to HPA axis dysregulation (Rosin et al., 2021).

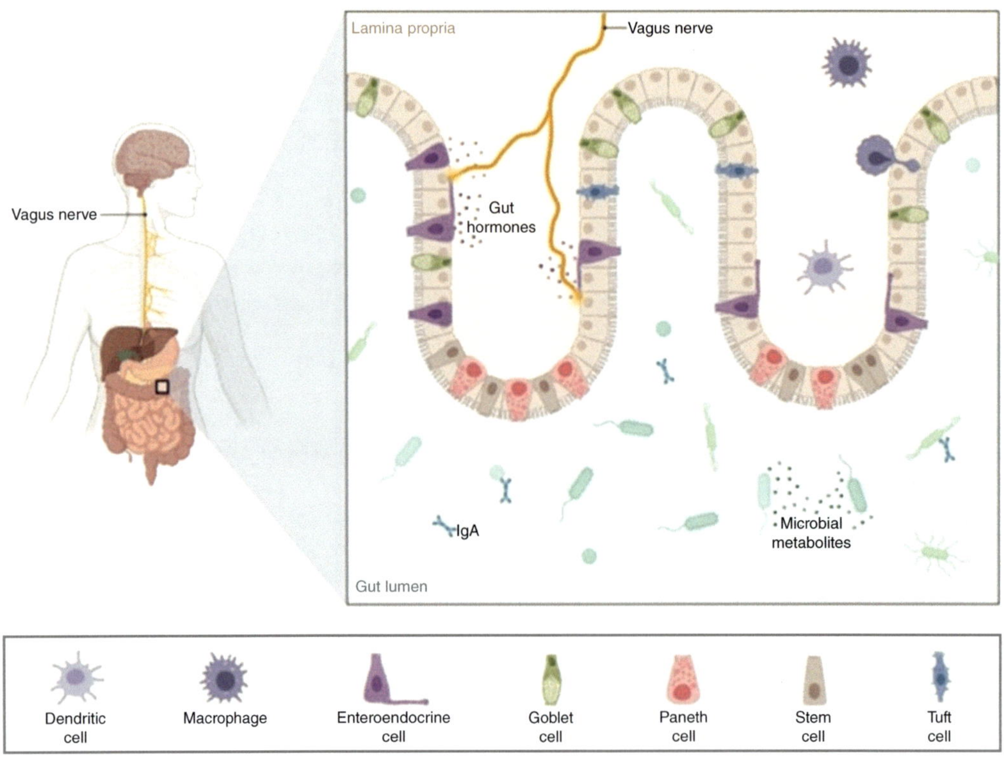

FIGURE 7.2 Gut–microbiome axis directly influencing the release of stress hormones.

Stress not only impacts the nervous system but also significantly alters immune function. Cortisol and catecholamines, released in response to stress, can modulate immune cell numbers and influence the production of pro-inflammatory cytokines. These stress hormones also affect microbial behaviour. For example, stress can promote the growth of certain pathogenic oral bacteria and increase their virulence, possibly explaining the observed link between psychological stress and periodontal disease (Ball and Darby, 2022). Interestingly, while the gut microbiome's relationship with the brain has been extensively studied, the oral microbiome's influence on systemic health and stress responses remains a relatively unexplored frontier.

Microbes, including bacteria within biofilms, are also capable of sensing environmental stressors, including those related to host responses. Bacteria residing inside the protective layers of biofilms can perceive signals produced by competing microorganisms and also host-associated stress hormones. This facilitates them to fine-tune their gene expression profiles, thereby increasing their chances of persistence, virulence, or competitive edge within hostile environments (Lories et al., 2020). These stress-adapted microbial populations may lead to the development of chronic inflammation, persistent and acute infections, and a collapse in maintenance of tissue integrity.

All of these findings focus on an intricate and dynamic association between the HPA axis, the microbiome, and microbial communities including biofilms. Stress hormones, which are crucial for handling host immune responses, also involuntarily aid in bacterial proliferation and perseverance in certain contexts. As research endures to unearth the distinctions of microbiome–endocrine–immune communications, it becomes crystal-clear that bacterial ecosystems, including oral as well as gut, play a central role in determining host immunity, health, and disease outcomes.

7.7 STRESS HORMONE-MEDIATED BIOFILM REGULATION

7.7.1 Impacts of Glucocorticoids

Bacterial biofilms are related to the metastasis of chronic rhinosinusitis alongside nasal polyposis (CRSwNP). Precisely, these hormones can deregulate microbial genes taking part in vital virulence-mediated pathways, such as iron sequestration, antimicrobial resistance, stress response, adhesion, and the formation of biofilms (Reidl and Monsó, 2015). Intranasal saline administration, as well as corticosteroids, are consistent treatment approaches. The biofilm-forming capacity of isolated bacterial strains was determined using a microtiter plate method, and the effects of therapeutic dosages of mometasone, fluticasone (a class of glucocorticoids), isotonic, and hypertonic saline on biofilm development were studied. In another study it was found that glucocorticoids suppressed the innate immune response, antimicrobial nitric oxide generation, and autophagy in mycobacteria-infected macrophages. Meanwhile, glucocorticoid therapy reduced the development and acidity of mycobacterial phagosomes in RAW 264.7 cells. As a result, we found that glucocorticoids increased the survival of intracellular mycobacteria in both primary macrophages and cell lines. Treatment with glucocorticoids reduced the activation of TBK1 kinase, which increases autophagosome maturation. Inhibition of TBK1 lowered nitric oxide generation, and autophagy-related genes were downregulated after glucocorticoid therapy, although the activation of the Akt/mTOR signalling pathway increased, which could explain autophagy inhibition during mycobacterial infection. The agonist rapamycin restored autophagy and eliminated glucocorticoid-mediated improvement of mycobacterial survival, implying that glucocorticoids inhibited anti-mycobacterial defence via autophagy (Wang et al., 2017). The effects of cortisol on *Flavobacterium columnare* biofilm production were investigated in another study. First, the kinetics of biofilm formation by one highly virulent (HV) and one low virulent (LV) *F. columnare* isolate with and without the stress hormone cortisol under microfluidic flow conditions were studied. This was done to confirm that *F. columnare* cells could develop biofilm in response to cortisol supplementation, as well as to assess the temporal and morphological variations between treatment groups. The particular study found that the cell aggregates resembling biofilms formed in both isolates within 7 hours of inoculation. As a result, cell clusters were sloughed away, followed by the reconstruction of bacterial cell aggregation, indicating a high spreading capacity. While the HV isolation demonstrated cell aggregates generated upstream at all time points, the LV isolate only showed this after cortisol treatment. Second, the transcriptional effect of genes (gldK, gldL, gldM, gldN, sprA, sprE, sprT, and porV) from the Type IX secretion system involved in gliding motility was investigated in planktonic and biofilm cells of an HV and LV isolate to which either no, a low (LD), or a high (HD) dose of cortisol was added. In LV isolate planktonic cells treated with HD-cortisol, gliding genes gldK, gldL, gldM, and gldN, as well as the protein secretion regulator porV, were significantly down-regulated. The LV isolate biofilm cells treated with HD-cortisol revealed a considerable overexpression of sprT, a gene that encodes mobile surface adhesion, which is critical for bacterial colonization. In the study with *S. aureus*, the particular strain was cultured in MBEC plates and treated with standard synthetic glucocorticoids. These were mixed with cerebrospinal fluid (CSF) soup after being dissolved in vehicle solvents. Concentrations (including therapeutic dosages) tested for budesonide ranged from 16 μg/200 μL to 2000 μg/200 μL, while fluticasone and mometasone ranged from 25 μg/200 μL to 400 μg/200 μL. The control pegs were given the same amount of the suitable solvent/CSF broth. Biofilms at 24 hours post-treatment were measured. There was a significant reduction in the biofilm (Declercq et al., 2019).

7.7.2 Dual Role of Corticosteroids and Cortisol in Biofilm Modulation

Corticosteroids and glucocorticoids have complex, and sometimes contrasting, effects on bacterial biofilm formation, depending on the type of bacteria and the local microenvironment. Some studies have demonstrated that corticosteroids can actively reduce biofilm formation in certain bacterial species. Biofilm production was greatly reduced by corticosteroids and saline solutions ($p < 0.01$ and $p < 0.05$, respectively), with fluticasone and isotonic nasal saline proving more effective. Fluticasone, mometasone, isotonic, and hypertonic nasal saline totally inhibited biofilm formation in 66, 50, 84, and 38% of bacterial strains, respectively. Biofilms generated by *Staphylococcus aureus*, *Pseudomonas aeruginosa*, and *Streptococcus pneumoniae* showed the highest density reduction compared to other bacterial species ($p < 0.01$, $p < 0.05$, $p < 0.05$, respectively). Corticosteroids and saline solutions had substantial antibiofilm effects based on bacterial biomass ($p < 0.05$), with a greater impact on biofilm formation with a higher amount. Topical steroids and nasal saline were found to be effective antibiofilm therapies in CRSwNP patients (Cirkovic et al., 2017). Likewise, other in vitro studies have showed the potentials of corticosteroids to unswervingly prevent formation of biofilms by *Staphylococcus aureus*, signifying that corticosteroids might hinder the signalling pathways of the bacterial or might also inhibit the extra polymeric biofilm matrix needed for maturation of biofilms.

On the other hand, cortisol is a stress-mediated glucocorticoid endogenously secreted by the host and can induce formation of biofilms in specific bacteria. Studies on *Streptococcus mutans*, which is a primary pathogen forming dental caries, exhibited that exposure to cortisol can promote the development of extremely vigorous biofilms. Therefore, this stress hormone seems to regulate microbial physiology, facilitating the survival of *S. mutans* in adverse conditions and encouraging its perseverance in dental plaques (Yahya et al., 2019).

Such distinct findings exemplify the context-dependent property of biofilm–steroid connections. Although corticosteroids may serve as an antibiofilm agent against certain pathogenic bacteria such as *S. aureus*, natural stress hormones, including cortisol, might aggravate formation of biofilms in others, including *S. mutans*. Parameters involving the class of corticosteroid being applied, species of bacteria, surrounding situations, and the physiological condition of the host altogether play a part in defining if the formation of biofilms is induced or suppressed.

7.7.3 Effects of Epinephrine on Biofilms

Epinephrine is a major stress hormone which plays a vital function in moderating microbial behaviours, such as formation of biofilms followed by adherence, virulence, and motility. Studies in the recent years demonstrated its complicated and species-specific activity on biofilms, having stark impacts found in pathogenic and commensal bacteria.

Epinephrine at physiologically pertinent concentrations (4.9×10^{-9} M) has been observed to modify the composition of the components of the biofilm matrix as well as the structure pointedly in a commensal skin bacterium *Micrococcus luteus* C01. Gannesen et al. (2022) established that exposure to epinephrine elevated the polysaccharide as well as extracellular DNA formation, thereby amending the lipid, protein, and carbohydrate profiles inside the biofilm matrix. Remarkably, the elongation factor Tu (Ef-Tu), which is a predominant extrapolymeric matrix-based protein, displayed a significant elevation in the profusion on treating with epinephrine. Moreover, few glucose-mannose polysaccharides and phospholipids were not present within the biofilm matrix in the presence of epinephrine, revelling that this hormone modifies not only the mass of the biofilms but also their biochemical construction.

In addition to these findings, another study, by Gannesen et al. (2021), showed that epinephrine exercises dualistic activities on the biofilms formed by *M. luteus* depending on the experimental model. For instance, in biofilms grown on the cubes made of polytetrafluorethylene (PTFE), epinephrine could elevate the entire biomass of biofilm and, subsequently, the metabolic activity. However, on glass fibre filters, it reduced colony-forming units (CFUs) while not affecting metabolic activity, and in glass-bottom plate assays, biofilm formation was notably suppressed. Transcriptomic analysis revealed that epinephrine down-regulated genes involved in stress response and biofilm regulation, including *groES* and lysine exporters, while upregulating genes associated with Fe–S cluster assembly and cyanide detoxification, suggesting a hormetic effect dependent on environmental factors and surfaces.

In stark contrast, epinephrine has been shown to enhance biofilm formation and virulence in *Pseudomonas aeruginosa*, a well-known opportunistic pathogen responsible for chronic infections, especially in immunocompromised patients and those with cystic fibrosis (CF) (Cambronel et al., 2019). Epinephrine-mediated treatment could strikingly increase the adherence, motility, and biofilm biomass of *P. aeruginosa* besides a raise in its pathogenic capacity, as observed in an in vivo *Galleria mellonella* larvae infection model. These observations proof that *P. aeruginosa* might influence host-derived catecholamines as ecological signals to supplement its biofilm-mediated persistence as well as virulence. Lei et al., 2023, further revealed that epinephrine could prominently elevate the growth level of *Mycobacterium tuberculosis* in macrophages as well as in serum media. Docking studies and computational analyses also demonstrated that the extra-cytoplasmic region of MprB can serve as a latent epinephrine receptor. Experimental investigations also showed that epinephrine exposure could invariably augment biofilm formation by *M. tuberculosis*. Such biofilms displayed modified composition of the matrix, increased antibiotic resistance, and enhanced tolerance to surrounding stress, signifying that the hormone significantly shapes the persistence and recalcitrance mechanisms of pathogens.

All of these studies together point out that epinephrine may not only act as an interkingdom signalling molecule, but also induce or suppress formation of biofilms on the basis of the species of bacteria and ecological context (Danilova et al., 2020). Although epinephrine can stimulate virulence and colonization behaviours in pathogenic bacteria, including *P. aeruginosa*, it employs intricate and inhibitory activities on commensal bacteria like *M. luteus*. A detailed explanation of such nuanced communications may help in paving the path towards the development of innovative anti-virulence or anti-biofilm therapeutic approaches via targeting stress hormone-mediated response pathways among clinically relevant bacterial strains.

7.7.4 Binary Effects of Epinephrine (Adrenalin) and Nor Epinephrine (Noradrenalin) on Biofilms

A recent study showed the activities of epinephrine (E) and norepinephrine (NE) on the growth of an enterotoxigenic *Escherichia coli* (ETEC) to 80%. The findings even indicated that NE and

E possessed negligeable impacts over the proliferation of 52038 isolates; in contrast, NE and E inhibited the development of 52011 isolates. The bacterial density in the ETEC+NE+α and ETEC+Epi+α groups was similar to that of the ETEC+α group after 12 and 24 hours. In the 52038, 52160, and 52130 isolates, the ETEC+Epi+β antagonist-added group had considerably lower bacterial density compared to the ETEC+Epi group. Epi may enhance development by acting as a β-adrenergic receptor antagonist (Niu et al., 2023). In another study, catecholamines increased *S. mutans* biofilm formation in orthodontic brackets ($p < 0.05$), but did not affect acid production (pH reduction) or the expression of biofilm-related genes (gtfB, gtfC, gbpA, gbpB, gbpC, gbpD, and brpA), aciduric (relA), or acidogenic properties (ldh) (Cortês et al., 2021). In another study four *Enterococcus faecalis* strains: *E. faecalis* MMH594 and *E. faecalis* V583, both pathogenic clinical strains; *E. faecalis* Symbioflor 1 clone DSM 16431, a pharmaceutical probiotic; and E. *faecalis* OB15, a probiotic strain previously isolated from *Tunisian rigouta*. Epi was discovered to influence the production of biofilm (biovolume and thickness) in *E. faecalis*, whether pathogens or probiotics. NE had a lesser influence on the production of biofilms by these bacteria (Cambronel et al., 2020).

Research has shown that the host stress hormone norepinephrine (NE) plays a significant role in modulating the behaviour of *Streptococcus pneumoniae*. Abundance of NE was observed to induce pneumococcal proliferation, augment formation of biofilms, and hence upregulate virulence gene expression profiles (Figure 7.3). Such findings indicated that NE might have influenced pathogenic attributes of *Pneumococcus* by providing suitable colonization and survival conditions, mainly at instances of host stress responses. This shows the multifaceted chemistry amongst the neuroendocrine parameters of the host cells and microbial pathogenicity (Sandrini et al., 2014). Duran-Pinedo et al. (2018) unveiled the way, in which cortisol, which is a predominant stress hormone, affects the buccal microbiota at the level of transcriptome. This finding further projected that exposure to cortisol may also modify gene expression inside the buccal microbiota, predominantly augmenting

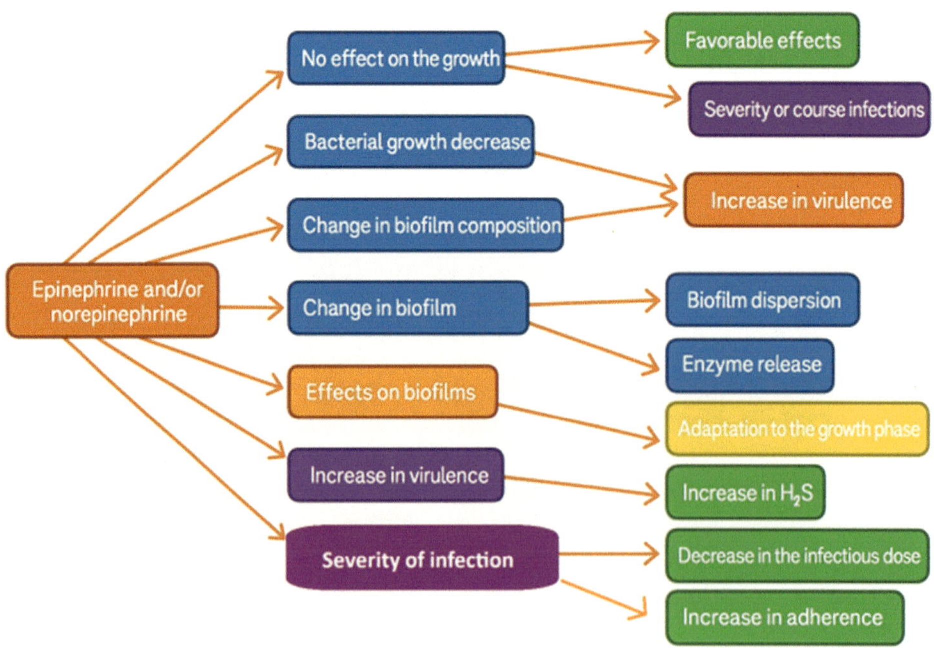

FIGURE 7.3 Schematic representation of binary effects of epinephrine (adrenalin) and nor epinephrine (noradrenalin) on biofilms.

pathways of virulence, iron sequestration, and formation of the biofilm. All these findings are indications suggesting that stress responses in host may result in swings in bacterial behaviour, possibly endorsing dysbiosis leading to disease progression, like periodontitis. It also throws light on the central association among and microbial-based oral infections and psychological stress.

7.8 CONCLUSION

To conclude, the association among the stress hormones, including NE and E, with microbial biofilms is extremely complicated and diverges remarkably according to the species of bacteria and the surrounding conditions. Although these catecholamines help in the formation of biofilms, virulence, and growth of bacteria, they can also suppress these procedures among other bacteria. The binary and antagonistic functions of these hormones accentuate the complicated association involved among stress responses by host and bacterial behaviour. A complete understanding of these dynamics is vital, since they pointedly implicate disease progression, antibiotic resistance, and discovery of innovative therapeutic approaches by targeting biofilm-mediated infections.

REFERENCES

Aguilera, G., Subburaju, S., Young, S., & Chen, J. (2008). The parvocellular vasopressinergic system and responsiveness of the hypothalamic pituitary adrenal axis during chronic stress. *Progress in Brain Research*, *170*, 29–39. https://doi.org/10.1016/S0079-6123(08)00403-2

Alborzkouh, P., Nabati, M., Zainali, M., Abed, Y., & Shahgholy Ghahfarokhi, F. (2015). A review of the effectiveness of stress management skills training on academic vitality and psychological well-being of college students. *Journal of Medicine and Life*, *8*(Spec Iss 4), 39–44.

Ball, J., & Darby, I. (2022). Mental health and periodontal and peri-implant diseases. *Periodontology*, *90*(1), 106–124. https://doi.org/10.1111/prd.12452

Botía, M., Escribano, D., Martínez-Subiela, S., Tvarijonaviciute, A., Tecles, F., López-Arjona, M., & Cerón, J. J. (2023). Different types of glucocorticoids to evaluate stress and welfare in animals and humans: General concepts and examples of combined use. *Metabolites*, *13*(1), 106. https://doi.org/10.3390/metabo13010106

Cambronel, M., Nilly, F., Mesguida, O., Boukerb, A. M., Racine, P. J., Baccouri, O., Borrel, V., Martel, J., Fécamp, F., Knowlton, R., Zimmermann, K., Domann, E., Rodrigues, S., Feuilloley, M., & Connil, N. (2020). Influence of catecholamines (epinephrine/norepinephrine) on biofilm formation and adhesion in pathogenic and probiotic strains of *enterococcus faecalis*. *Frontiers in Microbiology*, *11*, 1501. https://doi.org/10.3389/fmicb.2020.01501

Cambronel, M., Tortuel, D., Biaggini, K., Maillot, O., Taupin, L., Réhel, K., Rincé, I., Muller, C., Hardouin, J., Feuilloley, M., Rodrigues, S., & Connil, N. (2019). Epinephrine affects motility, and increases adhesion, biofilm and virulence of Pseudomonas aeruginosa H103. *Scientific Reports*, *9*(1), 20203. https://doi.org/10.1038/s41598-019-56666-7

Cameron J. L. (1997). Stress and behaviorally induced reproductive dysfunction in primates. *Seminars in Reproductive Endocrinology*, *15*(1), 37–45. https://doi.org/10.1055/s-2008-1067966

Chassin, C., Hornef, M. W., Bens, M., Lotz, M., Goujon, J. M., Vimont, S., Arlet, G., Hertig, A., Rondeau, E., & Vandewalle, A. (2007). Hormonal control of the renal immune response and antibacterial host defense by arginine vasopressin. *The Journal of Experimental Medicine*, *204*(12), 2837–2852. https://doi.org/10.1084/jem.20071032

Cirkovic, I., Pavlovic, B., Bozic, D. D., Jotic, A., Bakic, L., & Milovanovic, J. (2017). Antibiofilm effects of topical corticosteroids and intranasal saline in patients with chronic rhinosinusitis with nasal polyps depend on bacterial species and their biofilm-forming capacity. *European Archives of Oto-Rhino-Laryngology*, *274*(4), 1897–1903. https://doi.org/10.1007/s00405-017-4454-6

Cortês, I. T., Rosalen, P. L., Berto, L. A., Castro, M. L., Pedrini, D. L., Porto, A. N., Cogo-Müller, K., & Nobre Franco, G. C. (2021). Effect of adrenaline and noradrenaline on biofilm formation and virulence factors of Streptococcus mutans UA159. *Archives of Oral Biology*, *125*, 105091. https://doi.org/10.1016/j.archoralbio.2021.105091

Danilova, N. D., Solovyeva, T. V., Mart'yanov, S. V. et al. (2020). Stimulatory effect of epinephrine on biofilms of *micrococcus luteus* C01. *Microbiology*, *89*, 493–497. https://doi.org/10.1134/S0026261720040049

Declercq, A. M., Cai, W., Naranjo, E. et al. (2019). Evidence that the stress hormone cortisol regulates biofilm formation differently among *Flavobacterium columnare* isolates. *Veterinary Research*, *50*, 24. https://doi.org/10.1186/s13567-019-0641-3

Duran-Pinedo, A. E., Solbiati, J., & Frias-Lopez, J. (2018). The effect of the stress hormone cortisol on the meta-transcriptome of the oral microbiome. *npj Biofilms and Microbiomes*, *4*, 25. https://doi.org/10.1038/s41522-018-0068-z

Fareau, G. G., & Vassilopoulou-Sellin, R. (2007). Hypercortisolemia and infection. *Infectious Disease Clinics of North America*, *21*(3), 639–657. https://doi.org/10.1016/j.idc.2007.06.001

Galano-Frutos, J. J., Maity, R., Iguarbe, V., Aínsa, J. A., Velázquez-Campoy, A., Schaible, U. E., Mamat, U., & Sancho, J. (2024). L-Thyroxine and L-thyroxine-based antimicrobials against Streptococcus pneumoniae and other Gram-positive bacteria. *Heliyon*, *10*(7), e27982. https://doi.org/10.1016/j.heliyon.2024.e27982

Gannesen, A. V., Schelkunov, M. I., Geras'kina, O. V., Makarova, N. E., Sukhacheva, M. V., Danilova, N. D., Ovcharova, M. A., Mart'yanov, S. V., Pankratov, T. A., Muzychenko, D. S., Zhurina, M. V., Feofanov, A. V., Botchkova, E. A., & Plakunov, V. K. (2021). Epinephrine affects gene expression levels and has a complex effect on biofilm formation in *Micrococcus luteus* strain C01 isolated from human skin. *Biofilms*, *3*, 100058. https://doi.org/10.1016/j.bioflm.2021.100058

Gannesen, A. V., Ziganshin, R. H., Zdorovenko, E. L., Klimko, A. I., Ianutsevich, E. A., Danilova, O. A., Tereshina, V. M., Gorbachevskii, M. V., Ovcharova, M. A., Nevolina, E. D., Martyanov, S. V., Shashkov, A. S., Dmitrenok, A. S., Novikov, A. A., Zhurina, M. V., Botchkova, E. A., Toukach, P. V., & Plakunov, V. K. (2022). Epinephrine extensively changes the biofilm matrix composition in *Micrococcus luteus* C01 isolated from human skin. *Frontiers in Microbiology*, *13*, 1003942. https://doi.org/10.3389/fmicb.2022.1003942

Goldstein D. S. (1987). Stress-induced activation of the sympathetic nervous system. *Baillière's Clinical Endocrinology and Metabolism*, *1*(2), 253–278. https://doi.org/10.1016/s0950-351x(87)80063-0

Güner, B., & Kisadere, I. (2022). The efficacy of intravaginal antibiotic or their combination with different gonadotropins on vaginitis and fertility in ewes synchronized with progesterone impregnated intravaginal sponges. *Turkish Journal of Veterinary and Animal Sciences*, *46*(4), 4. https://doi.org/10.55730/1300-0128.4228

Helmreich, D. L., Parfitt, D. B., Lu, X. Y., Akil, H., & Watson, S. J. (2005). Relation between the hypothalamic-pituitary-thyroid (HPT) axis and the hypothalamic-pituitary-adrenal (HPA) axis during repeated stress. *Neuroendocrinology*, 81(3), 183–192.

Herman, J. P., Figueiredo, H., Mueller, N. K., Ulrich-Lai, Y., Ostrander, M. M., Choi, D. C., & Cullinan, W. E. (2003). Central mechanisms of stress integration: Hierarchical circuitry controlling hypothalamo-pituitary-adrenocortical responsiveness. *Frontiers in Neuroendocrinology*, *24*(3), 151–180. https://doi.org/10.1016/j.yfrne.2003.07.001

Howarth, E. J., O'Connor, D. B., Panagioti, M., Hodkinson, A., Wilding, S., & Johnson, J. (2020). Are stressful life events prospectively associated with increased suicidal ideation and behaviour? A systematic review and meta-analysis. *Journal of Affective Disorders*, *266*, 731–742. https://doi.org/10.1016/j.jad.2020.01.171

Itoh, N., Obata, K., Yanaihara, N., & Okamoto, H. (1983). Human preprovasoactive intestinal polypeptide contains a novel PHI-27-like peptide, PHM-27. *Nature*, *304*(5926), 547–549. https://doi.org/10.1038/304547a0

Kim, H. S., & Yosipovitch, G. (2020). The skin microbiota and itch: Is there a link? *Journal of Clinical Medicine*, *9*(4), 1190. https://doi.org/10.3390/jcm9041190

Lei, Y., Rahman, K., Cao, X., Yang, B., Zhou, W., Reheman, A., Cai, L., Wang, Y., Tyagi, R., Wang, Z., Chen, X., & Cao, G. (2023). Epinephrine stimulates *mycobacterium tuberculosis* growth and biofilm formation. *International Journal of Molecular Sciences*, *24*(24), 17370. https://doi.org/10.3390/ijms242417370

Leinweber, A., Weigert, M., & Kümmerli, R. (2018). The bacterium Pseudomonas aeruginosa senses and gradually responds to interspecific competition for iron. *Evolution*, *72*, 1515–1528. https://doi.org/10.1111/evo.13491

Lories, B., Roberfroid, S., Dieltjens, L., De Coster, D., Foster, K. R., & Steenackers, H. P. (2020). Biofilm bacteria use stress responses to detect and respond to competitors. *Current Biology: CB*, *30*(7), 1231–1244. https://doi.org/10.1016/j.cub.2020.01.065

Niu, L., Gao, M., Wen, S., Wang, F., Shangguan, H., Guo, Z., Zhang, R., & Ge, J. (2023). Effects of catecholamine stress hormones norepinephrine and epinephrine on growth, antimicrobial susceptibility, biofilm formation, and gene expressions of enterotoxigenic *Escherichia coli*. *International Journal of Molecular Sciences*, *24*(21), 15646. https://doi.org/10.3390/ijms242115646

Parijs, I., & Steenackers, H. P. (2018). Competitive inter-species interactions underlie the increased antimicrobial tolerance in multispecies brewery biofilms. *The ISME Journal*, *12*, 2061–2075. https://doi.org/10.1038/s41396-018-0146-5

Ranabir, S., & Reetu, K. (2011). Stress and hormones. *Indian Journal of Endocrinology and Metabolism*, *15*(1), 18–22. https://doi.org/10.4103/2230-8210.77573

Rather, M. A., Gupta, K., & Mandal, M. (2021). Microbial biofilm: Formation, architecture, antibiotic resistance, and control strategies. *Brazilian Journal of Microbiology: Publication of the Brazilian Society for Microbiology*, *52*(4), 1701–1718. https://doi.org/10.1007/s42770-021-00624-x

Reidl, J., & Monsó, E. (2015). Glucocorticoids and antibiotics, how do they get together? *EMBO Molecular Medicine*, *7*(8), 992–993. https://doi.org/10.15252/emmm.201505336

Rogerson, O., Wilding, S., Prudenzi, A., & O'Connor, D. B. (2024). Effectiveness of stress management interventions to change cortisol levels: A systematic review and meta-analysis. *Psychoneuroendocrinology*, *159*, 106415. https://doi.org/10.1016/j.psyneuen.2023.106415

Rosin, S., Xia, K., Azcarate-Peril, M. A., Carlson, A. L., Propper, C. B., Thompson, A. L., Grewen, K., & Knickmeyer, R. C. (2021). A preliminary study of gut microbiome variation and HPA axis reactivity in healthy infants. *Psychoneuroendocrinology*, *124*, 105046. https://doi.org/10.1016/j.psyneuen.2020.105046

Sandrini, S., Alghofaili, F., Freestone, P. et al. (2014). Host stress hormone norepinephrine stimulates pneumococcal growth, biofilm formation and virulence gene expression. *BMC Microbiology*, *14*, 180. https://doi.org/10.1186/1471-2180-14-180

Sarkodie, E. K., Zhou, S., Baidoo, S. A., & Chu, W. (2019). Influences of stress hormones on microbial infections. *Microbial Pathogenesis*, *131*, 270–276. https://doi.org/10.1016/j.micpath.2019.04.013

Verbrugghe, E., Boyen, F., Gaastra, W., Bekhuis, L., Leyman, B., Van Parys, A., Haesebrouck, F., & Pasmans, F. (2012). The complex interplay between stress and bacterial infections in animals. *Veterinary Microbiology*, *155*(2–4), 115–127.

Wang, J., Wang, R., Wang, H. et al. (2017). Glucocorticoids suppress antimicrobial autophagy and nitric oxide production and facilitate mycobacterial survival in macrophages. *Scientific Reports*, *7*, 982. https://doi.org/10.1038/s41598-017-01174-9

Webster, J. I., & Sternberg, E. M. (2004). Role of the hypothalamic-pituitary-adrenal axis, glucocorticoids and glucocorticoid receptors in toxic sequelae of exposure to bacterial and viral products. *The Journal of Endocrinology*, *181*(2), 207–221. https://doi.org/10.1677/joe.0.1810207

Yahya, A. A., Al-Haidar, A. H., & Al-Mizraqchi, A. S. (2019). The role of salivary cortisol and mutans streptococci in the development of early childhood caries. *Indian Journal of Public Health Research & Development*, *10*(10), 3081.

Yousuf, M., Ali, A., Khan, P., Anjum, F., Elasbali, A. M., Islam, A., Yadav, D. K., Shafie, A., Rizwanul Haque, Q. M., & Hassan, M. I. (2022). Insights into the antibacterial activity of prolactin-inducible protein against the standard and environmental MDR bacterial strains. *Microorganisms*, *10*(3), 597. https://doi.org/10.3390/microorganisms10030597

8 Phytohormones and Biofilm

Niladri Sarkar and Rina Rani Ray
Department of Biotechnology, Maulana Abul Kalam Azad University of Technology, Kolkata, India

8.1 INTRODUCTION

Biofilms, a complex group of microorganisms surrounded by a self-produced extracellular matrix, are widespread in nature. Biofilm appears on various biotic and abiotic surfaces, ranging from organic tissues to non-living objects. Both Gram-positive and Gram-negative bacteria are able to form biofilm, which is responsible for several pathogenic infections; however, biofilm may sometimes be considered beneficial. Biofilm formation is a dynamic and intricate process involving several stages such as microbial adherence in which bacteria, often in the planktonic stages, attaches to the different surfaces. This is followed by the production of extracellular polymeric substances (EPS), primarily composed of polysaccharides, proteins, and DNA. Due to the presence of sufficient nutrients, the bacterial cells form microcolonies by the help of EPS multiplication, signal transduction, and quorum sensing. The matured sessile cells then form multiple layers of biofilm which are dispersed across the new surface in a continuous process (Dutta et al., 2025). In recent years, the biofilm has become the subject of much research as it provides an array of contribution to several different system, including plant and animals' growth, development, pathogenic infections, food preservation etc. Nature is a reservoir of a wide range of bioactive compounds which, in a variety of interactions with biofilm either for induction or disruption, as shown in Figure 8.1. Plant hormones, or phytohormones, are among those which play a crucial role in biofilm.

Phytohormones can be described as a plant's chemical messenger synthesized from the different parts of the roots and shoots of plants. The bioactive hormones have several responsibilities, including root and shoot formation, growth, development, stem cell elongation, maturation, apical dominance, senescence, flowering, fruits ripening etc. Auxin and cytokinin are two of the preliminary plant hormones which are found in great abundance in all types of plants.

Additionally, gibberellin, abscisic acid, ethylene, jasmonic acid, melatonin, brassinosteroids, and salicylic acids are among the secondary phytohormones present at a lower concentration, playing an integrated role in seed dormancy, germination, flowering, the ripening of fruits, stress tolerance, and defense mechanisms (Pal et al., 2023). Consequently, chemically synthetic phytohormones, including auxin derivatives, zeatin, and 6-benzylaminopurine (BAP) etc., are used globally in the agricultural sector in order to achieve better yields. Horticulture and in vitro tissue culture are also facilitated by natural and synthetic hormones.

Indeed, studies have revealed intriguing interactions between phytohormones and biofilm-forming microorganisms; however, comprehensive and in-depth research is still needed to fully understand these dynamics. Several phytopathogens, such as *Agrobacterium tumefaciens, rhizobacteria, Pseudomonas spp. Bacillus subtilis, Staphylococcus aureus* etc., are widely studied, and can be affected by the phytohormones. Furthermost, some of the microbial pathogens are potentially able to synthesize their own auxin, cytokinin, abscisic acid, and also sometimes their derivatives. The self-contradictory role of phytohormones on bacterial pathogens and biofilm are species specific in nature where, a particular hormone showing their positive activity on growth of specific type of bacterial biofilm and for others are either non-responsive or having inhibitory effects. In this chapter,

 DOI: 10.1201/9781003519041-8

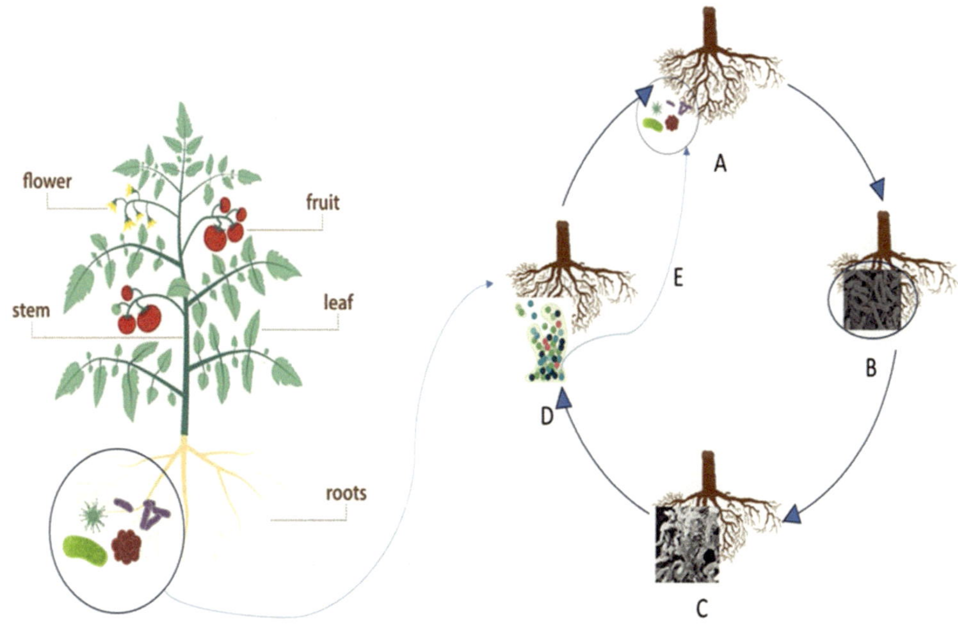

FIGURE 8.1 Diagram showing the sites of infection by phytopathogens and biofilm formation.

we will discuss the mode of action of different phytohormones on the inducing or inhibiting activities of microbial biofilm.

8.2 TYPES OF PHYTOHORMONES

Phytohormones, also known as plant hormones, are naturally occurring molecules that modulate the growth and development of plants. These signaling molecules, which are effective in lower concentrations, control several processes, from seed propagation and root formation to flowering, fruit ripening, and senescence (Fenn & Giovannoni, 2021). Phytohormones play the role of the primary chemical messenger within these plant bodies, resulting in responses to both congenital development and exterior signals. The action of plant hormones is particularly target-specific; it can range significantly, depending upon its concentration, maturation of the plants, and, most importantly, the presence of other hormones. The phytohormones can be classified based on the functional characteristics such as either growth promoter or growth elicitor. Auxin, cytokinin, gibberellin, abscisic acid, ethylene are five major classes of plant hormones found abundantly in the system, as shown in Figure 8.2. Their structural conformation is given in Figure 8.3.

8.2.1 Auxin

Auxins are Indole-based plant growth promoters, found in all types of plants in young developing tissues, such as shoot apical meristems, young leaves, and developing seeds. It comes from the Greek word 'Auxein', meaning 'to increase'. The father and son scientist duo, Charles Darwin and Francis Darwin, first observed the phototropic action of these hormone in grass coleoptiles in the late 19th century. However, the subsequent studies of Boysen-Jensen and Went unveiled the identity of the auxin hormones and then, later, Kenneth Thimann isolated the Indole acetic acid (IAA) as the most predominant form of auxin. Naturally occurring auxins are mostly derived from tryptophan (Trp) via complex pathways such as the Indole-3-pyruvic acid pathway (IPA), where

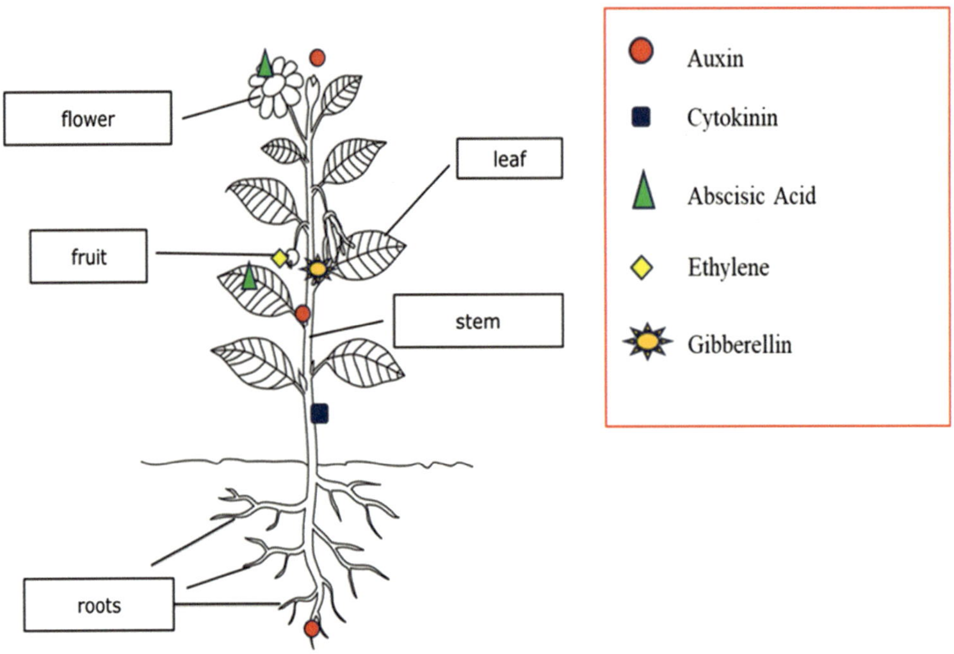

FIGURE 8.2 The diagrammatic representation of the source of different phytohormones.

tryptophan is transaminated into Indole-3-pyruvic acid. This is followed by the decarboxylation and oxidation into Indole acetic acid. In addition, tryptamine (TAM) and Indole-3-acetonitrile (IAN) pathways are among other following pathways involved in auxin biosynthesis. There are some tryptophan-independent pathways also present in the studies, but the exact mechanism is still not very clear. Indole-3 acetic acid (IAA), Indole-3 butyric acid (IBA) and 4-chloro Indole-3 acetic acid (4-Cl-IAA) are most abundantly found in natural sources, whereas there are several synthetic auxins: e.g. 2,4-Dichlorophenoxyacetic acid (2,4-D),1-Naphthaleneacetic acid (NAA), phenoxyacetic acid, phenoxypropionic acid, phenylbutyric acid, pyridinecarboxylic acids, quinolinecarboxylic acids, and quinclorac can all be generally used as herbicides, cell culture, tissue culture and parthenocarpy.

8.2.2 Gibberellin

Gibberellin is another crucial phytohormone found in the apical meristem of root and stem, leaves, and seed embryos. it can also be discovered in some fungi and bacteria. The first ever gibberellin hormone was identified by E. Kurosawa (1926) from the fungus *Gibberella fujikuroi*, an effective rice pathogen involved in random growth of the plant along with poor yield, commonly known as "Foolish Seedling Disease". The term gibberellin itself was coined by Teijiro Yabuta and Sumiki in 1935. Gibberellins, composed of tetracyclic diterpenoids, are based on a four-ring structure derived from a mostly 20-carbon isoprenoid (diterpene) skeleton. Gibberellic acid (GA) is one of the most common natural gibberellins, existing in nature alongside its 100 derivatives. GA12 is a well-known biosynthetic product. It is synthesized via stepwise complex procedures, beginning with the formation of geranylgeranyl diphosphate (GGPP) from isopentenyl diphosphate (IPP) and dimethylallyl diphosphate (DMAPP) within the plastid. This is then transported to endoplasmic reticulum (ER) and converted to the ent-kaurene and finally GA12 due to several enzymatic reactions. GA12 is the common precursor of all other gibberellic acids. GA3, GA4, GA5, GA6, GA7 are also among the popular gibberellic acids involved for plant growth development.

8.2.3 Cytokinin

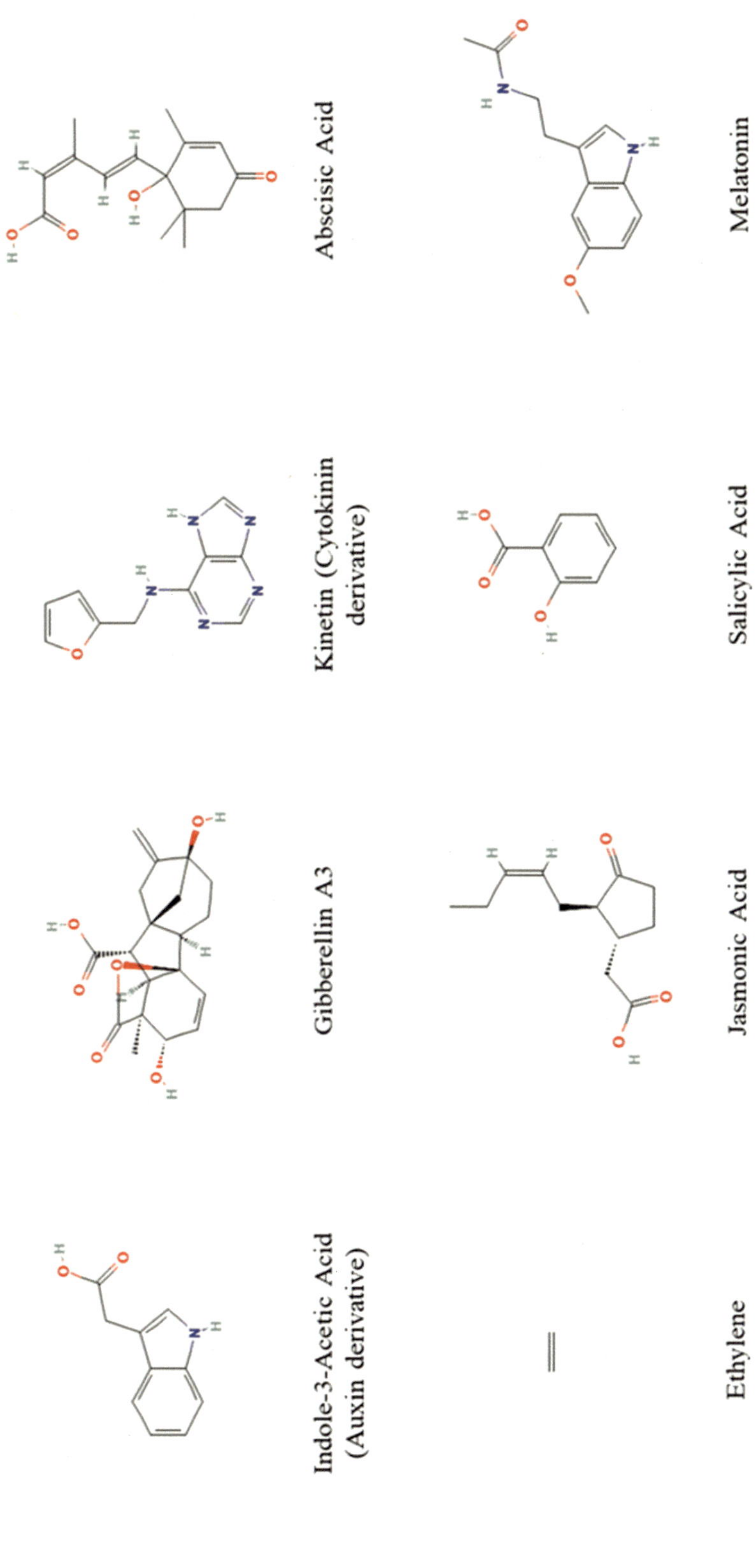

FIGURE 8.3 Representation of the chemical structure of phytohormones.

Of five primary phytohormones, cytokinin is one of the growth promoters derived from adenosine. The mechanism of the production of cytokinin begins from isopentenyladenosine-5'-monophosphate (iPMP), which is synthesized from adenosine-5′- monophosphate (AMP) and dimethylallyl pyrophosphate (DMAPP), and catalyzed by isopentenyl transferase (IPTs). This iPMP is modified in various forms of cytokinins, including zeatin, kinetin etc. The history of discovery of cytokinin is quite a lengthy one, being credited to Miller and Skoog in 1955 (Miller et al., 1955), who isolated kinetin from salmon sperm DNA. The precise functional activity of this plant hormone was observed by Mok in 1994 (Mok, 1994). Naturally occurring cytokinins are found in various sources such as roots, the endosperm of seeds, growing embryos, developing seeds, young fruits e.g. tomato, coconut water, and developing shoot buds. Cytokinin is classified as isoprenoid and aromatic type of cytokinin. The aromatic properties of cytokinin hormones, like 6-benzylaminopurine (BAP) and kinetin, can be attributed to the N-6 side chain linked to the adenine base. This aromatic side chain is a distinctive feature of this group of plant growth regulators.

8.2.4 Abscisic Acid

Abscisic acid (ABA) is considered to be a plant growth inhibitor, found in roots, stems, flowers, and seeds but most abundantly in mature leaves and stems. The presence of ABA was first observed in cotton fruits by Frederick T. Addicott and Larry A. Davis in 1963. Abscisic acid is responsible for the abscission of the mature leaves of the plants. ABA is a sesquiterpenoid phytohormone synthesized via the methylerythritol phosphate (MEP) pathway in plastids. The key regulatory step in ABA biosynthesis is catalyzed by 9-cis-epoxycarotenoid dioxygenase (NCED), which converts 9-cis-violaxanthin or 9-cis-neoxanthin into xanthoxin. Xanthoxin is then transported to the cytosol and further converted into ABA-aldehyde and eventually to ABA. ABA levels are tightly controlled through biosynthesis, catabolism (primarily via ABA 8′-hydroxylase), and transport. The expression of ABA biosynthesis genes, particularly NCED, is significantly upregulated under stress conditions.

8.2.5 Ethylene

Ethylene is one of the gaseous phytohormones synthesized from methionine via a multi-step pathway in which S-adenosylmethionine (SAM) is converted to 1-aminocyclopropane-1- carboxylic acid (ACC) by the catalysis of ACC synthase, followed by the ACC to ethylene formation through oxidation reaction. Ethylene is involved in fruit ripening, abscission, leaf senescence, stress tolerance etc. It is generally in the form of an aromatic simple unsaturated hydrocarbon found in higher plants such as roots, stems, fruits, flowers, tubers and seeds.

In addition to the primary phytohormones, there are several other plant hormones, including brassinosteroids, jasmonates, and salicylic acid. The essential role of brassinosteroids include cell division and the growth of the plant body, environmental stress tolerance, and some metabolic activity such as plant oxidation radicals and ethylene synthesis etc., whereas jasmonate and salicylic acid are two well-known phytohormones associated with plant defense mechanisms. Melatonin is another important documented phytohormone which involved in the activity of stress tolerance.

8.3 FUNCTIONS OF THE PHYTOHORMONES

Plants are very adaptive and tolerant of changes in the natural environment. Their ability to survive depends on their very well-designed arranged communication systems which coordinate growth, development, and responses to external stimuli. Phytohormones play a pivotal role as a chemical messenger in communication networks. Generally, hormones are produced in very specific locations within the plant, including apical meristematic tissues, leaves, stems and root tips etc and transported to other organs, where they react to receptors and trigger intracellular signaling cascades. These signaling pathways further modify the gene expression, resulting in changes in cellular

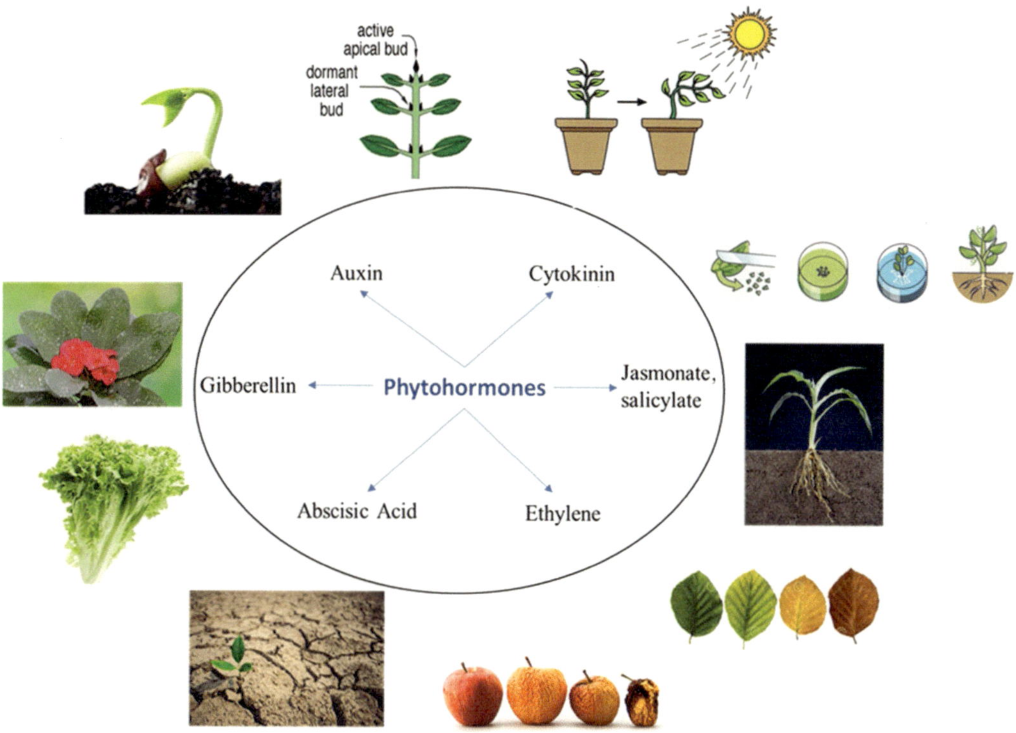

FIGURE 8.4 The schematic representation of the phytohormone's role in different physiological processes of plants.

activity and, consequently, changes in plant phenotype. The functions of plant hormones are often complex and involve interactions between different hormones, as well as environmental factors e.g. sunlight, water, nutrients, pathogens etc. as shown in Figure 8.4.

8.3.1 The Key Functions of Auxin

Auxin along with all derivatives including IAA, IBA, are important phytohormones that perform a range of functions: plant growth, developments and movements. Auxin distribution is highly sensitive for its functional activity. Auxin exhibits polar transport, as it falls preferentially from the apex towards the base of the plant in a cell-to-cell manner. This directional flow is largely dependent on two families of proteins: AUX1/LAX influx carriers that facilitate auxin entry into the cell; and PIN-FORMED (PIN) efflux carriers that facilitate auxin outgoing.

8.3.1.1 Cell Elongation

Auxin is considered to be the main growth promoter of the plant. According to the "acid growth hypothesis", auxin vivifies the activity of cell membrane H+-ATPases, resulting in the acidification of the cell wall. This acidification catalyses to activate the enzyme for the loosening of the cell wall; hence, the cell wall gets expanded in response to turgor pressure. The mechanism plays an important factor in stem and root elongation. Several studies have shown that the external application of auxin into plants may responsible for the augmentation of gene expression associated with cell wall extension such as CSLC4, CSLC5, EXPA4, EXPA1,XTH11 etc. for the production of primary and secondary cell wall components, structural proteins(EXPs), and peroxidases (Nemhauser et al., 2006).

8.3.1.2 Cell Division and Differentiation

Auxin and its activity on cell division in plants, especially during cambial activity and lateral root formation are often influential, in plants. It also influences cell differentiation, serving to the initiation of vascular tissue patterning and the growth of specific cell types. A study by Inze et al. showed that a lack of auxin in in vitro cell culture resulting in the cessation of cell division. Furthermore, the addition of auxin restores division after deprivation. Auxin is necessary but not sufficient to stimulate cell division in cultured cells or plant tissues. Another study has shown that auxin can control the patterns of cell division in the lateral root development by activating mitogen-activated protein kinase (MAPK) signaling through transmembrane kinases (TMKs), which are considered to be essential for lateral root development. TMK-mediated apocryphal signaling of auxin is imperative to the modulation of cell division patterns and connect auxin signaling to MAPK signaling (Huang et al., 2019). Additionally, auxin plays a role in the distinction of vascular tissues (xylem and phloem), resulting in the transport of water, nutrients, and signaling molecules throughout the plant.

8.3.1.3 Apical Dominance

Auxin produced generally in the apical bud of the plant can inhibit the growth and development of lateral buds, branching lower down the stem, termed as apical dominance. This phenomenon is well designed for investing the plant resources in upward growth. Auxin is transported from the shoot apex to the roots involved in this process.

8.3.1.4 Development of Plants

The presence of lower concentrations of auxin in the root promotes root growth and developments while higher auxin content in shoots is considered essential for lateral root formation and root architecture. This inclination of auxin is necessary for setting up the proper patterning in the root apical meristem. In addition to root development (Roychoudhry & Kepinski, 2021), auxin can play a predominant role in the development of fruits and leaves. Auxin and its derivatives can instigate seedless or parthenocarpic fruits such as tomato, cucumber, banana, pear, and watermelon (Qian et al., 2018) due to working on unpollinated ovaries. Chemically synthesized auxins e.g. 2,4-D and NAA, have potential for the growth and development of parthenocarpic fruit. 4-chlorophenoxyacetic acid (4-CPA) has been injected into ovaries in order to induce the seedless okra and grapes in experiments (Hossain et al., 2023; Sharif et al., 2022). Auxin is crucial for leaf initiation, patterning, and venation as auxin maxima help specify the position of leaf primordia on the shoot apical meristem.

8.3.1.5 Tropic Responses

Auxin is the leading phytohormone in terms of regulating the tropic movements of plants, including phototropism, gravitropism and hydrotropism. Differential auxin distribution, influenced by light and gravity, leads to asymmetric growth patterns, resulting in the plant bending towards the stimulus (Liscum et al., 2014).

8.3.2 The Key Functions of Gibberellin

8.3.2.1 Seed Germination

The role of the gibberellin hormone in the germination of seeds is to facilitate the breakdown of storage reserves and the elongation of the embryonic axis. During the period of seed dormancy, the content of active gibberellic acids (Gas) is considerably lower, and the DELLA proteins are stable, which inhibits germination. The application of exogenous GAs or the removal of the inhibitory effect of DELLAs by mutations can break seed dormancy and promote germination. The study investigates the role of externally applied gibberellic acid 3 (GA3) on seed germination and seedling growth of hemp—'Yunma 1' and 'Bamahuoma' under drought has shown the enhancement of seed germination percentage (Du et al., 2022).

8.3.2.2 Stem and Leaf Elongation

Like auxin, gibberellin has an impact on stem and leaf elongation, cell division, turgor pressure and cell differentiation, with levels being directly proportional to the growth rate of the plants. The gibberellic acid (GAs) supplementation can vivify stem elongation, enhancing the leaf size, leaf polarity and margin development. However, the inhibition of GA biosynthesis or signaling can reduce stem elongation, so-called dwarfism, along with shorter leaves (Sun et al., 2010). Several mechanisms are found in the process of gibberellin-associated elongation, including regulation of the expression of cell wall-loosening enzymes, e.g. expansions, stimulating cell division and expansion. The popular example of the role of GA in stem elongation has been shown in rice. Semi-dwarf rice varieties, such as those bred during the Green Revolution, induce mutations in genes linked to GA biosynthesis or signaling. These mutations can lower GA content or reduce sensitivity to GA, resulting in shorter, strapping stems that are less prone to falling off. As a result, the yield of the crop has increased significantly. Exogenous GA treatment has also been used in lettuce production to form a longer stem for easier harvesting. On the other site, in *Arabidopsis*, GA-deficient mutants have been found to present smaller leaves, along with minimized cell numbers and cell size. On the other hand, exaggeration of GA biosynthetic genes and the treatment of exogenous gibberellin leads to larger leaves.

8.3.2.3 Flowering

Anthesis plays an indispensable determinant of reproductive success in plants, facilitating seed production aimed ultimately at species survival. The metamorphosis of flowers is synchronized by several complex factors involving environmental factors, such as photoperiod (day length), warmth (vernalization), and nutrient availability, along with endogenous genetic and hormonal signals. Gibberellins and their role in flowering have received little attention to date. GAs regulate flowering in some specific plant species, especially in short-night plants and biennials following vernalization (Bao et al., 2020). The GAs assisted modulation in gene expression of transition in the period of vegetative to reproductive growth, promoting in flowering. However, the levels of auxin, abscisic acid and (sometimes) ethylene can interfere with the activity of gibberellin in case of flowering.

8.3.3 The Key Functions of Cytokinin

8.3.3.1 Root and Shoot Development

The most-studied role of cytokinin and its derivatives is to promote cell division in the shoot apical meristem (SAM) and lateral root formation. They activate the development of cells due to the cell cycle, primarily resulting in the growth of shoots and leaves. The interaction of auxin and cytokinin influences cell density and differentiation, proliferation in the context of callus formation and shoot regeneration. The ratio of auxin to CK often depends on the further differentiation of callus into roots or shoots (Sugimoto et al., 2010). Many studies have revealed that high cytokinin levels can inhibit primary root growth, whereas it promotes shoot formation. As a result, it is useful for de novo shoot organogenesis, an innovative in vitro shoot regeneration technology from several plant tissues (Raspor et al., 2021).

8.3.3.2 Senescence Regulation

Cytokinin is considered to delay senescence, the process of aging and programmed cell death in plants. Cytokinins can lengthen photosynthetic periods in leaves by decreasing the accumulation of sugars, proteins, and other cellular components linked to aging as well as by mitigating oxidative damage that leads to the damage of chlorophyll (Ahlawat et al., 2022, Wu et al., 2021). Exogenous application of cytokinin can reduce the chlorophyll, which promotes a delay in leaf senescence. In addition, cytokinin also promotes the assemblage of nutrients from source-sensing tissues to younger, maturing parts of the plant, facilitating the efficient allocation of supplements.

8.3.3.3 Stress Responses

Plants are exposed to changing environmental factors which then significantly show their impact in their growth, development, and, ultimately, survival. These stressed conditions can be classified as abiotic stresses (e.g., drought, salinity, extreme temperatures, nutrient deficiency) and biotic stresses (e.g., pathogen attacks, insect herbivory). Cytokinin can influence stomatal closure, which indirectly ceases the water loss in drought or saline water conditions. In this context, abscisic acid is the primary influencer in plants for stress management; however, studies have shown the complex regulation of cytokinin and abscisic acid in the survival of plants in adverse conditions (Zhao et al., 2021). Again, cytokinin can influence root structure and the formation of deep root growth during drought conditions, leading plants to collect deeper water resources. Inducing stress is the primary cause of formation of reactive oxygen species (ROS), introducing oxidative damage. Several studies have shown the activity of cytokinin against ROS production by stimulating antioxidant enzymes, such as superoxide dismutase (SOD), catalase (CAT), and peroxidase (POD).

8.3.4 The Key Functions of Abscisic Acid

8.3.4.1 Stomatal Closure

Abscisic acid (ABA) is a stress hormone that accumulates under different abiotic and biotic stresses and promote stomatal closure, which were opened in sunlight. During the challenging environment such as drought, salinity and elevated CO_2 conditions, ABA helps to retain the closure of leaf stomata, resulting in a decrease in both photosynthesis and transpiration. Hence, the plants can store water instead of suffering from dehydration. This mechanism involves the stimulation of ion channels, causing the efflux of potassium (K+) and chloride (Cl-) ions from guard cells; hence, turgor pressure decreases and there is a closure of stomatal pores.

8.3.4.2 Stress Tolerance

Abscisic acid (ABA) is the primary phytohormone involved in mediating plant responses to different biotic and abiotic stresses, including salinity, cold, and osmotic stress, lack of nutrients, pathogenic infections etc. ABA influences the expression of stress-responsive genes, encoding proteins such as LEA (late embryogenesis-abundant) proteins which protect cellular structures from dehydration and dehydrins. It along with glycine, proline, betaine stabilizes proteins and membranes, protects antioxidant enzymes that helps to scavenge ROS formation during stress. These solutes help maintain cell turgor and protect cellular structures from dehydration. Even ABA acts on seed dormancy, resulting in the inhibition of embryonic growth during adverse environments. Transcriptome studies unveiled that more than 50% of the genes aided by ABA are also associated with drought or salinity. Several experiments of molecular manipulation of ABA synthesis or signaling have been performed in different plants for improved stress tolerance, such as maize, rice, soybean, wheat, sorghum etc. Otherwise, the capacity for cold tolerance has also been shown in cucumber and alfalfa. Additionally, the newer transgenic crops induced the overexpression of the key ABA synthesis gene NCED and the ABA receptor gene PYR, which have been found to indicate improved drought tolerance (Wang et al., 2016).

8.4 KEY FUNCTIONS OF ETHYLENE

8.4.1 Fruit Ripening

Ethylene is a key regulator of fruit ripening in climacteric fruits, such as tomatoes, bananas, and apples. It triggers a cascade of events, including changes in fruit color, texture, and aroma, making the fruit palatable and attractive for seed dispersal. Ethylene-insensitive mutants exhibit delayed ripening, highlighting its essential role in this process.

8.4.2 Senescence

Ethylene promotes senescence in various plant organs, including leaves, flowers, and fruits. It induces the degradation of chlorophyll and other cellular components, leading to the eventual death of the senescing organ.

8.5 FUNCTIONAL OVERLAPPING OF PHYTOHORMONES

Plant growth is a convoluted system involving cell division, elongation, and differentiation, all of which are actively influenced by multiple hormones. In some cases, they even exhibit antagonistic activity, as shown in Figure 8.5. For example, auxin assists cell division and elongation, while gibberellic acid primarily stimulates shoot elongation. Again, the synergistic effect of auxin and gibberellin may promote growth by enhancing cell wall loosening and promoting the expression of genes associated with cell expansion. Many studies have shown that auxin can also catalyze GA biosynthesis for further development. Auxin and brassinosteroids (BRs) are two growth-promoting hormones which are useful for cell elongation. BRs can stimulate an auxin-signaling cascade by surging the expression of auxin transporters and receptors. Simultaneously, auxin facilitates the biosynthesis of brassinosteroids, inducing a positive feedback loop for growth promotion. Both cytokinin and auxin also play antagonistic roles in shoot and root development. Auxin encourages root growth, whereas cytokinin inhibits it (Hussain et al., 2021). The proportion of these two hormones is pivotal for determining apical dominance and lateral bud formation.

With regard to the other physiological parameters, the flowering of plants is regulated by mainly gibberellin hormones, but auxin and cytokinin have secondary involvement in the flowering of so-called 'long day' plants. Meanwhile fruit development, fruit coating, and fruit setting often depends on the auxin, cytokinin and ethylene, either particularly or synergistically at certain levels. Similarly,

	Germination	Growth & Maturation	Flowering	Fruit Development	Abscission	Seed Dormancy	Stress Tolerance
Auxin	−	+	X	+	−	+	+
Gibberellin	+	+	+	+	−	−	+
Cytokinin	+	+	+	+	X	−	+
Abscisic Acid	−	−	−	+	+	+	+
Ethylene	+	−	+	+	+	−	+
Salicylic Acid	X	+	+	+	+	X	+
Jasmonic Acid	X	+	+	+	X	−	+

FIGURE 8.5 Representation of the promoting or inhibiting properties of phytohormones in different processes of plants.

stress tolerance is also affected by the combined effect of abscisic acid, ethylene, jasmonate, salicylic acid and cytokinin.

8.6 CROSS-TALK BETWEEN PHYTOHORMONES AND BIOFILM

The impact of phytohormones on biofilm formation is critical and often species-specific. Several physiological factors, environmental stress, and presence of other hormones influences the activity of phytohormones on biofilm and plant–pathogen interaction.

8.6.1 The Role of Auxin as Bacterial Signal Molecules

Auxin is predominantly acceptable as a plant growth promoter, aiding the healthy growth of the stem and root of the plants. But the interesting fact which is rarely common to the population, is that auxin or Indole-3-Acetic Acid is imparting as a bacteria-signalling molecule for the attachment, growth and the development of the sessile cells, affecting the plant–pathogen interaction. Some of the microorganism, especially the pathogens associated with plants such as *Agrobacterium tumefaciens, Agrobacterium rhizogenes, Pseudomonas savastanoi pv., Savastanoi, P. syringae pathovars, rhizobacteria e.g.: Azospirillum, X. campestris, Pantoea agglomerans (formerly Erwinia herbicola), Dickeya spp. (formerly E. chrysanthemi)* etc. are well documented as far in literature as Indole-3-acetic acid producers. The phytopathogens generally inferring the growth and metabolic regulations of the plants by synthesizing the signalling molecules auxin (IAA, naturally occurring auxin commonly), which is again influencing for gene expressions in bacteria, resulting the enhanced pathogenic susceptibility, colonizing capability, and, hence, the formation of biofilm (Barash & Manulis-Sasson, 2009).

It is thought that the lower concentration of auxin may help in bacterial activity whereas the higher content of auxin within the system plays a role as an inhibitor to the growth of the bacteria was associated with crown gall tumor formation in plants through inducing tumor plasmid to the host cell. The plasmid DNA inscribed the synthesis of the Indole acetic acid, promoting tumor formation. The transfer is initiated by a regulatory system, VirA/G, which recognizes plant signal molecules and activates a regulon. Here, IAA immobilized vir gene expression by competing with acetosyringone, a phenolic compound which can inhibit the growth of plant-associated bacteria (Liu & Nester, 2006). The subsequent study shows the same result as the higher concentration of IAA suppresses the *Vir* and *Chv* regulon of the bacteria *A. tumefaciens* (Yuan et al., 2008). The action of auxin is still controversial and it needs more research. However, Indole is considered to be a signalling molecule and controlling parameters of quorum sensing, and, hence, biofilm formation, in many bacteria such as *Escherichia coli, Pseudomonas aeruginosa* and *Vibrio cholerae*. In *V. cholerae*, Indole plays a role as a promoting parameter of biofilm formation by influencing the gene expression involved in vibrio extra polysaccharide (EPS) production. EPS facilitated the growth and development of the biofilm by providing nutrients and extra safeguard to the pathogen. The signalling cascade, such as DksA protein along with regulators VpsR and CdgA, interfere in motility, ion production, siderophore activity etc. (Mueller et al., 2009). Additionally, Indole plays a crucial role in the development of biofilm of *E.coli* pathogen, due to the suppression of motility, and acid tolerance (e.g., hdeABD, gadABCEX) and signalling autoinducer 1(SdiA) production. It is observed that these proteins can directly interfere in the transcription of the SdiA and the biofilm characteristics. Indole promotes a flat colonization in *E.coli* and *P. aeruginosa* due to the reduction in motility (Lee et al., 2007). A study of Zhang and his coworkers has shown that the 26 tested Indole analogues, along with Indole-3-acetic acid, play a large role in controlling the swarming motility, biofilm generation in *Vibrio campbellii* in a shrimp brine model. The natural auxin Indole-3-acetonitrile was identified to be the most potential bioactive molecule against *V. campbellii*, and extremely efficient in protecting shrimp larvae at concentrations as low as 10 μM (Zhang et al., 2022). It is shown that auxin exhibits both promoting and inhibitory effects depending on the bacterial species, IAA concentration, environmental conditions, and interactions with other signaling molecules. Indeed, research is needed to fully elucidate the complex mechanisms underlying these interactions.

8.6.2 The Role of Gibberellin on Biofilm

The broader Gibberellic acid (GA) family is particularly involved in the growth of plants, in aspects such as stem development, delays in senescence, flowering, the development of fruits etc. GA3 is the most-studied gibberellic acid among the larger gibberellic acid family, found in plants, fungi, and bacteria. In addition to the common functional mechanism, gibberellin plays several other roles in the functions of microorganisms, including growth promotion and acting as signalling molecules. It is reported in many studies that gibberellin production is highly perceptible in soil microorganism and plant growth-promoting bacteria (PGPB) (Timofeeva et al., 2024). Rhizobacteria, providing total soil fertility, produce large amounts of gibberellin as a secondary metabolite. *Azotobacter chroococcum, Pseudomonas fluorescens, Pseudomonas aeruginosa, Azotobacter chroococcum, Azotobacter vinelandii, Azotobacter paspali, Azospirillum brasilense, Azospirillum lipoferum, Thiobacillus novellus growing autotrophically on thiosulfate, Pseudomonas monteilii, Rhizobium meliloti, Rhizobium trifolii, and Rhizobium leguminusarum* are among the heterotrophic and symbiotic bacteria which produce significant amounts of gibberellic acid during their growing periods (Chakraborty et al., 2024).

Although the exact mechanism of microbial gibberellic acid (GA) interaction with plants is still under investigation, it is believed that GA-producing or GA-responsive soil bacteria undergo physiological and genetic changes that can either promote colonization or act antagonistically as antimicrobial agents. Several studies have shown that GA can control the adhesion of the bacteria through interfering in motility. For instance, GA can affect the swarming motility of the *Pseudomonas aeruginosa* via the modulation of the flagellar biosynthesis and development which facilitate the pathogen as in attachment and growth, leading to biofilm formation (Kearns, 2010). Simultaneously, GA influences the production of adhesin protein which can inhibit the potential ability of pathogens to adhere to the plant surface. However, the dual nature of GA introduces conflicts of its functional quality and emphasizes the species-specific nature of such interactions.

8.6.3 The Role of Cytokinin on Biofilm

Cytokinins are a group of purine derivatives primarily known for their contribution in cell division, differentiation, and apical dominance. Interestingly, many bacterial species can synthesize and respond to cytokinins, suggesting that these hormones might act as inter-kingdom signaling molecules. Several studies reveal that there is a connection of cytokinin regulation and plant–pathogen interaction. The plant microbiome community assists plant growth and agricultural sustainability (Gupta et al., 2022). It is shown in various studies that the genes encoding enzymes and catalysing cytokinin biosynthesis, such as isopentenyl transferase (IPTs), have been isolated in various microorganisms. These include plant-associated bacteria like *Agrobacterium tumefaciens* and *Rhizobium spp., Bacillus subtilis*, and *Pseudomonas spp.* (Wei et al., 2023). However, cytokinin is mediated by a histidine kinase receptor in *Agrobacterium tumifaciens*. The receptor Chk, which converts to a response regulator ChpR via phosphorylation, results in transcriptional modification that influences the virulence factor of the pathogen (Nishimura et al., 2004). The mechanisms depend on the pathogen's species, the target, and the maturation stages. The potential activity of bacteria to both produce and respond to cytokinins given an insight of a pivotal role in cell-to-cell communication and adaptation to exterior factors, which promotes the formation of biofilms.

The hormones sometimes simply affect the pathogenic cell surface activity and the regulation of adhesion protein and motility, and also increase the production of EPS content which facilitates the attachment activity of the bacteria towards plants and the maturation after inserting themselves into systems. Several studies illustrate the involvement of cytokinin as a signalling molecule in the quorum-sensing networking system of bacteria. For instance, Gall formation in *Pantoea agglomerans pv. gypsophilae* is manipulated by hrp/hrc genes, phytohormones, and the quorum-sensing regulatory system. Interference in quorum sensing genes *pagI* and *pagR* minimized transcription levels of *hrp* regulatory genes. Indole-3-acetic acid (IAA) and cytokinin significantly forced the expression of quorum sensing and *hrp* regulatory genes, providing a clear understanding of influences of cytokinin on quorum-sensing mechanisms (Chalupowicz et al., 2009). Another study showed that the Gram-negative bacteria *Xanthomonas campestris*, which is involved in the black rot disease in crucifers, may sense plant stimuli such as cytokinin presence in leaves. The discovery of the bacterial receptor PcrK in Xanthomonas, which senses the plant hormone isopentenyladenine, highlights a key mechanism of interkingdom signaling between plants PcrK belongs to the histidine kinase superfamily and acts as a primary sensor in a major class of bacterial regulatory circuitries called "two-component systems" (Papon & Stock, 2019). The phosphodiesterase activity of PcrR promotes the expression of genes involved in resistance to oxidative stress, thereby enhancing bacterial resistance to host defenses (Kabbara et al., 2020).

8.6.4 The Role of Abscisic Acid on Biofilm

Abscisic acid (ABA), a derivative of sesquiterpenoid, plays an essential role in directing plant development and reciprocation to environmental stresses, such as drought, salinity, and cold. However, recent studies indicate that several microorganisms, including bacteria and fungi, are also potentially able to synthesize abscisic acid and derivatives. Furthermore, microorganisms possess ABA receptors and respond to external applied ABA, recommending a broader contribution for interspecies communication and microbial morphology. Generally, ABA can inhibit the biofilm by increasing the flagellar movement, motility, and negative action on transition from planktonic to sessile growth. Recent studies suggest that ABA can also derange the redox balance and metabolic pathways of microorganisms, resulting in their ability to thrive within a biofilm. These disturbances inhibit the growth, decreasing EPS production, and exhibiting susceptibility to environmental stresses, and, ultimately, the disruption of biofilm (Lievens et al., 2017). An experimental study shown that the subinhibitory concentration of ABA may induce biofilm formation of *Agrobacterium tumefaciens C58* and *Pseudomonas aeruginosa PAO1* while the concentration of doses exceed the level of subinhibitory concentration, the activity get conversed and the reduction of biofilm is being shown (Plyuta et al., 2013). Roussoellenic acid and pestabacillin B are two derivatives of abscisic acid isolated from *Rousella sp. (MFLUCC 17-2059)*, which are able to reduce by at least 34% the biofilm of *Staphylococcus aureus* (Phukhamsakda et al., 2018).

The exact influence of ABA on biofilm formation significantly depends on the microbial species, environmentally adverse conditions, and hormonal concentration. Conversely, in some cases, ABA promotes biofilm formation. These complex contradictory properties highlight the difficulties of ABA's role and the need for continuous researches. Generally. abscisic acid is popularly known to induce stress responses in microorganisms, enhancing the adaptability for their survival within the biofilm layer. Changes in gene expression associated with stress tolerance result in the biofilm development ensuring better stability and persistence (Brookbank et al., 2021). Otherwise, ABA can interfere in microbial metabolism, striking the utilization of nutrients within the biofilm and promoting its growth. A large number of other factors linked to biofilm formation, such as EPS production and quorum sensing, are also influenced by the action of abscisic acid. The indirect mechanism of ABA can improve the qualitative and quantitative levels of EPS and autoinducing signals; as a result, the quality of cell-to-cell communication becomes refined, promoting the biofilm formation and maturation.

8.6.5 The Role of Ethylene on Biofilm

Ethylene (C2H4), the gaseous plant hormone involved in fruit ripening, senescence, and stress responses, has recently been marked as an active chemical signaling molecule in various microbial systems. In addition to plants, bacteria and fungi are also able to produce ethylene. The role in biofilm growth and development is increasingly observed to be complex and context-dependent. Several bacterial and fungal species are potentially able to synthesize ethylene through various enzymatic pathways. The well-studied mechanism involves the enzyme 1-aminocyclopropane-1-carboxylate (ACC) deaminase, which reacts with ACC (a precursor to ethylene in plants) and breaks it down into ammonia and α-ketobutyrate (Gamalero et al., 2023). The primary enzyme is often isolated in plant-associated microorganisms, where ACC deaminase functions to promote plant growth by reducing ethylene levels and alleviating stress. The species-specific effect of ethylene may refer as a promoter or inhibitor of initial adhesion of microbial cells to any surfaces. In many studies, it significantly enhances the gene expression of adhesion molecules, stimulating the probability of surface colonization. Some studies state that ethylene helps in EPS production, leading to the formation of multilayer robust biofilm. *Azospirillum brasilense*, a nitrogen-fixing bacterium, possesses AzoEtr1, a functional ethylene receptor with high affinity for ethylene, a plant hormone and the receptor -hormone binding triggers metabolic changes in the bacterium with an accumulation of poly-hydroxybutyrate in cell. This experimental observation suggests that ethylene from host plants can change the colonization strategies of *A. brasilense* (Carlew et al., 2025).

8.7 CONTROVERSIAL EFFICACY OF PHYTOHORMONES ON MICROBIAL PATHOGENS

The role of phytohormones in biofilm formation is often more complicated and elaborate. Some phytohormones are positively assisted for biofilm formation in certain contexts, while others exhibit inhibitory effects. This heterogeneity is influenced by factors such as type and concentration of the phytohormone, the bacterial species involved, and the environmental factors. A clear understanding of the mechanisms involved in these interactions is important for effectively controlling biofilms. The intricate process of biofilm formation involves several stages, including initial attachment, micro-colony formation, maturation, and dispersal. These stages are tightly regulated by a network of environmental signals and cell-to-cell communication, predominantly through processes such as quorum sensing (QS). Conventionally, it is considered that QS signaling molecules produced by bacteria were the primary influencer of biofilm formation. However, recent research has revealed the involvement of other signaling molecules, autoinducers, and peptides. Plant hormones, an essential growth regulatory parameter, are interestingly being observed for their ability to influence bacterial behavior, including biofilm formation.

Phytohormones, including auxin, gibberellin, cytokinin, abscisic acid, ethylene, jasmonate, salicylates, have dual activities in terms of biofilm formation, where one hormone shows its positive or negative activity in a species-specific manner. For example, auxin is shown as a growth regulator for *Pseudomonas aeruginosa*, whereas in *Bacillus subtilis*, it plays an inhibitory effect (Tsukanova et al., 2017). However, it is seen in many studies that the phytohormones, in general, induce the biofilm of plant pathogens associated with plant growth such as *Rhizobacteria, agrobacterium spp.* etc. Other than that, some phytohormones are involved in the destruction of bacterial commensal communities and repopulate with different kinds of microorganisms.

8.8 CONCLUSION

The phytohormonal role on biofilm is a vigorous, new subject for research. Further investigation into the complex interaction between phytohormones and bacterial communities will surely lead to new discoveries into biofilm biology and the revolution of novel strategies for regulating these ubiquitous microbial structures to benefit both human beings and the environment.

REFERENCES

Ahlawat, Y., Li, S., Timilsena, P. R., Pliakoni, E. D., Brecht, J. K., & Liu, T. (2022). Identification of senescence-associated genes in broccoli (Brassica oleracea) following harvest. *Postharvest Biology and Technology*, 183, 111729. https://doi.org/10.1016/j.postharvbio.2021.111729

Bao, S., Hua, C., Shen, L., & Yu, H. (2020). New insights into gibberellin signaling in regulating flowering in Arabidopsis. *Journal of Integrative Plant Biology*, 62(1), 118–131. https://doi.org/10.1111/jipb.12892

Barash, I., & Manulis-Sasson, S. (2009). Recent evolution of bacterial pathogens: The gall-forming pantoea agglomerans case. *Annual Review of Phytopathology*, 47(1), 133–152. https://doi.org/10.1146/annurev-phyto-080508-081803

Brookbank, B. P., Patel, J., Gazzarrini, S., & Nambara, E. (2021). Role of basal aba in plant growth and development. *Genes*, 12(12), 1936. https://doi.org/10.3390/genes12121936

Carlew, T. S., Brenya, E., Ferdous, M., Banerjee, I., Donnelly, L., Heinze, E., King, J., Sexton, B., Lacey, R. F., Bakshi, A., & Alexandre, G., Binder, B. M. Ethylene signals through an ethylene receptor to modulate biofilm formation and root colonization in a beneficial plant-associated bacterium. *PLoS Genet*. 2025 Feb 7;21(2):e1011587. https://doi.org/10.1371/journal.pgen.1011587

Chakraborty, N., Halder, S., Keswani, C., Vaca, J., Ortiz, A., & Sansinenea, E. (2024). New aspects of the effects of climate change on interactions between plants and microbiomes: A review. *Journal of Basic Microbiology*, 64(10), e2400345. https://doi.org/10.1002/jobm.202400345

Chalupowicz, L., Barash, I., Panijel, M., Sessa, G., & Manulis-Sasson, S. (2009). Regulatory interactions between quorum-sensing, auxin, cytokinin, and the Hrp regulon in relation to gall formation and epiphytic fitness of pantoea agglomerans pv. Gypsophilae. *Molecular Plant–Microbe Interactions®*, 22(7), 849–856. https://doi.org/10.1094/MPMI-22-7-0849

Du, G., Zhang, H., Yang, Y., Zhao, Y., Tang, K., & Liu, F. (2022). Effects of gibberellin pre-treatment on seed germination and seedling physiology characteristics in industrial hemp under drought stress condition. *Life*, 12(11), 1907. https://doi.org/10.3390/life12111907

Dutta, B., Basu, D., Lahiri, D., Nag, M., & Ray, R. R. (2025). Antibacterial and antibiofilm activities of bacteriocin produced by a new strain of Enterococcus faecalis BDR22. *Naunyn-Schmiedeberg's Archives of Pharmacology*, 398(3), 2983–2999. https://doi.org/10.1007/s00210-024-03458-0

Fenn, M. A., & Giovannoni, J. J. (2021). Phytohormones in fruit development and maturation. *The Plant Journal*, 105(2), 446–458. https://doi.org/10.1111/tpj.15112

Gamalero, E., Lingua, G., & Glick, B. R. (2023). Ethylene, ACC, and the plant growth-promoting enzyme ACC deaminase. *Biology*, 12(8), 1043. https://doi.org/10.3390/biology12081043

Gupta, R., Elkabetz, D., Leibman-Markus, M., Sayas, T., Schneider, A., Jami, E., Kleiman, M., & Bar, M. (2022). Cytokinin drives assembly of the phyllosphere microbiome and promotes disease resistance through structural and chemical cues. *The ISME Journal*, 16(1), 122–137. https://doi.org/10.1038/s41396-021-01060-3

Hossain, A. S., Alenazi, M. M. A., Ahmed, A. S. A., Alrudayni, H. A., Haouala, F., Al-Hashimi, A., Kahrizi, D., & Alhudhaibi, A. M. (2023). Seedless okra pod, chlorophyll, vitamin C and mineral content development using plant growth regulator in vitro and in vivo culture. *Cellular and Molecular Biology (Noisy-le-Grand, France)*, 69(13), 53–58. https://doi.org/10.14715/cmb/2023.69.13.8

Huang, R., Zheng, R., He, J., Zhou, Z., Wang, J., Xiong, Y., & Xu, T. (2019). Noncanonical auxin signaling regulates cell division pattern during lateral root development. *Proceedings of the National Academy of Sciences*, 116(42), 21285–21290. https://doi.org/10.1073/pnas.1910916116

Hussain, S., Nanda, S., Zhang, J., Rehmani, M. I. A., Suleman, M., Li, G., & Hou, H. Auxin and Cytokinin Interplay during Leaf Morphogenesis and Phyllotaxy. *Plants (Basel)*. 2021 Aug 21;10(8):1732. https://doi.org/10.3390/plants10081732

Kabbara, S., Bidon, B., Kilani, J., Osman, M., Hamze, M., Stock, A. M., & Papon, N. (2020). Cytokinin sensing in bacteria. *Biomolecules*, 10(2), 186. https://doi.org/10.3390/biom10020186

Kearns, D. B. (2010). A field guide to bacterial swarming motility. *Nature Reviews Microbiology*, 8(9), 634–644. https://doi.org/10.1038/nrmicro2405

Lee, J., Jayaraman, A., & Wood, T. K. (2007). Indole is an inter-species biofilm signal mediated by SdiA. *BMC Microbiology*, 7(1), 42. https://doi.org/10.1186/1471-2180-7-42

Lievens, L., Pollier, J., Goossens, A., Beyaert, R., & Staal, J. (2017). Abscisic acid as pathogen effector and immune regulator. *Frontiers in Plant Science*, 8. https://doi.org/10.3389/fpls.2017.00587

Liscum, E., Askinosie, S. K., Leuchtman, D. L., Morrow, J., Willenburg, K. T., & Coats, D. R. (2014). Phototropism: Growing towards an understanding of plant movement. *The Plant Cell*, *26*(1), 38–55. https://doi.org/10.1105/tpc.113.119727

Liu, P., & Nester, E. W. (2006). Indoleacetic acid, a product of transferred DNA, inhibits vir gene expression and growth of Agrobacterium tumefaciens C58. *Proceedings of the National Academy of Sciences*, 103(12), 4658–4662. https://doi.org/10.1073/pnas.0600366103

Miller, C.O., Skoog, F., von Saltza, M. H., & Strong, M. (1955). Kinetin, a cell division factor from deoxyribonucleic acid. *Journal of the American Chemical Society* 77, 1329–1334.

Mok, M.C. (1994). Cytokinins and plant development: An overview. In D. W. S. Mok and M. C. Mok, (Ed.), *Cytokinins: Chemistry, Activity and Function*, (pp. 155–156). Boca Raton, FL: CRC Press.

Mueller, R. S., Beyhan, S., Saini, S. G., Yildiz, F. H., & Bartlett, D. H. (2009). Indole acts as an extracellular cue regulating gene expression in vibrio cholerae. *Journal of Bacteriology*, 191(11), 3504–3516. https://doi.org/10.1128/JB.01240-08

Nemhauser, J. L., Hong, F., & Chory, J. (2006). Different plant hormones regulate similar processes through largely nonoverlapping transcriptional responses. *Cell*, 126(3), 467–475. https://doi.org/10.1016/j.cell.2006.05.050

Nishimura, C., Ohashi, Y., Sato, S., Kato, T., Tabata, S., & Ueguchi, C. (2004). Histidine kinase homologs that act as cytokinin receptors possess overlapping functions in the regulation of shoot and root growth in arabidopsis. *The Plant Cell*, 16(6), 1365–1377. https://doi.org/10.1105/tpc.021477

Pal, P., Ansari, S. A., Jalil, S. U., & Ansari, M. I. (2023). Regulatory role of phytohormones in plant growth and development. In *Plant Hormones in Crop Improvement* (pp. 1–13). Elsevier. https://doi.org/10.1016/B978-0-323-91886-2.00016-1

Papon, N., & Stock, A. M. (2019). Two-component systems. *Current Biology*, 29(15), R724–R725. https://doi.org/10.1016/j.cub.2019.06.010

Phukhamsakda, C., Macabeo, A. P. G., Yuyama, K. T., Hyde, K. D., & Stadler, M. (2018). Biofilm inhibitory abscisic acid derivatives from the plant-associated dothideomycete fungus, roussoella sp. *Molecules*, 23(9), 2190. https://doi.org/10.3390/molecules23092190

Plyuta, V. A., Lipasova, V. A., Kuznetsov, A. E., & Khmel, I. A. (2013). Effect of salicylic, indole-3-acetic, gibberellic, and abscisic acids on biofilm formation by Agrobacterium tumefaciens C58 and Pseudomonas aeruginosa PAO1. *Applied Biochemistry and Microbiology*, 49(8), 706–710. https://doi.org/10.1134/S000368381308005X

Qian, C., Ren, N., Wang, J., Xu, Q., Chen, X., & Qi, X. (2018). Effects of exogenous application of CPPU, NAA and GA4+ 7 on parthenocarpy and fruit quality in cucumber (Cucumis sativus L.). *Food Chemistry*, 243, 410–413.

Raspor, M., Motyka, V., Kaleri, A. R., Ninković, S., Tubić, L., Cingel, A., & Ćosić, T. (2021). Integrating the roles for cytokinin and auxin in de novo shoot organogenesis: From hormone uptake to signaling outputs. *International Journal of Molecular Sciences*, 22(16), 8554. https://doi.org/10.3390/ijms22168554

Roychoudhry, S., & Kepinski, S. (2021). Auxin in root development. *Cold Spring Harbor Perspectives in Biology*, a039933. https://doi.org/10.1101/cshperspect.a039933

Sharif, R., Su, L., Chen, X., & Qi, X. (2022). Hormonal interactions underlying parthenocarpic fruit formation in horticultural crops. *Horticulture Research*, 9, uhab024. https://doi.org/10.1093/hr/uhab024

Sugimoto, K., Jiao, Y., & Meyerowitz, E. M. (2010). Arabidopsis regeneration from multiple tissues occurs via a root development pathway. *Developmental Cell*, 18(3), 463–471. https://doi.org/10.1016/j.devcel.2010.02.004

Sun, T. (2010). Gibberellin signal transduction in stem elongation & leaf growth. In P. J. Davies (Ed.), *Plant Hormones* (pp. 308–328). Netherlands: Springer. https://doi.org/10.1007/978-1-4020-2686-7_15

Timofeeva, A. M., Galyamova, M. R., & Sedykh, S. E. (2024). How do plant growth-promoting bacteria use plant hormones to regulate stress reactions? *Plants*, 13(17), 2371. https://doi.org/10.3390/plants13172371

Tsukanova, K. A., Chebotar, V. K., Meyer, J. J. M., & Bibikova, T. N. (2017). Effect of plant growth-promoting Rhizobacteria on plant hormone homeostasis. *South African Journal of Botany*, 113, 91–102. https://doi.org/10.1016/j.sajb.2017.07.007

Wang, Y., Yang, L., Chen, X., Ye, T., Zhong, B., Liu, R., Wu, Y., & Chan, Z. (2016). Major latex protein-like protein 43 (mlp43) functions as a positive regulator during abscisic acid responses and confers drought tolerance in arabidopsis thaliana. *Journal of Experimental Botany*, 67(1), 421–434. https://doi.org/10.1093/jxb/erv477

Wei, X., Moreno-Hagelsieb, G., Glick, B. R., & Doxey, A. C. (2023). Comparative analysis of adenylate isopentenyl transferase genes in plant growth-promoting bacteria and plant pathogenic bacteria. *Heliyon*, 9(3), e13955. https://doi.org/10.1016/j.heliyon.2023.e13955

Wu, W., Du, K., Kang, X., & Wei, H. (2021). The diverse roles of cytokinins in regulating leaf development. *Horticulture Research*, 8(1), 118. https://doi.org/10.1038/s41438-021-00558-3

Yuan, Z.-C., Liu, P., Saenkham, P., Kerr, K., & Nester, E. W. (2008). Transcriptome profiling and functional analysis of agrobacterium tumefaciens reveals a general conserved response to acidic conditions (Ph 5. 5) and a complex acid-mediated signaling involved in agrobacterium -plant interactions. *Journal of Bacteriology*, 190(2), 494–507. https://doi.org/10.1128/JB.01387-07

Zhang, S., Yang, Q., & Defoirdt, T. (2022). Indole analogues decreasing the virulence of Vibrio campbellii towards brine shrimp larvae. *Microbial Biotechnology*, 15(12), 2917–2928. https://doi.org/10.1111/1751-7915.14160

Zhao, B., Liu, Q., Wang, B., & Yuan, F. (2021). Roles of phytohormones and their signaling pathways in leaf development and stress responses. *Journal of Agricultural and Food Chemistry*, 69(12), 3566–3584. https://doi.org/10.1021/acs.jafc.0c07908

9 Calcium-Regulating Hormones and Biofilm

Sreejita Ghosh and Yashaswi Gupta
Department of Biotechnology, Maulana Abul Kalam Azad University of Technology, Kolkata, India

Amisha Singh
Department of Life Science, Maulana Abul Kalam Azad University of Technology, Kolkata, India

Rina Rani Ray
Department of Biotechnology, Maulana Abul Kalam Azad University of Technology, Kolkata, India

9.1 INTRODUCTION

Calcium is a vital mineral in the human body, playing significant roles in numerous physiological processes, involving the contraction of muscles, the transmission of nerve impulse, blood clotting, and bone mineralization. Calcium homeostasis is of importance in making the processes function well and is firmly regulated by a subtle hormone interplay. Recent research has even brought into focus the influence of calcium-regulating hormones on microbial biofilms, which are aggregates that bacteria form to defend themselves against unfavourable situations (Shoback et al., 2020).

Calcium-regulating hormones play a major role in functions which go beyond the outmoded realm of calcium homeostasis, disturbing the regulation of microbial biofilms. The conjunction of endocrinology and microbiology opens up promising avenues for future studies and therapeutic applications. By explicating the mechanisms through which these hormones impact biofilm formation and stability, it is possible to develop new strategies for the prevention and treatment of chronic infections, which can ultimately expand patient outcomes. This chapter aims to shed light on the intricate relationship between calcium-regulating hormones and biofilm formation, explaining their roles in microbial pathogenesis, immune modulation, and potential therapeutic strategies. Through the exploration of the mechanisms by which these hormones impact biofilm dynamics, new strategies can be developed to prevent and treat chronic infections, eventually refining patient outcomes.

9.2 HORMONES FOR BLOOD CALCIUM REGULATION

The major hormones taking part in calcium regulation are PTH, calcitonin and calcitriol (the active form of vitamin D) (Rapsinski et al., 2024).

- **PTH**: Produced by the parathyroid glands, PTH is secreted when there are levels of calcium in the blood. It elevates the level of calcium in the blood by stimulating bone resorption,

DOI: 10.1201/9781003519041-9

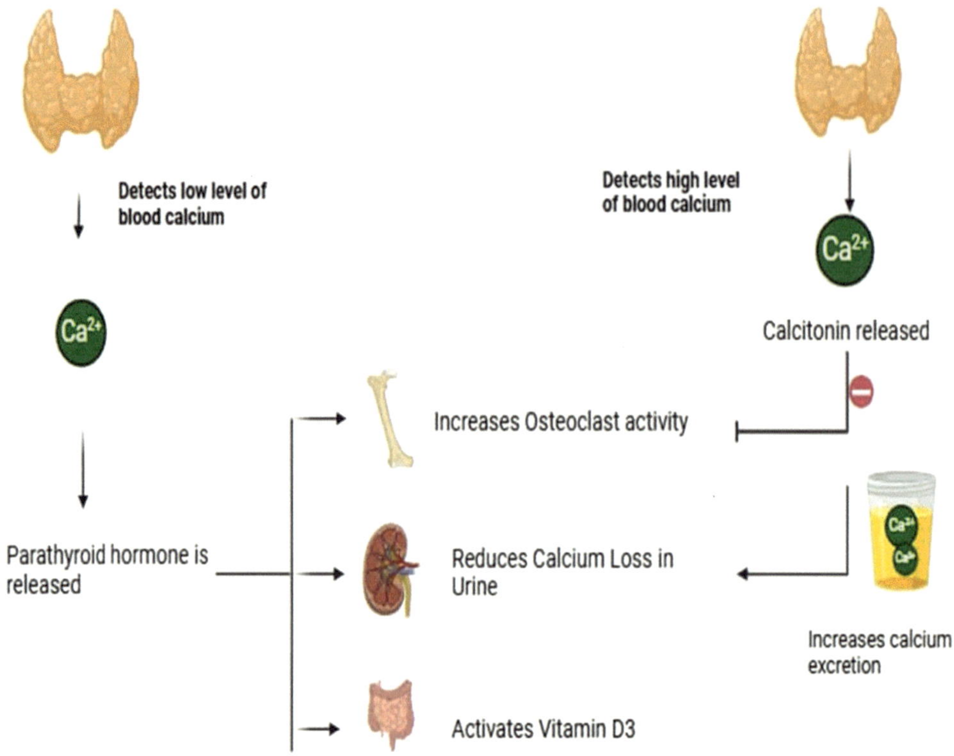

FIGURE 9.1 Calcium-regulating hormones in maintenance of calcium homoeostasis.

increasing renal calcium reabsorption, and promoting vitamin D activation within the kidneys. This, in turn, enhances the intestinal absorption of calcium.

- **Calcitriol**: Produced in the kidneys from its precursor forms, calcitriol enhances intestinal absorption of calcium and phosphorus, facilitating bone mineralization. It also modulates immune function and has been implicated in antimicrobial defense mechanisms.
- **Calcitonin**: Produced by the parafollicular cells of the thyroid gland, calcitonin reduces levels of calcium in the blood by hindering osteoclast activity, thereby reducing bone resorption. Its role in calcium homeostasis is considered less critical than that of PTH and calcitriol.
- **Vitamin D**: This is a secosteroid (pro) hormone, which has long been thought to be an important regulator of bone metabolism, as well as calcium and phosphorus homeostasis, via negative feedback with the parathyroid hormone.

These hormones are required to be at their optimal concentration for the maintenance of calcium homoeostasis and overall well-being, since patients with hypocalcemia are found to be more susceptible to various microbial infections, often forming biofilms (Figure 9.1).

9.3 CALCIUM-REGULATING HORMONES AND BACTERIAL GROWTH

9.3.1 Calcitonin and Bacterial Growth

The thyroid gland secretes the calcitonin hormone, which lowers blood calcium content to control them. This counteracts the action of PTH, which raises the levels of calcium in the blood. Thyroid is a tiny, butterfly-like gland situated underneath the skin at the anterior side of the neck. Calcitonin is produced and secreted by your thyroid's parafollicular cells, or C-cells. The symptoms of irregular calcitonin levels are either minimal or non-existent. The parafollicular cells (also called C cells) of the thyroid (or endostyle) among humans and other chordates within the ultimo-pharyngeal body secrete calcitonin, a peptide hormone containing 32 amino acids. Decreasing blood calcium (Ca2+), it counteracts the effects of the parathyroid hormone (PTH). Calcitonin, a hormone primarily involved in calcium homeostasis, has recently garnered attention for the role it plays in systemic infections. Its precursor, PCT, is emerging as a crucial biomarker for bacterial infections and sepsis. In contrast to traditional inflammatory markers including interleukin-6 (IL-6), and C-reactive protein (CRP), PCT is uniquely responsive to bacterial infections, making it a valuable diagnostic tool (Hausfater et al., 2007). Elevated PCT levels correlate with severe systemic infections, particularly those caused by bacterial, parasitic, or fungal pathogens, whereas viral infections generally do not induce a significant rise in PCT (Vijayan et al., 2017).

Several systemic diseases, including medullary thyroid carcinoma, are known to have elevated levels of calcitonin, its pro-peptide, procalcitonin (PCT), and other peptide products of the calcitonin gene. However, one clinical finding among humans is the observation of elevated procalcitonin levels without an increase in calcitonin in patients suffering from acute bacterial diseases (Assicot et al., 1993). A 116 amino acid-long peptide called PCT is converted into the mature hormone calcitonin, which is comprised of 32 amino acids, by post-translational proteolysis. PCT that is absent or low in the serum of healthy individuals, whether they are adults or infants, rises to high levels in patients suffering from meningitis, septicemia, or severe bacterial infections and then quickly falls off again following the right antibiotic treatment. Furthermore, PCT is low in individuals with inflammatory diseases or acute viral infections. Since every infectious disease is too complicated to be reduced to a single surrogate marker, the Food and Drug Administration (FDA) approved PCT in 2005, along with the determination of other laboratory values to evaluate the jeopardy of patients critically ill with sepsis. PCT is a precursor of the hormone calcitonin and is produced in the C cells of the thyroid. Other cells, such as liver or fat cells, also produce this prohormone under certain conditions, such as infections or surgeries (Becker et al., 2010).

It's unclear how PCTs are produced. After receiving intravenous interleukin-2 or TNF, patients with advanced cancer have shown a rapid and significant release of PCT (Assicot et al., 1993). PCT rises in the blood of healthy volunteers 3–4 hours after endotoxin injection, then quickly reaches a plateau at 6 hours and stays high for at least 24 hours. TNF and IL6 in the same patients reach baseline levels at 6 and 8 hours, respectively, after peaking at 90 minutes and 3 hours. Although there are variations in the PCT peaks' magnitudes amongst the patients, all subjects exhibit a noticeable increase.

In cases where the infection is localized or restricted to a solitary organ, with no systemic inflammatory reaction, the PCT is either low or slightly elevated. *Pseudomonas pseudomallei*, which is a Gram-negative bacillus, is the cause of melioidosis, a potentially serious infection. It has been discovered that PCT is a good prognostic indicator and a first-rate indicator of disease activity (low among patients with abscess and high during the septicemic stage) (Mouliou, 2023). One key factor in clinical diagnosis is the concentration of procalcitonin (PCT), which is used to guide antibiotic therapy, determine the severity of illness, and distinguish between sepsis and bacterial infections.

Septic shock is distinguished by significantly elevated procalcitonin levels, in contrast to the reduced levels observed in cardiogenic shock. Proinflammatory cytokines are also raised amongst both groups (Thanawongnuwech et al., 2004). Procalcitonin also facilitates the differential diagnosis of nonbacterial and bacterial causes of severe conditions. Since non-specific inflammation raises both CRP and IL-6, it is impossible to distinguish between the two groups in adult respiratory distress syndrome, whether or not there is a bacterial cause. On the other hand, PCT is only increased in patients with bacterial etiology; it does not overlap with the population of Acute Respiratory Distress Syndrome (ARDS) that have a toxic origin. In the case of biliary versus toxic aetiology, acute pancreatitis presents a very similar scenario: only PCT permits discrimination, not CRP or IL6.

Increased inflammatory activity is indicated when the PCT concentration rises above specific thresholds. Compared to severe bacterial infections and sepsis, the blood's PCT concentration likely rises less in cases of localized viral infections, bacterial infections, non-infectious inflammatory diseases like autoimmune processes, or following trauma and surgery. However, the PCT concentration ranges for the various inflammatory conditions overlap to such an extent that it can occasionally be challenging to identify the decision limits that allow for the diagnostic benefits of PCT determination. To date, larger studies have been conducted to determine such decision limits for the use of PCT for the diagnosis of clinical pictures like sepsis and lower respiratory infections.

The literature has extensively discussed the application and utility of PCT in the form of a biomarker among two primary fields: as a guide for managing antibiotic therapy and as a primary marker of sepsis (differentiating between non-bacterial and bacterial etiologies). PCT has gained attention in recent decades as a biomarker that can help determine when to stop antibiotic treatment, especially in patients who are very sick. It is safe and effective when used alone or in combination with other biomarkers like presepsin, CRP and alpha 2 macroglobulin (A2M). In comparison to other biomarkers like CRP, and white blood cells (WBC), PCT seems to be especially vulnerable to toxins from bacteria, and infections with viruses cause it to be down-regulated.

The pathophysiology behind PCT elevation during bacterial infections is complex and involves multiple inflammatory mediators:

1. Endotoxin and Cytokine Activation:
   a. During bacterial infections, endotoxins produced by Gram-negative bacteria and peptidoglycans produced by Gram-positive bacteria stimulate proinflammatory cytokines, including, IL-1β, IL-6 and tumor necrosis factor-alpha (TNF-α) (Kragsbjerg et al., 1998).
   b. This triggers systemic PCT production, not just from thyroid C cells but also from extra-thyroidal sources, such as hepatocytes and adipocytes (Hausfater et al., 2007).
2. Differentiation Between Bacterial and Viral Infections:
   a. Viral infections trigger an interferon-gamma (IFN-γ) response that reduces synthesis of PCT (Kiya et al., 2024).
   b. This clarifies the reason behind PCT remaining low in case of viral infections, making it a beneficial differential diagnostic marker.

9.3.2 PTH and Bacterial Growth

PTH is important for calcium homeostasis through regulating phosphate and calcium metabolism in kidneys, bones, and the intestine. Recent studies proposed that PTH not only impacts bone remodelling but also interrelates with the gut microbiome, manipulating bacterial growth as well as immune responses (Yu et al., 2020). This section explores the association between PTH and bacterial growth, predominantly in the context of gut microbiota and systemic infections.

PTH influences bone metabolism, partially via its interactions with the gut microbiota. A study by Coskun et al. (2024) analysed the gut microbial profiles of patients suffering from primary hyperparathyroidism (PHPT). The study included 9 PHPT patients and 9 healthy persons, who were matched by gender, age, and body mass index. Gut microbiota composition was evaluated by 16S rRNA gene sequencing, at baseline as well as 1 month following parathyroidectomy in the PHPT group. The study stated that the overall microbial diversity (alpha and beta diversity) did not fluctuate pointedly between the PHPT group and healthy controls, nor did it change post-surgery. On the other hand, particular groups of bacteria showed discrete patterns. Patients having PHPT showed a higher abundance of *Subdoligranulum*, whereas *Alloprevotella*, *Ruminococcus*, *Clostridium sensu stricto_1*, and *Phascolarctobacterium* were notably reduced compared to controls ($p < 0.001$). After parathyroidectomy, *Subdoligranulum* levels dropped, but *Alloprevotella* and *Ruminococcus* levels augmented significantly ($p < 0.001$). Furthermore, bone mineral density (BMD) measurements indicated that PHPT patients had significantly lowered total femoral and lumbar BMD as compared to the healthy persons ($p < 0.05$). Correlation analysis at baseline demonstrated that *Alloprevotella* abundance was positively in association with serum phosphorus, but *Subdoligranulum* was linked to higher total lumbar BMD. In the interim, *Clostridium sensu stricto_1* demonstrated a negative correlation with serum calcium but a positive correlation with femoral neck BMD. Following surgery, *Alloprevotella* remained positively associated with baseline phosphorus levels, and *Phascolarctobacterium* correlated positively with distal radius BMD.

PTH is a polypeptide hormone, which is produced by the parathyroid gland when levels of calcium in serum are low. It leads to an upsurge in calcium reabsorption in the kidneys, kindles osteoclast-mediated bone resorption, and improves intestinal calcium absorption through the activation of vitamin D (1,25-dihydroxyvitamin D) (Mihai & Farndon, 2000). PTH regulation is vital for the maintenance of calcium equilibrium and averting disorders involving hyperparathyroidism and osteoporosis.

The intestinal microbiota is significant in maintaining bone metabolism via its interplay with PTH. Studies have revealed that PTH-mediated bone loss is the result of microbiota-associated mechanisms, particularly via the expansion of pro-inflammatory T cells inside the gut (Yu et al., 2020). In hyperparathyroidism models, PTH augments the production of tumour necrosis factor (TNF)-producing Th and T cells, contributing to elevated osteoclast activity and bone resorption.

In addition, the microbiome impacts the accessibility of short-chain fatty acids (SCFAs) modulating PTH activity and its effects on bones. The gut–bone axis is gradually recognized as a vital regulator of bone homeostasis, with bacterial metabolites shaping systemic effects of PTH.

Apart from its impacts on bone, PTH has been implicated in curbing bacterial growth during systemic infections. Studies showed that hyperparathyroidism modulates immune responses, potentially upsetting the body's ability to battle bacterial infections. PTH-mediated fluctuations in gut microbiota composition may impact bacterial colonization and systemic inflammation (Yu et al., 2020).

Moreover, calcium homeostasis plays an important part in bacterial pathogenesis. Several bacteria need calcium for formation of biofilm, enzyme activation, and the production of virulence factors. PTH, by controlling calcium levels, could circuitously impact bacterial growth dynamics, predominantly in calcium-dependent pathogens such as *P. aeruginosa* and *S. aureus*.

Understanding the linkage between PTH and bacterial growth has vital clinical consequences. Improving the gut microbiota via probiotics or dietary involvements may serve as a potential approach to alleviate PTH-mediated bone loss and bacterial dysbiosis. Further research is needed to explore how PTH modulates bacterial growth in different physiological and pathological contexts, including osteoporosis, chronic kidney disease, and sepsis.

PTH is an indispensable calciotrophic hormone, which plays a crucial part in regulating the development of the postnatal skeleton. Continuous infusion of PTH (cPTH) in animals simulates primary hyperparathyroidism, a disorder marked by chronic and continuous hyperproduction of PTH by the parathyroid glands. One common reason for fractures and osteoporosis is primary hyperparathyroidism. Primary hyperparathyroidism is a diverse condition, though, as few patients experience bone damage whereas others do not. Whether or not a body loses bone does not depend on whether hypercalcemia is present or absent; the absence of hypercalcemia refers to the norm calcemic variant.

Whether target cells are continuously exposed to PTH or sporadically will largely determine whether PTH causes bone loss or promotes the acquisition of bone mass. PTH causes net bone loss when it is produced or infused in excess and continuously, as it promotes bone resorption and, to a lesser extent, bone formation.

Although IL-17 and TNF secreted by BM T cells cause bone loss in inflammatory conditions, interaction with common bacterial products, like flagellin and lipopolysaccharide (LPS), causes the small intestine to produce a large number of TNF-producing T cells. In the intestinal lamina propria of mice, Th cells are also produced in large quantities in response to particular microbiota components. The main cause of intestinal Th cell production in healthy mice is segmented filamentous bacteria (SFB), commensal Gram-positive bacteria which form spores and strongly stimulate Th cell differentiation. Infections with a number of extracellular pathogens, including *Citrobacter rodentium* and *Candida albicans*, can also activate Th cells (Umar et al., 2018).

The human microbiome is a diverse ecosystem comprised of trillions of native bacteria, fungi, and viruses that live on the exposed surfaces of mammals. The most abundant microbiome, the gut microbiota, is now understood to control the action of the chief endocrine bone regulators as well as postnatal skeletal development. One hormone that regulates bone, PTH, needs components of the gut microbiota to have both anabolic and catabolic effects on bone. Much research is being carried out on how the gut microbiota controls the skeletal response to PTH.

9.3.3 Vitamin D3 (Cholecaliciferol) and Bacterial Growth

Vitamin D plays an essential role in triggering an immune response against invasive pathogens (Ismailova & White, 2021). Immune cells contain VDR and 1α-hydroxylase, which are essential for converting vitamin D into its active form, 1,25 dihydroxy vitamin D. By strengthening the innate immune system, vitamin D contributes to protective immunity. Toll-like receptors (TLRs) of macrophages could identify LPS, a mimic for bacterial infections. Defensin and cathelicidin are instances of antimicrobial peptides (AMPs), which are produced when TLRs get activated. AMPs exhibit strong antimicrobial activity and co-localize inside phagosomes that contain ingested bacteria, rupturing bacterial cell membranes (Lei et al., 2019). Furthermore, vitamin D's immunomodulatory potential in host defenses is highlighted by its inductive and suppressive effects on immune system participants (Lei et al., 2019). By assembling an indication that vitamin D3 reduces biofilm-mediated infections through immunomodulation and expression of cathelicidin, the advantages of vitamin D supplementation in preventing and reducing infections caused by organisms that produce biofilms can be proved.

Concerns regarding sunburn and skin cancer, coupled with indoor lifestyles, religious practices, and extreme weather conditions, have led to an increased interest in vitamin D supplementation. Oral vitamin D exists in two distinct forms: vitamin D3 (cholecalciferol) and vitamin D2 (ergocalciferol). Vitamin D3 is predominantly present in fatty fish, fish oils, egg yolks, liver, fortified foods, and various dietary sources. Mushrooms cultivated under ultraviolet light exhibit the highest concentrations of Vitamin D2 (Judprasong et al., 2023). The physiologically active form of vitamin D is 1,25-Dihydroxyvitamin D_3, or calcitriol. This is vital for the mineralization of bones, calcium and phosphate homeostasis, and a number of physiological functions, like cellular variation and immunological response. Vitamin D binding protein (VDBP), which is a serum glycoprotein that has only one site for binding all vitamin D metabolites, binds ingested vitamin D (Bouillon et al., 2020). First, cytochrome P450 (CYP) 2R1 hydroxylates vitamin D at the C-25th position inside the liver to create 25-hydroxy vitamin D, also known as calcidiol. Second, 1,25-dihydroxy vitamin D, the functional form, which mediates genomic activities, is produced within the kidney by CYP27B1 adding a hydroxyl group at C-1 (Pop et al., 2022). Renal hydroxylases are controlled by PTH, fibroblast growth factor 23, and hypocalcemia. Because of its strong affinity for 25-hydroxy vitamin D, VDBP helps people with vitamin D deficiency by raising the amount of 25-hydroxy vitamin D in the blood (Bouillon et al., 2020). Vitamin D deficiency increases the risk of developing serious infections because it is associated with decreased levels of AMP (such as cathelicidin), elevated levels of inflammatory cytokines, and the formation of biofilms (Figure 9.2). There is also evidence linking many infections linked to biofilms to lower vitamin D levels. As a modifiable risk factor, deficiency of vitamin D may be prevented and diseases may be lessened with vitamin D supplements that restore normal levels. Lastly, for patients with infections caused by microbes that produce biofilms, we recommend routine testing for biofilm production and serum 25-hydroxy vitamin D monitoring.

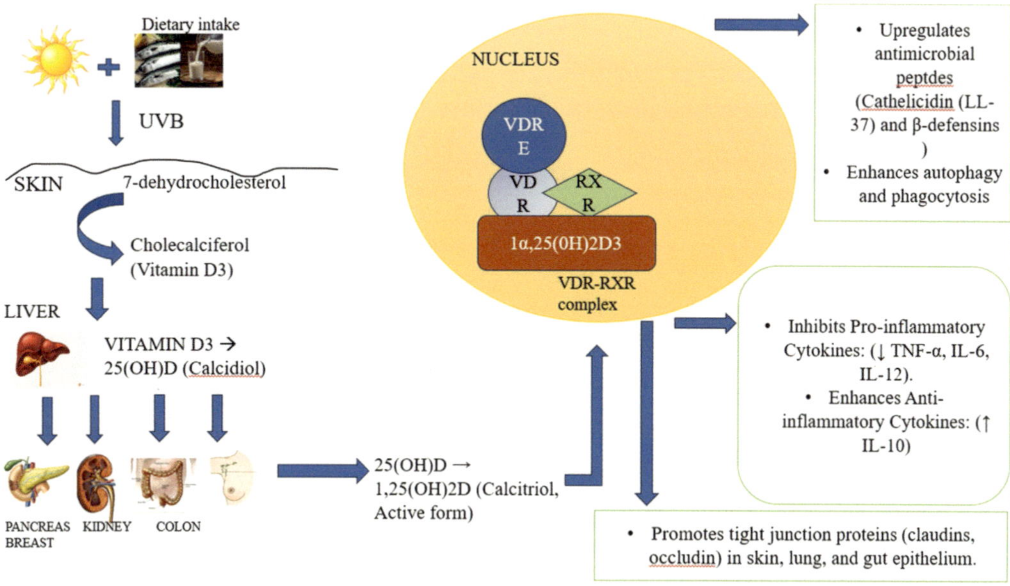

FIGURE 9.2 Vitamin D3 and its role in immune modulation.

Vitamin D3 (cholecalciferol) is widely recognized for its activity in bone metabolism and calcium homeostasis. However, emerging research suggests that it also exhibits antimicrobial properties against various bacterial pathogens. Numerous in vitro studies have demonstrated the ability of vitamin D3 to inhibit bacterial growth and enhance host immune responses, making it a potential adjunct in treating bacterial infections.

Numerous species of bacteria, including *Streptococcus pyogenes*, *Staphylococcus aureus*, *Escherichia coli*, and *Klebsiella pneumoniae*, exhibit significant growth inhibition when exposed to high concentrations of vitamin D3 (Guevara et al., 2020). Studies have reported that at doses of 50,000–90,000 IU/mL, vitamin D3 can either completely eliminate bacterial growth or significantly reduce bacterial proliferation. The association among low levels of vitamin D and increased vulnerability to Gram-positive bacterial infections, counting invasive pneumococcal infections, *Neisseria meningitidis* disease, and group A Streptococcus infection, suggests a possible therapeutic function of vitamin D supplementation for preventing bacterial diseases (Guevara et al., 2020).

Macrophage-induced immunity is vital in bacterial removal. However, studies on cultured macrophages challenged with 1,25-dihydroxyvitamin D3 ($1,25(OH)_2D_3$) showed weakened action against *Listeria* species because of the suppression of IFNγ-associated macrophage activation. This suppression results in abridged reactive oxygen species (ROS) production, which are indispensable for killing bacteria (Baisa et al., 2019). Conversely, Baisa et al. (2019) stated that the deficiency of vitamin D receptor (VDR) resulted in the delayed removal of *Listeria monocytogenes*, highlighting the intricate immunomodulatory function of vitamin D3.

9.3.3.1 Impact of Vitamin D3 on Opportunistic and Pathogenic Bacteria

P. aeruginosa, an opportunistic pathogen predominantly distressing individuals with cystic fibrosis (CF), offers substantial treatment difficulties. While largely extracellular, it has the capability to infiltrate host cells. Antal et al. (2011) established that vitamin D3 improves innate immunity by regulating the expression of the AMPs and moderating cytokine responses. Besides, vitamin D3 was revealed to upregulate VDR expression among macrophages infested with *P. aeruginosa*, signifying a latent immunomodulatory activity.

Oxidative burst, a perilous defence mechanism, is mostly mediated due to NADPH oxidase, which produces reactive nitrogen and oxygen species such as nitric oxide (NO) and hydrogen peroxide (H_2O_2). Vitamin D3 has been found to impact oxidative burst responses, contributing to bacterial eradication (Antal et al., 2011).

9.3.3.2 Vitamin D3 and Central Nervous System (CNS) Infections

Djukic et al. (2015) proved that vitamin D3 could protect the CNS against bacterial infections. Their investigation showed that vitamin D3-deficit mice had increased mortality rates and persistent presence of *E. coli* in the brain, whereas those with tolerable vitamin D3 levels displayed better pathogen removal. In addition, vitamin D3 was allied with amplified IL-10 and diminished IL-6 concentrations, demonstrating its anti-inflammatory activities (Aranow, 2011). This indicates that vitamin D3 may aid in reducing inflammation-based neuronal damage in case of bacterial meningitis.

9.3.3.3 Vitamin D3 and Streptococcal Infections

Saputo et al. (2018) examined the antibacterial effects of vitamin D3 against *Streptococcus mutans*, which is a predominant pathogen leading to the formation of dental caries. More in vitro studies evaluated the action of vitamin D3 against *Streptococcus sobrinus* and *S. mutans*, showing important antibacterial features. The minimum bactericidal concentration (MBC) and minimum inhibitory concentration (MIC) values were indicative of bacterial growth suppression by vitamin D3. In addition to this, scanning electron microscopy (SEM) images depicted that vitamin D3 could interact with peptidoglycan of the cell wall of bacteria, causing membrane contraction, surface irregularity, and, ultimately, cell death.

Even though vitamin D3 could display robust antibacterial activities, the exact molecular mechanisms behind the bactericidal effects are yet to be totally understood. Further research needs to be done regarding how vitamin D3 can act on bacterial membrane ultrastructure, metabolic pathways, and its probable synergistic functions with available antibiotics (Golpour et al., 2019). Having a deeper insight in these mechanisms can boost the development of novel antibacterial therapies, mainly in resisting antibiotic-resistant pathogens.

9.4 CALCIUM-REGULATING HORMONES AND BIOFILM

Biofilms can be defined as clusters of bacterial cells embedded in a self-produced extra polysaccharide (EPS) matrix, attaching to biotic or abiotic surfaces and showing resistance against antibiotics and host immune responses. They are responsible for causing numerous chronic and acute diseases, such as medical device-associated infections, chronic wounds, and respiratory tract infections (Nishikawa & Kobayashi, 2021). The antibiotic resistance of biofilms poses momentous challenges in clinical settings, resulting in persistent chronic infections and incurring huge healthcare expenses.

Emerging studies proved that calcium-regulating hormones can influence biofilm formation and stability. Calcitriol can modulate immune response, increasing the antimicrobial effects of the immune cells such as macrophages. This variation may impact the capability of the immune system to respond against biofilm-associated infections. PTH can impact calcium levels in the gut, thereby altering the composition of the gut microbiota. Modifications in microbiota composition can induce biofilm formation, since few bacterial species are more susceptible to biofilm development. Although the direct impacts of calcitonin on biofilms are not entirely understood, the way it regulates bone metabolism also indicates possible indirect effects. Through manipulation of bone turnover and calcium release, calcitonin might have affected the local surroundings encountered by bacteria, strongly influencing their capability of biofilm formation. In few scenarios, biofilm-associated infections can be even detected by assessing the markers like enhanced C-reactive protein (CRP), blood leukocyte counts, procalcitonin levels, and interleukin-6 (IL-6), which are generally linked to infectious and inflammatory diseases.

For chronic gastritis, *H. pylori* biofilms may serve as a protective sheath against detrimental external agents. Apart from its protective activity, this biofilm also acclimatizes its surface attributes in response to specific compounds, like calcitonin, helping the bacteria to become accustomed to the varying surrounding (Gaddy et al., 2015). Furthermore, the biofilm architecture limits the infiltration of ROS, explaining the reason why immune cells, like phagocytes, repeatedly struggle to completely eradicate bacteria within the biofilms (Costerton et al., 1999).

Bacterial pathogens can be broadly divided into several factors contributing to infection development, via infiltration, adherence, the impairment of host cells, dodging host immunity, and propagation. The synthesis and secretion of chemicals that harm host cells enzymatically or non-enzymatically frequently aids bacterial invasion (Ribet and Cossart, 2015). It is known that Ca^{2+} regulates a variety of secreted enzymes in bacterial pathogens (King et al., 2020).

One of the most effective ways in which bacterial pathogens can take over a host is by producing toxins. Ca^{2+} affects how many toxins are produced, secreted, and function in various pathogens. A complex signaling landscape for invasive pathogens is created by variations in Ca^{2+} levels within a host and fluctuations during the course of the disease. Thus, it is crucial for pathogens to sense host Ca^{2+} because this allows them to effectively adapt to their surroundings by inducing the expression of genes that confer resistance and virulence. In part, bacteria use two-component regulatory systems (TCS) to achieve this. This often involves a response regulator and a sensor kinase. Typically, the sensor region faces the periplasm, and the sensor kinase is entrenched in the internal membrane. When the signal is bound, the sensor kinase auto-phosphorylates and then phosphorylates the response regulator, which usually controls the transcription of a group of target genes (Raisz, 1981).

The inability to regenerate, primarily brought on by the ongoing infection caused by multispecies biofilms, prevents chronic wounds from healing. The impacts of biofilm-mediated wound infections and how they interact with the host are still not entirely understood. Our goal was to investigate physiologically-induced functional biofilms in vitro and their possible activities on in vivo regeneration and health. *Enterococcus faecalis*, *S. aureus*, and *P. aeruginosa* were thus seeded in collagen-associated scaffolds for dermal regeneration. Scaffolds contained viable biofilms that were resistant to antibiotics, and their bacterial loads after 24 hours varied based on the original inoculum. Scaffolds were then placed onto the mice's entire skin wounds, along with daily monitoring and antibiotic therapy. All of the mice lived, but their health was compromised; they showed signs of fever and weight loss. Scaffold histomorphology revealed significant variation in samples and within groups after ten days. Colony-forming units indicated whether the wounds were strongly, mildly, or not infected, and *P. aeruginosa* was more frequently identified. Leucocyte infiltration increased interleukin-10 and interferon-γ in scaffolds, splenic weight and size growth, and elevated systemic procalcitonin concentrations were all caused by biofilm infection (Cárdenas-Calderón et al., 2022).

Furthermore, pro-calcitonin levels were pointedly more for biofilm-infected groups in comparison to controls, with concentrations of proteins in plasma within stated ranges for chronic-infected patients, signifying that, apart from unfavourable impacts over wellbeing, other systemic activities, which mimic symptoms of the patients, were also present (Cárdenas-Calderón et al., 2022). In the context of osteonecrosis, microbial biofilms have been implicated in jawbone necrosis secondary to bisphosphonate therapy. Studies have identified microbial biofilms directly associated with osteonecrotic lesions. Interestingly, vitamin D3 (VD3) has been revealed to play an inhibitory role in formation of biofilm by fungi. The capacity of VD3 to constrain hyphal development and prevent biofilms in a variety of Candida species (e.g., *C. glabrata*, *C. albicans*, *C. krusei*, *C. parapsilosis*, and *C. tropicalis*) has been reported. Additionally, hydrophobicity of fungi diminished after treatment with VD3, preventing the proliferation of *Candida* in a dose-dependent way. Kherad et al. (2023) investigated the capacity of VD3 as an effective antifungal as well as antibiofilm agent against

Candida species. The findings demonstrated how VD3 could inhibit *Candida* growth disrupted their biofilm development, which are mainly accountable for persistent infections and treatment challenges. It can be inferred from here that VD3 can act as a capable adjunct or alternative approach to manage *Candida*-associated infections, opening up a plethora of antifungal therapies. Lutfi et al. (2024) found that vitamin D, as well as vitamin K1, brought about a remarkably decreased formation of biofilms among Gram-negative bacteria, thereby hinting at a potential application in combating biofilm-associated infections caused by these bacteria. Hence, Vitamin D and K1 can act as potential drug candidate to improve the available antibacterial therapeutic strategies, especially when antibiotic resistance is a booming challenge in clinical settings globally. It even improvises the antibiofilm activities of different antibiotics against *P. aeruginosa* infections. Biswas et al. (2025) observed that VD3 could potentially improve both antibiofilm and antimicrobial activities of thymoquinone as well as streptomycin against *P. aeruginosa*. It can be further concluded that VD3, when applied in synergy with these agents, could significantly resist *P. aeruginosa* infections and disrupt its biofilm network. Therefore, VD3 can be applied as a valuable adjunct to treat biofilm-associated infections caused *P. aeruginosa*.

9.5 IMPORTANCE OF THE ANTIBIOFILM ACTIVITIES POSSESSED BY CALCIUM-REGULATING HORMONES

The association between calcium-regulating hormones and their prospective antibiofilm activities can be understood using three different approaches:

- Therapeutic approaches: Calcitriol of PTH level moderation may serve as a therapeutic approach for host immunity enhancement to resist biofilm-associated diseases.
- Infection control: Maintenance of optimum calcium homeostasis could suppress dysbiosis and reduce the biofilm-associated infection risks, mainly among vulnerable populations (elderly or immunosuppressed individuals).
- Diagnostic markers: Continuous monitoring calcium-regulating hormones biomarks early detection of biofilm-associated infections, allowing for timely intervention.

9.6 CONCLUSION

Calcium is not only an indispensable mineral required for bone strength, but also regulates the immune responses through microbial behaviour modulation. Calcium-regulating hormones, like calcitonin, PTH, and VD3, could not only maintain systemic calcium balance but also aid the host immunity in defending against pathogens. All these hormones indirectly regulate the calcium-enriched niches exploited by bacteria at times of infection and biofilm propagation. Biofilms, having an intricate and solid architectural backbone, could survive in situations where the presence of calcium facilitates bacterial attachment and the virulence factor production. A high level of extracellular calcium can influence bacterial toxins, enzymes and signalling pathways (quorum sensing), which are essential for biofilm growth and proliferation. Furthermore, biofilm-associated infections delay wound healing via inducing chronic inflammation, which is a result of modified host calcium homeostasis. VD3, one of the major calcium-regulating hormones, plays a unique double role. It not only stimulates calcium absorption and bone strength, but also prevents biofilm formation, control host immune responses, and augment AMP (cathelicidin) production. These attributes can open up new therapeutic boulevards for treating biofilm-associated infections (osteomyelitis, chronic wounds, or oral pathologies involving osteonecrosis of the jaw). The modulation of systemic calcium, in conjunction with the localized bacterial calcium-sensing mechanism, may compromise biofilm integrity and enhance treatment strategies for chronic infections. Future perspectives can aim in exploring endocrine–microbial interplay for novel strategy development, combining hormonal therapy with traditional antimicrobial therapies.

REFERENCES

Antal, A. S., Dombrowski, Y., Koglin, S., Ruzicka, T., & Schauber, J. (2011). Impact of vitamin D3 on cutaneous immunity and antimicrobial peptide expression. *Dermato-Endocrinology*, *3*(1), 18–22. https://doi.org/10.4161/derm.3.1.14616

Aranow, C. (2011). Vitamin D and the immune system. *Journal of Investigative Medicine*, 59(8), 881–886. https://doi.org/10.2310/JIM.0b013e31821b8755

Assicot, M., Gendrel, D., Carsin, H., Raymond, J., Guilbaud, J., & Bohuon, C. (1993). High serum procalcitonin concentrations in patients with sepsis and infection. *The Lancet*, 341(8844), 515–518. https://doi.org/10.1016/0140-6736(93)90277-n

Baisa, G. A., Plum, L., Marling, S., Seeman, J., & DeLuca, H. F. (2019). Vitamin D is not required for adaptive immunity to listeria. *Physiological Reports*, *7*(16), e14209. https://doi.org/10.14814/phy2.14209

Becker, K. L., Snider, R., & Nylen, E. S. (2010). Procalcitonin in sepsis and systemic inflammation: a harmful biomarker and a therapeutic target. *British Journal of Pharmacology*, *159*(2), 253–264. https://doi.org/10.1111/j.1476-5381.2009.00433.x

Biswas, P., Bose, S., & Chakraborty, S. (2025). Vitamin D3 potentiates antimicrobial and antibiofilm activities of streptomycin and thymoquinone against Pseudomonas aeruginosa. *World Journal of Microbiology & Biotechnology*, *41*(3), 104. https://doi.org/10.1007/s11274-025-04304-y

Bouillon, R., Schuit, F., Antonio, L., & Rastinejad, F. (2020). Vitamin D binding protein: A historic overview. *Frontiers in Endocrinology*, *10*, 910. https://doi.org/10.3389/fendo.2019.00910

Cárdenas-Calderón, C., Veloso-Giménez, V., González, T. et al. (2022). Development of an implantable three-dimensional model of a functional pathogenic multispecies biofilm to study infected wounds. *Scientific Reports*, *12*, 21846. https://doi.org/10.1038/s41598-022-25569-5

Coskun, M., Babayeva, A., Barlas, T., Muhittin Yalcin, M., Akturk, M., Balos Toruner, F., Ayhan Karakoc, M., Karakan, T., Cindoruk, M., Yetkin, I., & Eroglu Altinova, A. (2024). Relationship between gut microbiome and bone deficits in primary hyperparathyroidism: A proof-of-concept pilot study. *Journal of Investigative Medicine: The officIal Publication of the American Federation for Clinical Research*, *72*(6), 541–552. https://doi.org/10.1177/10815589241251695

Costerton, J. W., Stewart, P. S., & Greenberg, E. P. (1999). Bacterial biofilms: A common cause of persistent infections. *Science*, *284*, 1318–1322. https://doi.org/10.1126/science.284.5418.1318

Djukic, M., Sostmann, N., Bertsch, T., Mecke, M., Nessler, S., Manig, A., Hanisch, U. K., Triebel, J., Bollheimer, L. C., Sieber, C., & Nau, R. (2015). Vitamin D deficiency decreases survival of bacterial meningoencephalitis in mice. *Journal of Neuroinflammation*, *12*, 208. https://doi.org/10.1186/s12974-014-0208-1

Gaddy, J. A., Radin, J. N., Cullen, T. W., Chazin, W. J., Skaar, E. P., Trent, M. S., & Algood, H. M. (2015). Helicobacter pylori resists the antimicrobial activity of calprotectin via lipid a modification and associated biofilm formation. *mBio*, *6*(6), e01349-15. https://doi.org/10.1128/mBio.01349-15

Golpour, A., Bereswill, S., & Heimesaat, M. M. (2019). Antimicrobial and immune-modulatory effects of vitamin D provide promising antibiotics-independent approaches to tackle bacterial infections - Lessons learnt from a literature survey. *European Journal of Microbiology & Immunology*, *9*(3), 80–87. https://doi.org/10.1556/1886.2019.00014

Guevara, M. A., Lu, J., Moore, R. E., Chambers, S. A., Eastman, A. J., Francis, J. D., Noble, K. N., Doster, R. S., Osteen, K. G., Damo, S. M., Manning, S. D., Aronoff, D. M., Halasa, N. B., Townsend, S. D., & Gaddy, J. A. (2020). Vitamin D and streptococci: The interface of nutrition, host immune response, and antimicrobial activity in response to infection. *ACS Infectious Diseases*, *6*(12), 3131–3140. https://doi.org/10.1021/acsinfecdis.0c00666

Hausfater, P., Juillien, G., Madonna-Py, B., Haroche, J., Bernard, M., & Riou, B. (2007). Serum procalcitonin measurement as diagnostic and prognostic marker in febrile adult patients presenting to the emergency department. *Critical Care (London, England)*, *11*(3), R60. https://doi.org/10.1186/cc5926

Ismailova, A., & White, J. H. (2021). Vitamin D, infections, and immunity. *Reviews in Endocrine and Metabolic Disorders*, 22(1), 1–3. https://doi.org/10.1007/s11154-021-09679-5

Judprasong, K., Chheng, S., Chimkerd, C., Jittinandana, S., Tangsuphoom, N., & Sridonpai, P. (2023). Effect of ultraviolet irradiation on vitamin D in commonly consumed mushrooms in Thailand. *Foods*, *12*(19), 3632. https://doi.org/10.3390/foods12193632

Kherad, Z., Yazdanpanah, S., Saadat, F., Pakshir, K., & Zomorodian, K. (2023). Vitamin D_3: A promising antifungal and antibiofilm agent against *Candida* species. *Current Medical Mycology*, *9*(2), 17–22.

King, M. M., Kayastha, B. B., Franklin, M. J., & Patrauchan, M. A. (2020). Calcium regulation of bacterial virulence. *Advances in Experimental Medicine and Biology*, *1131*, 827–855. https://doi.org/10.1007/978-3-030-12457-1_33

Kiya, G. T., Asefa, E. T., Abebe, G., & Mekonnen, Z. (2024). Procalcitonin guided antibiotic stewardship. *Biomarker Insights*, *19*, 11772719241298197. https://doi.org/10.1177/11772719241298197

Kragsbjerg, P., Söderquist, B., Holmberg, H., Vikerfors, T., & Danielsson, D. (1998). Production of tumor necrosis factor-alpha and interleukin-6 in whole blood stimulated by live gram-negative and gram-positive bacteria. *Clinical Microbiology and Infection: The Official Publication of the European Society of Clinical Microbiology and Infectious Diseases*, *4*(3), 129–134. https://doi.org/10.1111/j.1469-0691.1998.tb00375.x

Lei, J., Sun, L., Huang, S., Zhu, C., Li, P., He, J., Mackey, V., Coy, D. H., & He, Q. (2019). The antimicrobial peptides and their potential clinical applications. *American Journal of Translational Research*, *11*(7), 3919–3931.

Lutfi, L.L., Shaaban, M. I. & Elshaer, S. L. (2024). Vitamin D and vitamin K1 as novel inhibitors of biofilm in Gram-negative bacteria. *BMC Microbiology 24*, 173. https://doi.org/10.1186/s12866-024-03293-6

Mihai, R., & Farndon, J. R. (2000). Parathyroid disease and calcium metabolism. *British Journal of Anaesthesia*, *85*(1), 29–43.

Mouliou, D. S. (2023). C-reactive protein: Pathophysiology, diagnosis, false test results and a novel diagnostic algorithm for clinicians. *Diseases*, *11*(4), 132. https://doi.org/10.3390/diseases11040132

Nishikawa, M., & Kobayashi, K. (2021). Calcium prevents biofilm dispersion in bacillus subtilis. *Journal of Bacteriology*, *203*(14), e0011421. https://doi.org/10.1128/JB.00114-21

Pop, T. L., Sîrbe, C., Benţa, G., Mititelu, A., & Grama, A. (2022). The role of vitamin D and vitamin D binding protein in chronic liver diseases. *International Journal of Molecular Sciences*, *23*(18), 10705. https://doi.org/10.3390/ijms231810705

Raisz L. G. (1981). Calcium regulation. *Clinical Biochemistry*, *14*(5), 209–212. https://doi.org/10.1016/s0009-9120(81)90904-8

Rapsinski, G. J., Michaels, L. A., Hill, M., Yarrington, K. D., Haas, A. L., D'Amico, E. J., Armbruster, C. R., Zemke, A., Limoli, D., & Bomberger, J. M. (2024). Pseudomonas aeruginosa senses and responds to epithelial potassium flux via Kdp operon to promote biofilm. *PLoS Pathogens*, *20*(5), e1011453.

Ribet, D., & Cossart, P. (2015). How bacterial pathogens colonize their hosts and invade deeper tissues. *Microbes and Infection*, *17*(3), 173–183. https://doi.org/10.1016/j.micinf.2015.01.004

Saputo, S., Faustoferri, R. C., & Quivey Jr, R. G. (2018). Vitamin D compounds are bactericidal against Streptococcus mutans and target the bacitracin-associated efflux system. *Antimicrobial Agents and Chemotherapy*, *62*(1), 10–1128.

Shoback, D., Rosen, C. J., Black, D. M., Cheung, A. M., Murad, M. H., & Eastell, R. (2020). Pharmacological management of osteoporosis in postmenopausal women: An endocrine society guideline update. *The Journal of Clinical Endocrinology and Metabolism*, *105*(3), dgaa048. https://doi.org/10.1210/clinem/dgaa048

Thanawongnuwech, R., Thacker, B., Halbur, P., & Thacker, E. L. (2004). Increased production of proinflammatory cytokines following infection with porcine reproductive and respiratory syndrome virus and mycoplasma hyopneumoniae. *Clinical and Diagnostic Laboratory Immunology*, *11*(5), 901–908. https://doi.org/10.1128/CDLI.11.5.901-908.2004

Umar, M., Sastry, K. S., & Chouchane, A. I. (2018). Role of vitamin D beyond the skeletal function: A review of the molecular and clinical studies. *International Journal of Molecular Sciences*, 19(6), 1618. https://doi.org/10.3390/ijms19061618

Vijayan, A. L., Vanimaya, N., Ravindran, S. et al. (2017). Procalcitonin: a promising diagnostic marker for sepsis and antibiotic therapy. *Journal of Intensive Care*, *5*, 51. https://doi.org/10.1186/s40560-017-0246-8

Yu, M., Malik Tyagi, A., Li, J. Y. et al. (2020). PTH induces bone loss via microbial-dependent expansion of intestinal TNF^{+} T cells and Th cells. *Nature Communications*, *11*, 468. https://doi.org/10.1038/s41467-019-14148-4

10 Sugar-Regulating Hormones and Biofilm

Sreejita Ghosh, Bandita Dutta, Yashaswi Gupta, Samriddhi Kumari, and Rina Rani Ray
Department of Biotechnology, Maulana Abul Kalam Azad University of Technology, Kolkata, India

10.1 INTRODUCTION

Sugar-regulating hormones play a major role in the maintenance of metabolic equilibrium inside the human body, striking an intricate balance of the rate of glucose uptake followed by its storage and utilization as per requirement. Hormones, including insulin, glucagon, epinephrine, cortisol and Growth Hormone (GH), are the pivotal hormones in this regulatory process which ensures that the homoeostasis process is strictly regulated. Over recent years, however, it has been found that these hormones extend their activity even beyond the human physiology and into the world of microbes (Patel et al., 2021). Researches have indicated that modifications in human sugar-regulating hormones can affect microbial aggregation, their virulence, and, subsequently, biofilm formation (Plotkin et al., 2023).

Biofilms, being the bacterial cell clusters enclosed in a self-produced extrapolymeric matrix, play a crucial role in almost all bacterial infections. The bacterial cells residing within the biofilm matrix are not easily eradicated by the available antimicrobial therapies. The biofilm matrices also do not possess susceptibilities towards various antibiotics; hence such cells are called antibiotic-resistant (Sharma et al., 2019). These biofilms are responsible for causing a wide range of chronic and persistent infections among the host, thereby spurring a cross-talk between microbial behaviour and host endocrine signalling. Bacterial biofilm formation is extremely adaptable to the environment and is generally regulated by the microbial response quorum-sensing (QS) compounds, including mammalian hormones like insulin. Insulin is also known to be an interkingdom QS molecule, modulating biofilm formation and playing a key role in potentiating the secondary spread of systemic infections (Plotkin et al., 2023). In other words, sugar-regulating hormones work as a rheostat in modulation of biofilm formation inside the host body with a subsequent spreading of infections.

This chapter delves into the synergy of microbiology and endocrinology by assessing how the host sugar-regulating hormones affect the biofilm dynamics. The chapter details the indirect and direct mechanisms via which these hormones can influence bacterial growth, QS molecule production and signalling and, ultimately, the maturation of biofilms.

10.2 BLOOD SUGAR AND THE IMPORTANCE OF ITS MAINTENANCE IN THE BODY

Blood sugar is the level of glucose concentration present within the bloodstream. This substance, glucose, is one of the principal energy sources for almost all cells in the body, being particularly important for red blood cells (RBCs), brain, and muscles. It is primarily obtained from dietary carbohydrates and is distributed through the circulatory process to the organs and tissues, where it is either utilized immediately for energy generation or stored for later use. The maintenance of

DOI: 10.1201/9781003519041-10

TABLE 10.1
Physiological Implication of Blood Glucose Regulation in the Human Body

Aspect	Role and Importance	References
Supply of energy to the cells	Glucose serves as the main source of energy for cellular respiration. The brain is immensely dependent on glucose for proper functioning. Any type of fluctuations in glucose level may lead to impairment of brain function, dizziness and unconsciousness	Ritter (2017)
Diabetes and other complication prevention	Chronic hyperglycaemia results in damage of the blood vessels, nerves, eyes, heart and kidneys. Stable glucose levels aids in prevention of diabetes-associated complications	Rawal et al. (2012)
Hormonal balance and metabolic regulation	Glucagon and insulin maintain glucose homoeostasis. Proper levels of glucose ensure healthy protein biosynthesis, lipid metabolism and immune function	Tao and Cheng (2023)
Muscle and physical performances	Muscles utilize glucose during physical activities. Stable blood sugar supports recovery and endurance, whereas hypoglycaemia causes weakness and fatigue	Pataky et al. (2021)
Immune functions	Balanced glucose levels ensure proper immune function. Hyperglycaemia impairs blood cell activities increasing infection susceptibility	Berbudi et al. (2020)
Cognitive and neurological health	Poor glucose homoeostasis is related to cognitive impairment and increased chances of stroke. Stable levels of glucose support long-term brain health and proper functioning	Shomali et al. (2021)

optimum blood sugar levels (70–110 mg/dl) in the fasting state is vital for the maintenance of physiological stability. The human body is dependent on an intricate hormonal balance for the regulation of glucose production, its uptake and storage. Disturbance in this balance may lead to disorders such as hypoglycaemia (low blood sugar levels) or hyperglycaemia (high blood sugar levels), both of which have acute and chronic health consequences. The importance of blood sugar level maintenance is described in Table 10.1.

10.3 HORMONES REGULATING BLOOD SUGAR

Blood glucose regulation is not just a passive consequence of digestion followed by metabolism, but an increasingly dynamic system controlled by the endocrine system. Homoeostasis of the level of glucose in blood is strictly controlled by a complex hormonal network maintaining energy balance as well as proper metabolism. Hormones serve as biochemical messengers, having the capacity to detect fluctuations in blood sugar levels by coordinating proper responses amongst all the organs like adipose tissues, muscles, and the pancreas. This network of hormones not only aids in immediate modulations of glucose availability but also regulates long-term energy balance and proper metabolic adaptation in situations of fasting, stress, and illness. The two major hormones involved in the regulation of blood glucose homoeostasis are insulin and glucagon. In addition to this, several other hormones, such as epinephrine, cortisol and GH, also play a part in regulating blood sugar levels (Table 10.2).

10.3.1 Glucagon

Glucagon, a 29-amino acid peptide hormone, is produced from proglucagon, which is expressed in various tissues, such as the brain, pancreas, and intestine, where it is subsequently transformed into glucagon through a process of proteolysis. Pancreatic alpha cells in the islet of Langerhans release

TABLE 10.2
Sources and Functions of Major Blood Sugar-Regulating Hormones

Hormones	Source	Roles	References
Glucagon	α-cells of pancreas	Increases blood glucose level by triggering glycogenolysis and gluconeogenesis in the liver	Zeigerer et al. (2021)
Insulin	ß-cells of pancreas (islets of Langerhans)	Decreases blood glucose levels by promotion of glycogenesis, cellular uptake of glucose and lipogenesis	Rawal et al. (2012)
Cortisol	Adrenal cortex	Promotion of gluconeogenesis and reduction of glucose usage in situations of prolonged fasting or stress	De Nys et al. (2022)
Epinephrine	Adrenal medulla	Raises glucose levels during stress conditions by stimulating breakdown of glycogen and inhibition of insulin	Albayrak et al. (2011)
GH	Anterior pituitary	Antagonistic to insulin, stimulates lipolysis and reserves glucose for important organs	Zhang et al. (2021)

glucagon, which is essential for maintaining glucose homeostasis by encouraging hepatic glucose synthesis. Glucagon's main role is to elevate blood glucose levels by prompting the liver to decompose glycogen and release glucose into the blood. Additionally, it assists in releasing energy reserves by facilitating fat breakdown (lipolysis) and the turnover of amino acids. According to Müller et al. (2017), glucagon may also indirectly regulate its secretion by inducing beta cells to release insulin.

10.3.2 Insulin

Insulin controls blood sugar levels by enabling glucose to be absorbed from the blood into cells, where it can either be utilized for energy or stored for later use. Additionally, it contributes to general metabolism by affecting the breakdown and storage processes of fats, proteins, and carbohydrates. Glucokinase senses the entry of glucose into the β cell and phosphorylates it into glucose-6-phosphate (G6P), thereby generating ATP. The opening of voltage-dependent calcium channels and membrane depolarization through the closure of K+-ATP-dependent channels increase intracellular calcium levels, thereby resulting in pulsatile insulin release (Soria et al., 2004). In response to environmental stimuli and the interaction of other hormones, nutritional and non-nutrient secretagogues regulate the synthesis and secretion of insulin. The nutrient secretagogues, such as glucose, appear to stimulate the β cell to secrete insulin through the elevation of intracellular ATP and the closure of K^+-ATP channels. Insulin release is also augmented by the enhanced production of cyclic AMP and other cellular energy intermediates. Insulin action is not required for glucose, fructose, mannose, or galactose to pass into the β cell. Non-nutrient secretagogues may act through peptide hormones, cationic amino acids, or brain stimulation such as cholinergic and adrenergic mechanisms.

10.3.3 Epinephrine (Adrenaline)

The primary functions of adrenaline, primarily secreted by the adrenal glands' medulla, are to raise blood glucose levels and cardiac output. Epinephrine, commonly referred to as adrenaline, is crucial for the management of blood sugar levels as it can either raise or lower glucose concentrations. It prompts the liver to generate additional glucose via glycogenolysis and gluconeogenesis. Additionally, it hinders the absorption and utilization of glucose by tissues, especially in muscle and fat, which diminishes the body's capacity to use glucose and can inhibit insulin release from the pancreas, further leading to elevated blood glucose levels.

10.3.4 Cortisol

Glucocorticoids, such as cortisol, play a crucial role in managing blood sugar levels, though their impact can be intricate and, at times, contradictory. These hormones stimulate the liver to generate more glucose, even when blood sugar levels are already high. This mechanism includes enhancing the expression of enzymes that contribute to glucose production. Glucocorticoids disrupt the effects of insulin, the hormone that enables glucose to enter cells for energy use. As a result, there can be an accumulation of glucose in the bloodstream, even while cells are in need of it. Nevertheless, they also affect the release of insulin and glucagon, hormones that directly manage blood sugar levels.

10.3.5 Growth Hormone

Human growth hormone (GH) is a crucial hormone which is released by the pituitary gland, affecting blood sugar levels through various direct and indirect mechanisms. Although this can lead to elevated blood sugar, the body typically adjusts by producing more insulin, and GH is also involved in regulating blood sugar balance during instances of low blood sugar. GH can raise blood glucose levels by promoting glycogenolysis and gluconeogenesis in the liver and can lead to insulin resistance, which means that cells become less sensitive to insulin, resulting in increased blood sugar. Additionally, GH contributes to glucose counter-regulation, a mechanism that helps to prevent and correct low blood sugar levels.

10.4 SIGNIFICANCE OF BIOFILM FORMATION IN BODY

Patients with hyperglycemia are susceptible to various chronic diseases. Chronic inflammatory diseases may appear because of the persistent and robust environment that bacterial biofilms contribute to harmful microbes inside the body. Biofilms are complex bacterial colonies that can develop in the human body, chronic diseases, and medical equipment, among other places. Biofilms present a significant therapeutic and diagnostic issue and are identified by a safety model of extracellular polymeric substance that envelop the bacteria (Zhao et al., 2023). When it comes to persistent infection, biofilms differ from free-floating bacteria in a few key ways. Significant treatment consequences of these alterations include tissue damage, chronic inflammation, elevated antibiotic resistance, and difficulties with diagnosis. The biofilm-bound microbial cells are able to evade immune system recognition. The bacteria are shielded from immune cells by the extracellular matrix that envelops them, making it more difficult for the body to eradicate the infection. Chronic inflammation is the outcome of the immune system's continuous attempts to fight the infection, which prolongs the inflammatory response.

Biofilms are often resistant to antibiotics. Some bacteria within the biofilm go into a dormant condition, which reduces their sensitivity to drugs, and the shape of the biofilm can impede the penetration of antibiotics. Antibiotic resistance exacerbates chronic inflammation while the infection continues (Uruén et al., 2020).

Dysbiosis of gut microbiome leads to the development and proliferation of liver diseases and colorectal cancer (CRC). In the case of CRC, microbiome changes make the intestinal barrier susceptible by initiating tumor development and inflammation. Major signaling pathways involve ß-catenin, E-cadherin, MYD88, TLR-4, NF κß, RAS, SMO, p38 and MAPK. Bacteria such as *Enterococcus faecalis*, *F. nucleatum*, *Bacteroides fragilis*, *Escherichia coli*, *Helicobacter pylori*, and *Streptococcus bovis* are mostly responsible for host DNA damage promoting oncogenesis. Of these, *F. nucleatum* has been found to infiltrate the tumor cells among most of the CRC patients (Kostic et al., 2012).

The gut–liver axis via which gut-associated metabolites and microorganisms reach to the liver through the portal vein also leads to liver diseases. The bacteria responsible for liver diseases include Proteobacteria, Firmicutes and Fusobacteria.

10.5 BIOFILM IN IMPLANT

Implant-associated microbial infection is a primary cause of biomaterial implant failure and also a substantial threat. Medical equipment such as cochlear implants, total joint prostheses, cardiac pacemakers, prosthetic heart valves, renal dialysis membranes, and different kinds of intravenous catheters are being used increasingly in practically every area of medicine. Over the past two decades or sos, biofilms from a variety of medical devices have been thoroughly investigated. Research has also clarified the materials' susceptibility to bacterial adhesion and biofilm formation for certain devices, including contact lenses and urinary catheters. The main implants that are susceptible to infection are: Cardiovascular (heart valves, ventricular assist devices, coronary catheters); intravascular (mostly central venous catheters); and neurosurgical (nervous stimulators implanted, ventricular shunts). In addition to following hygienic guidelines when implanting medical devices, a new and promising approach is the creation of materials that may be impervious to microbial adherence and colonization. Such spoilage of implants by biofilm is more prominent in those patients with high levels of blood sugar (Shang & Gao, 2021).

10.6 HYPERGLYCEMIA AND BACTERIAL BIOFILM

By encouraging bacterial pathogenicity and biofilm formation, conditions of high blood sugar, or hyperglycemia, can induce infections even faster, including the formation of biofilms. This may result in delays in the healing of wounds or continuous infections. Sugars provide bacteria with an easily available energy source, which enhances bacterial growth and the formation of biofilms.

Hyperglycemia, innate immune cell dysfunction, and infection with antimicrobial-resistant bacterial strains are among the diabetes-related complications that may worsen bacterial infections. Hyperglycemia, which is characterized by a rise in blood and tissue glucose concentration, is one of the marked symptoms of diabetes. Many infectious bacteria choose glucose as their carbon source, and hyperglycemia causes a rise in the number of the growth and virulence of bacteria. Certain bacterial virulence functioning, such as tissue attachment and biofilm formation, which are known to contribute to the seriousness of infection, are facilitated by hyperglycemia (Darwitz et al., 2024).

10.7 SPECIFIC BACTERIA INVOLVED IN BIOFILM FORMATION

Both abiotic and biotic surfaces are capable of supporting the formation of biofilm by bacteria under natural and therapeutical circumstances. Both types of bacteria create extracellular matrixes of their own, which consolidate the bacterial groups within biofilm. As a leading infection-causing organism in biofilms, *S. aureus* occurs in abundant numbers. By inhibiting the host immune system and antibiotics from targeting bacteria, the development of biofilm has the potential to make bacteria resistant to external threats. Consequently, biofilm infections are now challenging to treat clinically. An innovative and effective treatment method must be developed in order to meet this urgent need. A deep understanding of the mechanisms that control the development and control of bacterial biofilms can provide valuable information into preventing antibiotic resistance caused by bacterial biofilms (Peng et al., 2022).

Reports of staphylococcal species yielding higher amounts of biofilms on increased glucose concentrations have been noted. Since the pathogenicity of organisms involved in causing implant infections is attributed to the production of biofilms, the finding holds clinical importance. Given that hyperglycaemia is related to a higher risk of nosocomial infections, it is difficult to determine the relationship between microbial pathogenicity and the severity of abnormal glucose levels, whether they are low or high. There is a significant correlation between the amount of bacterial biofilm and the concentration of glucose.

10.8 ROLE OF SUGAR-CONTROLLING HORMONES ON BIOFILM

10.8.1 Insulin and Biofilm

As a quorum chemical signaling molecule, human-derived insulin controls the growth of *E. coli* biofilms on uroepithelial cells, catheter material, plastic, and glass. Human insulin regulates *E. coli* biofilm formation and has global rheostat-like control over all phases, from adhesion to motility, the population growth of cells, and the maturation and establishment of biofilms (Plotkin et al., 2023). The structural changes to cell walls are equivalent to alterations that have been demonstrated to increase virulence. The changes appear to be glucose uptake and insulin binding to the cell, which agrees with observations of insulin's interaction with mammalian cells. However, insulin acts in a covalent manner. The virulence factors associated with colonization are up-regulated when insulin is combined with glucose. When glucose is absent, insulin appears to act as an indicator, inducing the planktonic form by acting as a chemo-repellent that induces movement away from surfaces. Insulin, on the other hand, has a bivalent effect. Insulin appears to act as a warning signal when glucose is not present and encourages the planktonic lifestyle by acting as a chemo-repellent that induces movement away from surfaces. But when glucose is present, the encouragement of surface adherence—the initial step in biofilm formation—enhances chemoattraction (Figure 10.1).

Additionally, insulin alone has no influence on the extent of biofilm development or cell growth in the host's physiological conditions, but both are significantly elevated when glucose is present. Insulin regulates the development of biofilms under a variety of environmental conditions. This is insulin concentration-specific. It is fascinating to observe that galactose and insulin produce the most biofilm per cell, regardless of temperature or any other nutrient. Consequently, it must be most detrimental to the efficacy of antibiotics. In cystic fibrosis-associated pneumonia and chronic wound infections, *Pseudomonas aeruginosa* is significant in biofilm formation. *Staphylococcus aureus*, *P. aeruginosa*, *Enterococcus faecalis*, coagulase-negative staphylococci, *Proteus* spp., and anaerobic bacteria have been isolated from chronic wounds. *P. aeruginosa* is the most prevalent Gram-negative. It is also a significant pathogen in patients with chronic lung infections and is the first cause of morbidity and mortality in cystic fibrosis. In contrast to planktonic bacteria, biofilm bacteria both can evade the innate and adaptive immune system and are also protected from phagocytosis

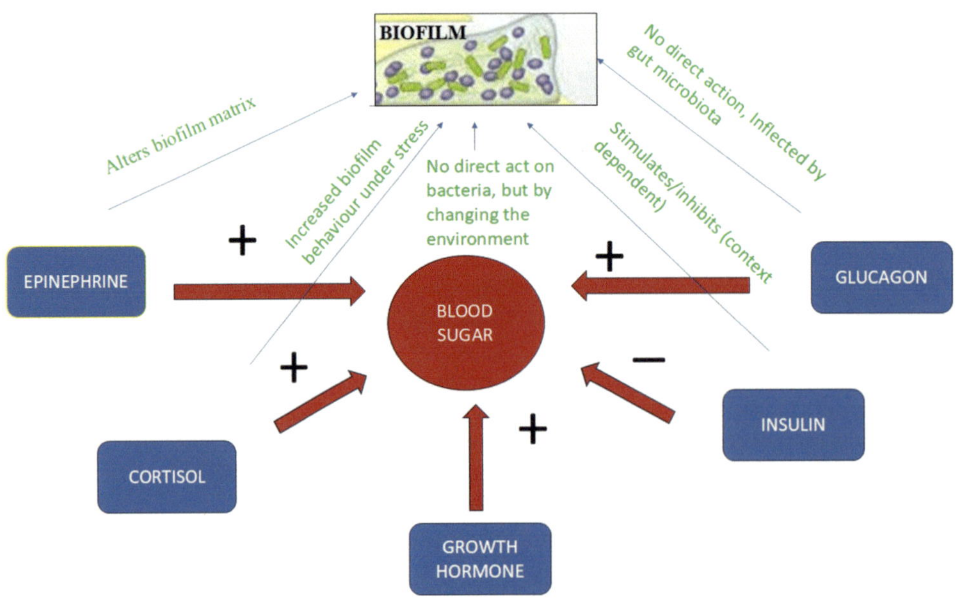

FIGURE 10.1 Effect of sugar-controlling hormones on biofilm.

and antibacterial therapy. Catheter-associated urinary tract infections (UTIs) account for 80% of all medical care-related UTIs. As a whole, the mortality risk of catheterized patients with a UTI is three times that of otherwise infected non-catheterized patients. Of those who are catheterized, patients with diabetes mellitus are at a higher risk of dying from catheter-associated infections compared to those without (Patel et al., 2021). Type 2 diabetes mellitus is a common disease. Insulin resistance in the periphery is found in type 2 diabetics. Hyperinsulinemia is the consequence of increased insulin secretion by their bodies in reaction. Depending on their disease state and the degree of glycemic control, this excess insulin is excreted in the urine both with and without glucose. In comparison with individuals with type 1 diabetes mellitus or normal metabolism, these individuals also have an increased rate of UTIs. Earlier studies have proved that insulin and glucose affect the activity of *E. coli*, such as the formation of biofilms. A substance known as biofilm, composed of a matrix, protects bacteria from environmental stress and antibiotics. Insulin and glucose increase the attachment of *E. coli* to rubber and its favorable surface charge, enabling greater bacterial association with negatively charged silicon-coated catheters. Insulin and glucose also enhance the formation of biofilm on rubber and silicon catheter segments (Plotkin et al., 2023). Microorganisms referred to as biofilms play a very crucial role in numerous aspects of research. The ability of bacteria to establish biofilms is something that can be tested in the presence of a variety of host chemicals, which can have further influences upon biofilm-forming capabilities. Patients take chemicals, referred to as synthetic hormones, in order to compensate for their hormone deficiency. One of the most widely used synthetic hormones is insulin, which is administered to diabetic patients. A number of insulin solutions on the market have antibacterial properties and can create biofilms against the bacterial species *E. coli* and *S. aureus*. There are no antibacterial activities of insulin detemir. In all concentrations tested (from 10% to 10–8%), *S. aureus* with a diluted suspension of insulin detemir had an increased biofilm formation, but the other species did not show any change of biofilm formation after adding insulin (Novalić & Bektić 2020). The biofilm formation ability of *S. aureus* and *Pseudomonas* is higher than *E. coli* and *Klebsiella* in diabetic and nondiabetic populations. This implies that symptomatic infections cannot be eliminated by antibiotics and raise the threat of UTI and bacteriuria (bacteria in the urine) by 3–6% every day. 2.5 U/ml of insulin enhanced the growth rate of the bacterial isolates. It has been shown that insulin alters the growth kinetics of Gram-positive and Gram-negative bacteria and infers that nature has a variety of responses to this hormone. 0.1% glucose, insulin–glucose combination altered growth in a manner that enhanced the growth rate (Plotkin et al., 2023). Studies have indicated that insulin directly affects the growth kinetics of *E. coli* (Abed et al., 2021), which is among the most common pathogens in diabetic patients that can involve all organs and systems; it is responsible for 70% of diabetic patients with emphysematous urinary tract infections and 40% of all diabetic patients with emphysematous cholecystitis. Many experiments have proven that diabetic patients are more prone to microbial infections due to their hyperglycaemic physiological condition, which is responsible for phagocytic dysfunction. Human hormone insulin, even at low levels, not only produced a vigorous stimulating effect on the growth of bacterial cells, but also profoundly influenced *E. coli* metabolic activity and served as an autoinducer of biofilm formation (Figure 10.2). Because insulin therapy has the potential to induce expression of the infectious factor aspartyl proteinase, an enzyme which can also serve as a signal molecule for biofilm formation and quorum sensing, it has the potential to facilitate the transmission of *E. coli* infection among individuals with diabetes. On the contrary, surprisingly, regardless of temperature or the presence of other nutrients, galactose with insulin produces the most biofilm growth per cell and thus is considered to be detrimental.

10.8.2 Glucagon and Biofilm

Glucagon and biofilms share a somewhat intricate connection, as glucagon is mainly recognized for its function in controlling blood sugar levels, while biofilms consist of groups of microorganisms that adhere to surfaces. Although a direct causal relationship between glucagon and biofilm

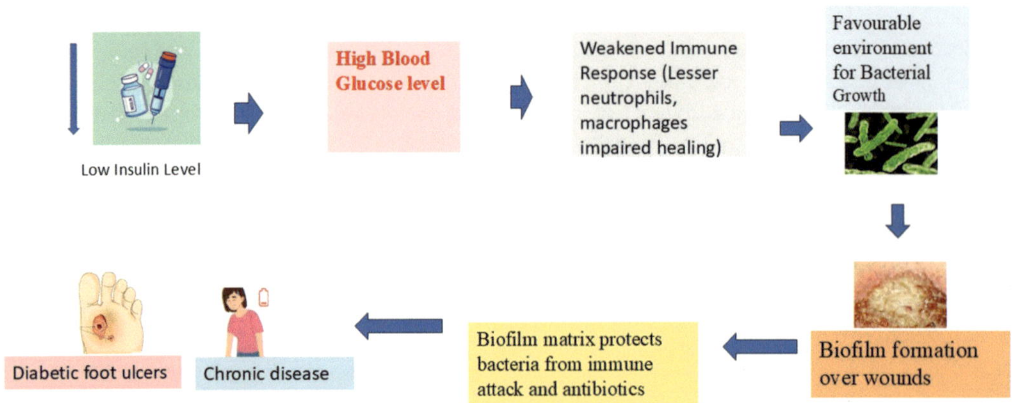

FIGURE 10.2 Role of low insulin level on development of biofilm-associated person.

development has not been established, some studies indicate that glucagon could affect processes linked to biofilms, especially regarding inflammation and immune responses. Research indicates that glucagon may influence immune responses, and inflammation can contribute to the formation and maintenance of biofilms. For instance, glucagon has the ability to reduce the migration of neutrophils, which are immune cells involved in combating infections.

10.8.3 Epinephrine and Biofilm

Epinephrine can profoundly change the biofilm matrix structure in *Micrococcus luteus* C01 residing on human skin (Gannesen et al., 2022). The changes arising due to human hormones affects the commensal microbiome and biofilms forming microorganisms have been a subject of many studies. Gannesen et al. (2022) reported that, in the presence of epinephrine, there was a modification of biofilm matrix by phospholipid changes within the matrix. These changes suggest that epinephrine can modify the physical properties and structural stability of biofilms, which affects the survivability-bound host microorganism and the host–pathogen relationship. Their findings elucidate, in a sense, the role of neuroendocrine signals in determining the bacterial physiology, suggesting notably how stress inflicted by a host can influence parameters of pathogenic microorganisms.

Cambronel et al. (2019) studied the well-known pathogen *P. aeruginosa*. It was found that in a *Galleria mellonella* (wax moth larvae) infection model, exposure to epinephrine heightened bacterial adhesion, biofilm formation, and virulence in vivo. Epinephrine was found to promote the production of factors that assist in the biofilm development processes, such as initial surface adhesion and cell aggregation. This indicates that hormonal changes due to stress in the host can directly fuel the omnipotent pathogenicity of *P. aeruginosa* for more aggressive infections, especially in the weakly defended patients.

Microbial endocrinology is exemplified by host epinephrine's interaction with bacterial signalling (Danilova et al., 2020). Bacteria have sensors of catecholamines, for example, two-component systems and adrenergic-like receptors. Such responses can initiate a cascade of changes in gene expression with alterations in biofilm maturation, quorum sensing, and biofilm antibiotic tolerance occurring. This inter-kingdom cross-talk emphasizes the importance of host factors such as stress and disease in understanding the pathogenicity of bacteria.

10.8.4 Cortisol and Biofilm

Over the past few years, significant research has focused on characterizing the complex microbial communities inhabiting the human body. Despite the advancement, the mechanisms underlying host–microbe interactions that maintain a stable microbiota remain enigmatic. One key factor that can imbalance the host–microbe interactions is host stress, which can also influence a range of systemic and localized diseases. While acute stress may predispose the immune system to respond to infections, chronic stress has been linked to inflammatory diseases such as rheumatoid arthritis, diabetes, and periodontitis. In addition, stress can not only affect the structure of commensal microbiota, but also further influence disease progression.

The most accepted theory explaining how hormones influence microbial infections affecting the immune system. Under stress conditions, the central nervous system triggers the hypothalamus to release corticotropin-releasing hormone and arginine vasopressin, which stimulate adrenocorticotropic release from the pituitary. This cascade leads to the production of cortisol by the adrenal cortex. Notably, the presence of stress hormones such as adrenaline and cortisol has been correlated with increased vulnerability to infections. For instance, a lower dose of *Clostridium* is required to induce infection in the presence of adrenaline, implying that microorganisms directly respond to stress hormones, under the field of microbial endocrinology.

Microbes that have evolved mechanisms to detect host hormones can potentially benefit evolutionarily by adapting gene expression in response to environmental changes. New evidence even suggests that the microbiome can regulate host hormone levels. For instance, *Ruminococcus*, a gut microbe in piglets, has been proposed to modulate brain levels of n-acetylaspartate via cortisol regulation.

Although most of the studies focusing on the interaction between stress hormones and bacteria have been directed towards the gut microbiome, a few recent studies have shown that stress hormones may also affect oral pathogens. Within the oral cavity, glucocorticoids like cortisol suppress immune functions, including secretory immunoglobulin production and neutrophil activity, which play a critical role in the defence mechanism against periodontal infections. Elevated levels of cortisol have been detected in the saliva and serum of individual patients with severe periodontal disease, suggesting a direct linkage between stress and oral health (Duran-Pinado et al., 2018).

10.8.4.1 Cortisol Regulates Biofilm Formation of *Flavobacterium Columnare*

Living organisms to survive should maintain homeostasis. Yet this equilibrium is constantly being challenged by both internal and external stimuli, the latter of which are termed as stressors—stimuli that elicit a stress response. In fish, the hypothalamic–pituitary–interrenal (HPI) axis becomes activated by stress, triggering the release of glucocorticoids, particularly cortisol, into the bloodstream. Cortisol not only circulates through the blood and tissues but is also excreted through mucus, urine, and feces. Significantly, stress is not fundamentally harmful; it may be considered as "eustress" when it is of benefit to the animal or "distress", which, when it becomes chronic or overly intense, potentially disrupts physiological function. Chronic stress is widely recognized as a key factor in the outbreak of diseases in aquaculture, including those caused by *Flavobacterium columnare*, the pathogen responsible for columnaris disease. This condition results in significant clinical symptoms such as gill lesions, skin damage, and mouth rot, particularly in species such as carp, catfish, and trout. Notably, susceptibility to columnaris disease varies among individual fish, even when exposed to low-virulence strains of *F. columnare*. This variation has been linked to individual coping styles and different cortisol responses following stress, such as during experimental immersion challenges where handling and bathing procedures induce cortisol elevation. These differences may influence the degree to which fish are vulnerable to infection.

Recent studies imply a more direct action of cortisol, on not only the host but also the pathogen. Studies indicate that alteration of the morphology of *F. columnare* colonies is cortisol dose-dependent; higher cortisol levels lead to colonies with reduced rhizoid structure, a morphology correlated with decreased gliding motility and, by extension, virulence. Gliding motility is critical for colonization and virulence, is mediated by the type IX secretion system (T9SS), which contains crucial components like GldK, GldL, GldM, GldN, SprA, SprE, SprT, and PorV. These components promote the secretion of surface adhesins like SprB and RemA, which are essential for locomotion over surfaces. Mutations in these genes suppress gliding and attachment, indicating their importance.

Infections begin with the development of biofilms on the gill filaments, initiated when bacteria lose motility, adhere, and form sessile colonies. Since cortisol has an impact on gliding motility, it may also influence biofilm formation and gene expression responsible for motility and adhesion (Declercq et al., 2019).

10.8.4.2 Surface Translocation of *P. Gingivalis* by Cortisol

The pathogenicity of the oral microbial community is regulated by host susceptibility, microbial community composition, and interbacterial interactions. Periodontal disease, like periodontitis, occurs when host–microbe equilibrium shifts into dysbiosis, a transition affected by factors, including genetics, diet, smoking, and most importantly, stress. Stress is a recognized risk factor for several diseases, including periodontal disease, due to its ability to disrupt the human microbiome. Research indicates that stress hormones like cortisol can alter the gene expression profiles of oral microbes, resulting in disease progression.

The primary glucocorticoid in humans, cortisol, is increased in both saliva and serum in individuals with severe periodontal disease. In the oral cavity, a high level of glucocorticoids suppresses immunity by reducing secretory immunoglobulin production and neutrophil function, thereby disrupting microbial homeostasis. The major pathogen linked to periodontitis is *Porphyromonas gingivalis*, a Gram-negative anaerobe strongly associated with oral disease and systemic conditions such as diabetes, Alzheimer's disease, cardiovascular disease, and rheumatoid arthritis.

P. gingivalis utilizes a type IX secretion system (T9SS), a specific protein export pathway found in members of the *Bacteroidota* phylum, to secrete more than 30 proteins, along with virulence factors like gingipains. In other *Bacteroidota* species, T9SS also regulate gliding motility, although *P. gingivalis* lacks this common motility. Instead, it exhibits surface translocation on the interface between soft agar and a solid surface, a T9SS and fimbrial proteins-dependent process.

P. gingivalis differs from most bacteria in having a distinctive metabolism; it is extremely proteolytic and asaccharolytic, depending on proteins rather than sugars as energy sources. Additionally, it combines with amino acid fermentation to utilize pyruvate and lactate for energy production. This metabolic process also regulates the fimbrial adhesins expression, which are essential for surface adhesion and movement.

Fimbriae in *P. gingivalis* are categorized as either major or minor. Major fimbriae are encoded by the *fim* gene cluster (including *fimA*, *fimB*, *fimC/D/E*), while minor fimbriae are derived from the *mfa* cluster (including *mfa1*, *mfa2*, and tip proteins *mfa3/4/5*). Of these, Mfa5 is characterized by having a C-terminal domain (CTD), directing it through the T9SS, and a von Willebrand factor type A (vWF) domain, which facilitates protein–protein interactions and adhesion. In other pathogens, vWF-containing proteins at the tip of the fimbrials are essential for attachment to the host cell. In *P. gingivalis*, deletion of the *mfa5* gene disrupts the surface translocation, highlighting its function in motility and colonization (Kim et al., 2022).

10.8.5 Growth Hormone and Biofilm

There is no well-established or direct association between GH and biofilm formation, but certain studies showed that few hormones like those related to the GH/IGF-1 axis can affect bacterial adhesion and formation of biofilms, thereby affecting the gut microbiome and other microbial communities (Møller et al., 1991). However, more detailed researches are required to confirm the involvement of GH in the formation of biofilms in human bodies.

10.9 CONCLUSION

"Sugar is a silent killer." This means that the presence of excessive glucose in blood not only damages the systemic efficacy expediting death of a person, but also promotes biofilm formation by opportunistic pathogens within various body parts and implants, if any. Of all the sugar-regulating hormones, insulin has the ability to both encourage and hinder biofilm formation, which varies based on glucose concentration and the type of bacteria involved. In certain instances, the combination of insulin and glucose can reduce biofilm formation; in contrast, in other situations, it may increase it. Although glucagon does not play any role on biofilm, cortisol and epinephrine can impact biofilm formation and destruction depending on the bacteria involved. Hence, depending on the nature of bacteria, the hormone may either stimulate or eradicate biofilm.

REFERENCES

Abed, W. G., Al-Shawk, R. S., & Jassim, K. A. (2021). The role of insulin level on the biofilm-forming capacity in diabetes-related urinary tract infection. *Mustansiriya Medical Journal*, *20*(2), 66–70. https://doi.org/10.4103/mj.mj_12_21

Albayrak, Y., Saglam, M. B., Yildirim, K., Karatay, S., Polat, B., Uslu, T., Suleyman, H., & Akcay, F. (2011). Effects of epinephrine and cortisol on the analgesic activity of metyrosine in rats. *Archives of Pharmacal Research*, *34*(9), 1519–1525. https://doi.org/10.1007/s12272-011-0914-5

Berbudi, A., Rahmadika, N., Tjahjadi, A. I., & Ruslami, R. (2020). Type 2 diabetes and its impact on the immune system. *Current Diabetes Reviews*, *16*(5), 442–449. https://doi.org/10.2174/1573399815666191024085838

Cambronel, M., Tortuel, D., Biaggini, K., Maillot, O., Taupin, L., Réhel, K., Rincé, I., Muller, C., Hardouin, J., Feuilloley, M., Rodrigues, S., & Connil, N. (2019). Epinephrine affects motility, and increases adhesion, biofilm and virulence of Pseudomonas aeruginosa H103. *Scientific Reports*, *9*(1), 20203. https://doi.org/10.1038/s41598-019-56666-7

Danilova, N. D., Solovyeva, T. V., Mart'Yanov, S. V., Zhurina, M. V., & Gannesen, A. V. (2020). Stimulatory effect of epinephrine on biofilms of micrococcus luteus C01. *Microbiology*, *89*, 493–497. https://doi.org/10.1134/S0026261720040049

Darwitz, B. P., Genito, C. J., & Thurlow, L. R. (2024). Triple threat: How diabetes results in worsened bacterial infections. *Infection and Immunity*, *92*(9), e0050923. https://doi.org/10.1128/iai.00509-23

De Nys, L., Anderson, K., Ofosu, E. F., Ryde, G. C., Connelly, J., & Whittaker, A. C. (2022). The effects of physical activity on cortisol and sleep: A systematic review and meta-analysis. *Psychoneuroendocrinology*, *143*, 105843. https://doi.org/10.1016/j.psyneuen.2022.105843

Declercq, A. M., Cai, W., Naranjo, E., Thongda, W., Eeckhaut, V., Bauwens, E., Arias, C., De La Fuente, L., Beck, B. H., Lange, M. D., Peatman, E., Haesebrouck, F., Aerts, J., & Decostere, A. (2019). Evidence that the stress hormone cortisol regulates biofilm formation differently among Flavobacterium columnare isolates. *Veterinary Research*, *50*(1), 24. https://doi.org/10.1186/s13567-019-0641-3

Duran-Pinedo, A. E., Solbiati, J., & Frias-Lopez, J. (2018). The effect of the stress hormone cortisol on the metatranscriptome of the oral microbiome. *Npj Biofilms and Microbiomes*, *4*(1), 25.

Gannesen, A. V., Ziganshin, R. H., Zdorovenko, E. L., Klimko, A. I., Ianutsevich, E. A., Danilova, O. A., Tereshina, V. M., Gorbachevskii, M. V., Ovcharova, M. A., Nevolina, E. D., Martyanov, S. V., Shashkov, A. S., Dmitrenok, A. S., Novikov, A. A., Zhurina, M. V., Botchkova, E. A., Toukach, P. V., & Plakunov, V. K. (2022). Epinephrine extensively changes the biofilm matrix composition in *Micrococcus luteus* C01 isolated from human skin. *Frontiers in Microbiology*, *13*, 1003942. https://doi.org/10.3389/fmicb.2022.1003942

Kim, H. M., Rothenberger, C. M., & Davey, M. E. (2022). Cortisol promotes surface translocation of *porphyromonas gingivalis*. *Pathogens (Basel, Switzerland)*, *11*(9), 982. https://doi.org/10.3390/pathogens11090982

Kostic, A. D., Gevers, D., Pedamallu, C. S., Michaud, M., Duke, F., Earl, A. M., Ojesina, A. I., Jung, J., Bass, A. J., Tabernero, J., Baselga, J., Liu, C., Shivdasani, R. A., Ogino, S., Birren, B. W., Huttenhower, C., Garrett, W. S., & Meyerson, M. (2012). Genomic analysis identifies association of Fusobacterium with colorectal carcinoma. *Genome Research*, *22*(2), 292–298. https://doi.org/10.1101/gr.126573.111

Møller, N., Jørgensen, J. O., Abildgård, N., Orskov, L., Schmitz, O., & Christiansen, J. S. (1991). Effects of growth hormone on glucose metabolism. *Hormone Research*, *36*(Suppl 1), 32–35.

Müller, T. D., Finan, B., Clemmensen, C., DiMarchi, R. D., & Tschöp, M. H. (2017). The new biology and pharmacology of glucagon. *Physiological Reviews*, *97*(2), 721–766. https://doi.org/10.1152/physrev.00025.2016

Novalić, J., & Bektić, B. (2020). Effect of commercially available synthetic insulin on the biofilm formation in S. *aureus* and E. *coli* Bacterial Strains. In A. Badnjevic, R. Škrbić, L. Gurbeta Pokvić (Eds.), *CMBEBIH 2019. CMBEBIH 2019. IFMBE Proceedings*, Vol 73. Cham: Springer. https://doi.org/10.1007/978-3-030-17971-7_112

Pataky, M. W., Young, W. F., & Nair, K. S. (2021). Hormonal and metabolic changes of aging and the influence of lifestyle modifications. *Mayo Clinic Proceedings*, *96*(3), 788–814. https://doi.org/10.1016/j.mayocp.2020.07.033

Patel, N., Curtis, J. C., & Plotkin, B. J. (2021). Insulin regulation of *escherichia coli* abiotic biofilm formation: Effect of nutrients and growth conditions. *Antibiotics (Basel, Switzerland)*, *10*(11), 1349. https://doi.org/10.3390/antibiotics10111349

Peng, Q., Tang, X., Dong, W., Sun, N., & Yuan, W. (2022). A review of biofilm formation of *staphylococcus aureus* and its regulation mechanism. *Antibiotics (Basel, Switzerland)*, *12*(1), 12. https://doi.org/10.3390/antibiotics12010012

Plotkin, B. J., Halkyard, S., Spoolstra, E., Micklo, A., Kaminski, A., Sigar, I. M., & Konaklieva, M. I. (2023). The role of the insulin/glucose ratio in the regulation of pathogen biofilm formation. *Biology*, *12*(11), 1432. https://doi.org/10.3390/biology12111432

Rawal, L. B., Tapp, R. J., Williams, E. D., Chan, C., Yasin, S., & Oldenburg, B. (2012). Prevention of type 2 diabetes and its complications in developing countries: A review. *International Journal of Behavioral Medicine*, *19*(2), 121–133. https://doi.org/10.1007/s12529-011-9162-9

Ritter, S. (2017). Monitoring and maintenance of brain glucose supply: Importance of hindbrain catecholamine neurons in this multifaceted task. *Appetite and Food Intake*, 177–204. https://doi.org/10.1201/9781315120171-9

Shang, R., & Gao, L. (2021). Impact of hyperglycaemia on the rate of implant failure and peri-implant parameters in patients with type 2 diabetes mellitus: Systematic review and meta-analysis. *The Journal of the American Dental Association*, *152*(3), 189–201. https://doi.org/10.1016/j.adaj.2020.11.015

Sharma, D., Misba, L., & Khan, A. U. (2019). Antibiotics versus biofilm: An emerging battleground in microbial communities. *Antimicrobial Resistance and Infection Control*, *8*, 76. https://doi.org/10.1186/s13756-019-0533-3

Shomali, N., Mahmoudi, J., Mahmoodpoor, A., Zamiri, R. E., Akbari, M., Xu, H., & Shotorbani, S. S. (2021). Harmful effects of high amounts of glucose on the immune system: An updated review. *Biotechnology and Applied Biochemistry*, *68*(2), 404–410. https://doi.org/10.1002/bab.1938

Soria, B., Quesada, I., Ropero, A. B., Pertusa, J. A., Martín, F., & Nadal, A. (2004). Novel players in pancreatic islet signaling: from membrane receptors to nuclear channels. *Diabetes*, *53*(Suppl 1), S86–S91. https://doi.org/10.2337/diabetes.53.2007.s86

Tao, Z., & Cheng, Z. (2023). Hormonal regulation of metabolism-recent lessons learned from insulin and estrogen. *Clinical Science (London, England: 1979)*, *137*(6), 415–434. https://doi.org/10.1042/CS20210519

Uruén, C., Chopo-Escuin, G., Tommassen, J., Mainar-Jaime, R. C., & Arenas, J. (2020). Biofilms as promoters of bacterial antibiotic resistance and tolerance. *Antibiotics (Basel, Switzerland)*, *10*(1), 3. https://doi.org/10.3390/antibiotics10010003

Zeigerer, A., Sekar, R., Kleinert, M., Nason, S., Habegger, K. M., & Müller, T. D. (2021). Glucagon's metabolic action in health and disease. *Comprehensive Physiology*, *11*(2), 1759–1783. https://doi.org/10.1002/cphy.c200013

Zhang, C., Tian, R., Dreifus, E. M., Hashemi Shahraki, A., Holt, G., Cai, R., Griswold, A., Bejarano, P., Jackson, R., Schally, A. V., & Mirsaeidi, M. (2021). Activity of the growth hormone-releasing hormone antagonist MIA602 and its underlying mechanisms of action in sarcoidosis-like granuloma. *Clinical & Translational Immunology*, *10*(7), e1310. https://doi.org/10.1002/cti2.1310

Zhao, A., Sun, J., & Liu, Y. (2023). Understanding bacterial biofilms: From definition to treatment strategies. *Frontiers in Cellular and Infection Microbiology*, *13*, 1137947. https://doi.org/10.3389/fcimb.2023.1137947

11 Sex Hormones and Biofilm

Poulami Sarkar, Niladri Sarkar, and Sumit Nath
Department of Biotechnology, Maulana Abul Kalam Azad University of Technology, Kolkata, India

Swarnadeep Sarkar
Department of Life Science, Maulana Abul Kalam Azad University of Technology, Kolkata, India

Rina Rani Ray
Department of Biotechnology, Maulana Abul Kalam Azad University of Technology, Kolkata, India

11.1 INTRODUCTION

Sex hormones are steroid hormones which are indispensable for sexual dimorphism, that is, the development of primary and secondary sexual characteristics. They are mainly characterized into three major types: androgens, estrogens, and progestogens. These hormones act as a secondary messenger for the complex interplay of physiological processes throughout the human lifespan, from embryonic development to aging. Reduced or over-production of sex hormones and the regulation of signaling can lead to severe health issues. The production of sex hormones is essentially regulated by the hypothalamic–pituitary–gonadal (HPG) axis. The hypothalamus discharges gonadotropin-releasing hormone (GnRH), which manipulates the secretion of the pituitary gland to release luteinizing hormone (LH) and follicle-stimulating hormone (FSH). LH activates the production of testosterone in males and theca cell activity in females, while FSH stimulates spermatogenesis in males and granulosa cell function in females. Sex hormones, in turn, exhibit a negative feedback mechanism on the hypothalamus and pituitary gland, regulating the HPG axis and thereby maintaining hormonal balance. With the exception of those afflicted by genetic disorder, hormonal regulation depends on the age, lifestyle, environment and composition of microbiota existing in different parts of the body, the formation of biofilm and the extracellular polymeric substances.

Biofilm is a multilayer three-dimensional structural formation of bacteria that can adhere to living and non-living surfaces such as uterus, vagina, prostate glands like any other organs and medical equipment etc. It is a complex structural blanket of microorganisms formed in order to facilitate their intercommunication for transfer of nutrients and genes and to deal with environmental stress (Ghosh et al., 2022). Bacterial exopolysaccharides or extracellular polymeric substances (EPS) help in the formation of several layers of biofilm, offering protection from environmental stress and antibiotic treatments as it prevents the penetration of external molecules to the biofilm layer. The generation and development of biofilm occurs in a cyclic process, in which, initially, the planktonic bacterial cells become attached to the biotic and abiotic surfaces and uptake an ample amount of nutrients to mature and develop, forming several layers of biofilm. After the formation of multiple layers of biofilm, the sessile cells are dispersed and seek out a new surface. Recent studies have

DOI: 10.1201/9781003519041-11

shown the effects of sex hormones on biofilm-bound recalcitrant microbial cells. On the other hand, biofilm formation with consequent antibiotic resistance leads to the inception of various diseases, including sexually transmitted diseases, in which the growth and formation of the biofilm is regulated by the composition of the microbiome, hormone profiles and immune status.

All steroids demonstrate a concentration-specific dual functionality; they not only promote the expression of traits linked to virulence, such as biofilm development, but they also offer initial defense against chronic infections associated with biofilms (Patt et al., 2018).

This chapter describes the relationship between sex steroids and biofilm, with special attention being given to the effects of these hormones on biofilm-mediated important diseases.

11.2 HUMAN SEX HORMONES

Sex steroid hormones (SSHs) are the major biochemicals synthesized in the human body that are essential for the growth and maintenance of the human skeletal system in both males and females. SSHs regulate a variety of activities in the circulatory, immunological, muscular and nervous systems. They also influence the prenatal and postnatal development of different brain areas, including those linked with critical physiological, behavioral, cognitive and emotional functions. This action can be mediated by both intracellular and transmembrane receptors. There are three principal categories of SSHs, including androgens, estrogens and progestogens, as shown in Figure 11.1. Androgens are recognized as primary "male sex hormones" due to their masculinizing effects, whereas estrogens and progestins are known as primary "female sex hormones" (Pillerová et al., 2021).

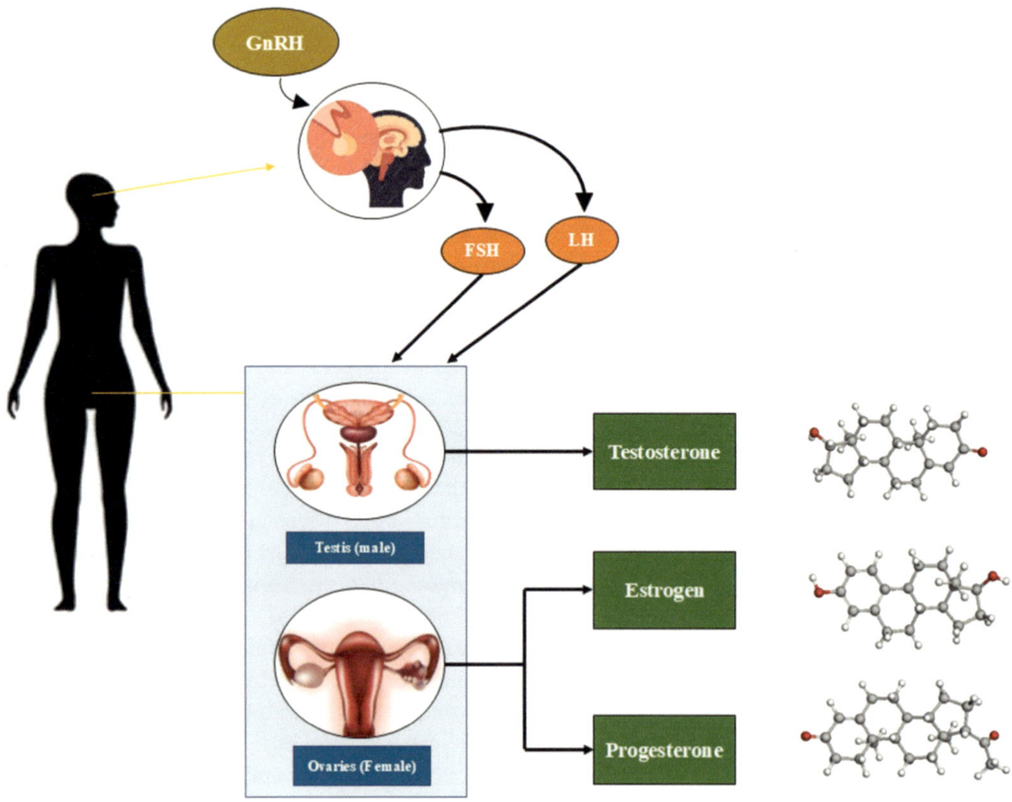

FIGURE 11.1 A diagrammatic representation of sex hormones sourced from their endocrine glands.

11.2.1 Testosterone

Gonadotropin-releasing hormone (GnRH), secreted by the hypothalamus stimulates the release of exuding of luteinizing hormone (LH) and follicle-stimulating hormone (FSH) from anterior pituitary. In males, LH can bind to the receptors present in the Leydig cells of testis, activating the conversion of cholesterol into testosterone. Testosterone, the main male sex steroids, control the primary and secondary sexual characteristics, including the formation of sperm, development of muscle mass, body and facial hair, deep voice, strong bones, and so on. FSH, primarily working on Sertoli cells of the seminiferous tubules (Smith et al., 2021), promotes spermatogenesis and helps in binding testosterone with Androgen Binding Protein (ABP).

11.2.2 Estrogen

The ovaries are the primary origin of estrogen in females during their reproductive periods. Ovaries containing follicles, one of which matures to become a Graafian follicle under the impact of FSH released from the pituitary. This hormone, along with LH, regulates ovulation. Being stimulated by the LH, androstenedione is synthesized inside the theca cells from cholesterol, which has been taken up to the granulocyte cells of ovaries and is converted to estrone (E1) and then to estradiol (E2). Estrone and estradiol are the most important and copious estrogens in premenopausal women. In addition, adrenal glands, and peripheral tissues, especially adipose tissue, skin, and brain, can orchestrate estrogen through the help of aromatase. This enzyme can modify androgens or androstenedione, to estrogens. During pregnancy, the placenta also plays a crucial role in the production of estrogen. The role of estrogen in the female body is to maintain the menstrual cycle and the development of reproductive organs such as uterus and ovaries. The secondary sexual characteristics, such as well-developed breasts, fatty hips, fat deposition in adipose tissues, and softening of skin, are influenced by estrogen (Simpson, 2003).

11.2.3 Progesterone

Progesterone is a dominant hormone derived from multiple organs, playing a crucial role in influencing overall levels at different phases of the life cycle and reproduction. The corpus luteum is the key source during the luteal phase and early pregnancy, while the placenta contributes as the controlling producer in later gestation. On the other hand, adrenal glands and testes in males are also considered secondary sources of progesterone. Unpopularly, progesterone is present in males in much smaller amounts as a precursor of testosterone synthesis. Progesterone plays a vital role in the development of several female secondary sexual characteristics as well as inhibiting lactation before parturition (Nagy et al., 2021). Moreover, the blood released during menstruation and capillary growth of the uterus wall are also functioned by progesterone.

11.3 CHEMISTRY OF SEX HORMONES

Steroidogenesis, the manufacture of all steroid hormones, including sex steroids (androgens, estrogens, and progesterone), begins with a common precursor molecule, cholesterol, and involves complex biochemical pathways in mitochondria. Cholesterol, a 27-carbon compound with a cyclopentanoperhydrophenanthrene ring, is the main component for the synthesis of all kinds of sex steroid hormones. Its synthesis is regulated by the sterol regulatory element binding transcription factor 1 (Srebf 1) and 2 (Srebf 2) genes, as shown in Figure 11.2. The resulting sex steroids are androgens (C 19) and estrogens (C18).

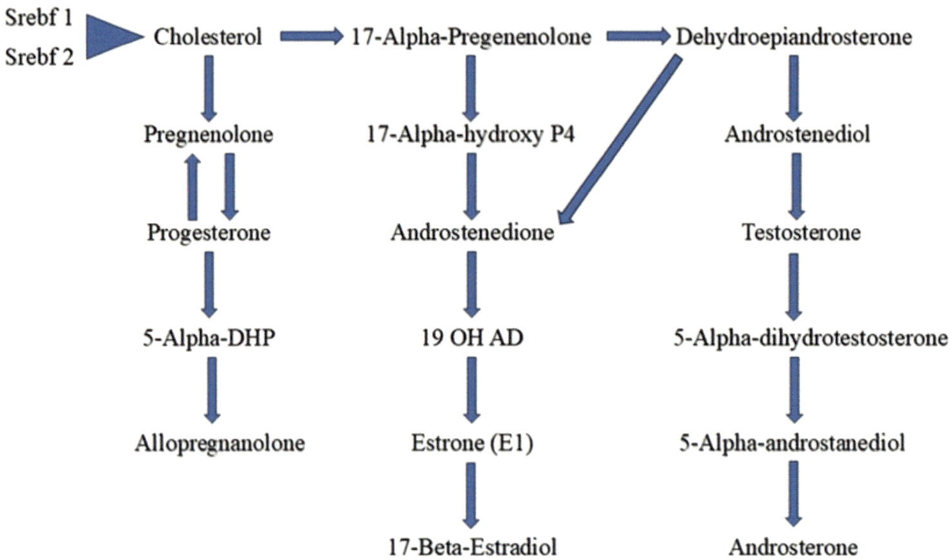

FIGURE 11.2 A flow chart representation of biosynthesis of sex hormones in human.

11.4 SEX STEROIDS AND MICROBIOTA

In recent studies, there has been a growing appreciation that microbiota plays an important role in modulating metabolism, sex steroid (hormone) production, and the immune system. Though it has a potent role, some sex steroids, such as estrogen (estrone, estradiol, and estriol), are vulnerable to rapid degradation in the liver (Santos-Marcos et al., 2023). Sex steroid estrogen plays a direct role on the gut microbiota metabolism through steroid receptors, particularly the estrogen receptor beta. Estrogen usually results in conformational changes when it interacts with nuclear receptors. The central and peripheral changes related to irritable bowel syndrome (IBS) has been regulated by the immune regulation and permeability of the intestinal barrier. According to epidemiological research, IBS is more common in women, which suggests that sex hormones might play a role in its etiology. It has also been noted that the length of GI transit varies with the menstrual cycle, pregnancy, and the postpartum period. Faecal microbiota abundance and α-diversity were linked to total urine estrogen levels in both male and postmenopausal females. Faecal *Clostridia taxa* and some *Ruminococcaceae*, which produce β-glucuronidase, were also linked to non-ovarian estrogens. Faecal estrogens and faecal β-glucuronidase had an inverse relationship. On the other hand, neither the faecal microbiota nor the enzymes were associated with premenopausal female estrogen levels. It has been demonstrated that the gut microbiota and immune cells are directly impacted by estrogen and testosterone. The production of IL-12 and IFN-γ by dendritic cells is influenced by β-estradiol. As a result, pro-inflammatory cytokine pathways are therefore activated. Polyclonal B cells are activated and B cell survival is prolonged by estradiol. Gut microbiota migrate into the lamina propria as a result of the ensuing pro-inflammatory milieu and altered intestinal gut permeability, which further encourages inflammatory processes. There has been research on the connection between microbiome and androgen. Unconjugated and glucuronidated testosterone levels were assessed in an animal investigation (Yoon and Kim, 2021). Faecal levels of unconjugated dihydrotestosterone (DHT) were 70 times greater than serum levels in young adult males. Mice without germs had very low quantities of free DHT in their distal colon, but substantial levels of glucuronidated testosterone and DHT. They came to the conclusion that intestinal metabolism and androgen glucuronidation are influenced by the gut microbiota. Testosterone inhibits the growth of T cells in males

(Colldén et al., 2019). Unlike estradiol, testosterone does not change the intestinal barrier. There is a correlation between gut microbiota and blood testosterone levels. Additionally, male castration reversed the sex difference, confirming the impact of androgens on microbiota. A disruption in the gut microbiota is linked to androgen excess, which is exemplified by polycystic ovarian syndrome (PCOS). A high-fat diet (HFD) combined with 5α-Dihydrotestosterone (DHT) may reduce the variety of gut bacteria. According to a human investigation, *Streptococcus*, *Escherichia/Shigella*, and *Bacteroides* were the co-abundance bacterial groups that were elevated in PCOS. They were shown to be positively correlated with body mass index and testosterone, but conversely related to ghrelin. On the other hand, PCOS caused a decrease in Akkermansia and *Ruminococcaceae*, which also had the opposite effects on body weight, sex hormone, and brain–gut peptides (Liu et al., 2017).

The gut microbiota has a contribution in modifying host steroids, but more research is required in this field. Progesterone is found to enrich the microbiota, especially inducing the growth of Bifidobacterium, in the late stages of pregnancy (Nuriel-Ohayon et al., 2019). The gut microbiome composition could significantly predict plasma progesterone levels. New research has shown that several gut microorganisms can synthesize the neuro steroid allopregnanolone (a derivative of progesterone) (Greenhill, 2024).

On the other hand, research suggests a complex interplay between testosterone and the gut microbiome, with testosterone levels potentially influencing gut microbiome composition and diversity, and gut microbes potentially impacting testosterone metabolism and levels. It is found that the microbial diversity of microbiota is associated with the testosterone level of a man, although certain families from the Actinobacteria, Proteobacteria, Firmicutes, and Verrucomicrobia phyla were negatively correlated with testosterone level (Moadi et al., 2024).

11.5 SEX HORMONES AND DIMORPHISM IN MICROBIOTA

The human gut is considered to be one of the most suitable places for trillions of bacteria, archaea, fungi, viruses, and other microorganisms, collectively forming a system of gut microbiota. The different dynamic nature of the microbiota is attributed to the age, diet, genetic aberration, and sexual characteristics of the host. Even the amount of gut microbiota present differ from male to female. It is observed in various studies that there are some sex differences in overall gut microbial profiles between males and females (McGee and Huttenhower, 2021).

The compositional differences of the gut microbiota are even involved in the alteration of functionality, including sexual metabolic pathways and immune responses (Zhang et al., 2021). Dimorphism, especially in the involvement of the sex steroids, plays a pivotal role for understanding the personalized hormonal status and immunity status. A gender preference by the microbiota is also observed. For instance, males contain a lower amount of Bacteroidetes compared to females while an abundance of Firmicutes is observed in males. *Lactobacillus*, *Bifidobacterium*, and *Prevotella*, some of the other genera also showing differences in quantities depending upon sex differences (Kim et al., 2020).

Moving further into integrative variations, the metabolism associated with sex-specificity has been shown as a differential mechanism due to microbiota present, including bile acid metabolism, the degradation of amino acids, short chain fatty acid (SCFA) formation etc. Several studies reveal the correlation of bile acid metabolism with sex dimorphism and gut microbiome composition. The secondary bile acid production and inhibition of bile production is regulated by the gut microbiota in the liver while, simultaneously, the expression system and bile acid levels can inhibit the overgrowth of microbiota. Several important expression systems, such as fibroblast growth factor 15 (FGF15) and cholesterol 7α-hydroxylase (CYP7A1) produced in the ileum and in the liver, respectively, via farnesoid X receptor (FXR) dependent pathway can suppress the bile acid production in liver. Interestingly, the lower levels of bile acid again promotes the nature of the gut microbiome composition whereas the FXR reduces the colonization of the bacteria (Golonka et al., 2024). Apart from playing a pivotal role in reproduction, human sex hormones, primarily estrogen and testosterone, are also involved in bone metabolism and immune functions. Testosterone levels and functionalities are, to some degree, passively dependent on the composition of gut microbiota as these sex steroids are synthesized from bile acid, which is quantitatively and structurally modified by the microbiota. Furthermore, the interference of testosterone in metabolic pathways, such as carbohydrate and lipid metabolism, intestinal motility and immune responses, can indirectly improve the accessibility of nutrients for gut microbes. Androgen receptor (AR), which is activated by testosterone, may stimulate the immunosuppressive activity, and facilitates the modification of an environment to make it suitable for the gut microbial growth and proliferation, specifically for some bacterial taxa such as *Sutterella* and *Ruminococcus*. Some recent studies have shown how the supplementation of testosterone in mice in the case of castration to plays an active role in shaping the gut microbial landscape (Cohen et al., 2021). Table 11.1 describes how the levels of sex steroids affect the microbiome composition.

As sex hormones, estrogens are potentially involved in directing the bacterial growth, biofilm formation and sometimes the virulence factor is also introduced directly. Bacterial metabolism is regulated by the Estrogen Receptor-β, (ERβ), which is supposed to depend on the sex hormone levels. Moreover, it has long been recognized that, for a long time, estradiol and progesterone uptake facilitates an increase in the vitamin K level. This promotes the growth and multiplication of the anaerobic bacteria *Prevotella Intermedius*, which causes a series of periodontal diseases due to higher levels of estradiol, resulting in the increase of protein and polysaccharides which both introduces and promotes biofilm formation. *Dipeptidyl peptidase IV (DPPIV)*, a protease secreted by the bacteria *Prevotella intermedius*, acts as a virulence factor and plays a vital role in the destruction of periodontal tissues (de Freitas et al., 2016). The ability of the estradiol to modulate protease activity has been ascertained. Again, gut microbiome composition is directly associated with the substrate levels and energy generation which is mediated by the sex hormone concentration. A large number of gut microbiota produces GUSB enzymes that act as a catalytic molecule in the production of gluconic acid from sex steroids, which are conjugated in the liver and release carbon molecules which facilitate the growth of the intestinal bacteria as an energy source. Various research findings show that activation of the sex hormone receptors (SHR) through sex steroids can also affect the gut microbiome composition, as shown in Figure 11.3. The composition, organization, and function is strongly dependent on the gastrointestinal cellular barrier, composition, permeability etc. Estrogen receptor -β (ER-β) is the abundant estrogen receptor in the human gut epithelial cell, which is

TABLE 11.1

Effect of Hyper and Hypo Synthesis of Sex Hormones on Bacterial Abundance of Gut Microbiota ↑: Increased Secretion, ↓: Decreased Secretion

Hormones	Effects on Gut	Effects on Gut Microbiota	References
Estrogen ↑ (Hyperestrogenism)	Increase growth rate of beneficial bacteria, enhance microbial diversity, anti-inflammatory effect, gut barrier improvement.	*Lactobacillus, Bifidobacterium*	**Baker et al. (2017)**
Estrogen ↓ (Hypoestrogenism)	Decrease microbial diversity, propagation of pathogenic bacteria, weekend gut barrier, reduced short chain fatty acid (SCFA) production.	*Proteobacteria, Clostridium*	**Baker et al. (2017)**
Testosterone ↑ (Hyperandrogenism)	Increase of firmicutes, reduction in diversity, enhance metabolic production, anti-inflammatory effect.	Bacteroidetes, firmicutes	**Moadi et al. (2024)**
Testosterone ↓ (Hypoandrogenism)	Reduce beneficial bacteria, induce pro-inflammatory effect, reduce microbial diversity, reduction of short chain fatty acid (SCFA) production.	Proteobacteria, Enterobacteria	**Moadi et al. (2024)**

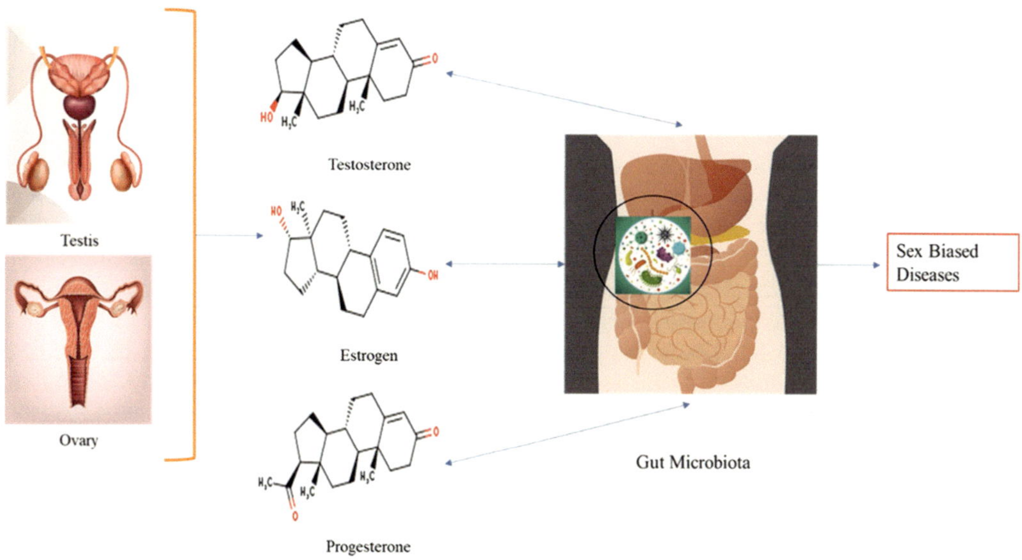

FIGURE 11.3 Schematic representation of the cross-talk between sex hormones (SH) and gut microbiota (GM) and the sex-associated diseases.

considered to be an important factor in maintaining the construction and organization of the epithelial cell. The experimental findings regarding the ER-β knockout mice indicate that their gut microbiome composition is more pronounced compared to that of the wild type mice, as the knockout exhibited several pre-pathogenic traits, including damage to cell-to-cell tight junctions, which significantly impacts the permeability of the gut epithelial layer.Systemic inflammation is caused by the default of immune cells located in the gastrointestinal tract associated with gut cellular lymphoid tissues. The illness caused by diseases like Crohn's disease (CD), ulcerative colitis (UC), multiple sclerosis (MS), and rheumatoid arthritis (RA) (Forbes et al., 2018), which are directly related to the immune mediated inflammation, is reflected by the gut microbiome composition.

Since the males and females are not equally prone to the inflammatory diseases, gut microbiome composition may also differ according to sex orientation. Estrogen and testosterone activity on immune response is completely conflicting. Estrogen is able to create a hyperactive immune environment which facilitates inflammation whereas testosterone always suppresses the T cell proliferation and immune responses so that an anti-inflammation condition is created. Similarly, progesterone eliminates the pro-inflammation properties due to the synthesis of anti-inflammatory cytokines (Polikarpova et al., 2019). Anti-inflammation is assisted by the short chain fatty acid (SCFAs) related to the gut microbiome composition. In addition, butyrate is the major metabolic product of the gut microbiota such as *Ruminococcaceae, Eubacterium, Clostridia*, and *Firmicutes* etc. is found to have an inhibitory role in inflammation as it promotes the suppression of NF-κB activity and the transcription factors of inflammation and modulates the negative mechanism in innate immune responses.

11.6 EFFECTS OF SEX STEROIDS ON GROWTH AND BIOFILM FORMATION

Hormones are bioactive chemicals with low molecular weight, heterogeneous non-nutrient substances which are produced in lower amounts and act as an intercellular signaling molecule. Sex hormones, cholesterol-derived molecules, are responsible for the regulation of various physiological processes in the human body. Bacteria principally utilize the sex steroid hormones as an alternative to vitamin K, gradually leading to induction and an increase in the virulence factor through the formation of biofilm. Biofilms are formed by a distinctly hydrated mucous layer consisting of proteins, polysaccharides and nucleic acid and collectively called an extracellular polymeric substance

(EPS). Almost 80% of the pathogenic microorganisms form biofilm and reside on the mucosal surface of the epithelia, enhancing resistance against antibiotics through biofilm-induced tolerance. Considering the disadvantages of biofilm formation, the sex steroids can be used and observed for their impacts on biofilm formation (Cornejo Ulloa et al., 2023). Recent studies reveal various effects on gut microbiota exerted by the sex hormones. Progesterone reduced the biofilm development of tested Gram-positive bacteria, while boosting biofilm formation in Gram-negative bacteria. It increased the adhesion of *Limosilactobacillus reuteri* to Caco-2/HT-29 cell line coculture. and *B. fragilis* adhesion to the co-cultured cell lines (Wasfi et al., 2023). On the other hand, estradiol (estrogen derivative) had an antibiofilm effect against *L. reuteri* (Hammouda et al., 2023). Sex steroid estradiol showed a variable impact on biofilm formation in patients suffering from cystic fibrosis (CF) caused due to *Pseudomonas aeruginosa* infection. Another study unveiled that progesterone and estradiol showed variation in the growth of biofilm of *Staphylococcus aureus*, on different kinds of strains such as *methicillin-resistant Staphylococcus aureus* (MRSA) and *methicillin-susceptible S. aureus* (MSSA), respectively. The lower concentration of progesterone and the higher concentration of estradiol (estrogen derivatives) can enhance the growth and biofilm formation of MRSA while the higher concentration of progesterone may inhibit the growth of *methicillin-susceptible S. aureus* (MSSA) (Kalaycı-Yüksek et al., 2023).

Actinobacillus seminis, the causal agent of epididymitis, showed a significant reduction of biofilm formation (nearly 32%) when exposed to estradiol while testosterone showed no effect in this case (Ramírez-Paz-y-Puente et al., 2023) The effect of sexual hormones on the proliferation and production of virulence factors in different microbes appears to be important for its colonization and persistence in the host.

11.7 PROBABLE MODE OF ACTION OF SEX STEROIDS ON GROWTH OF BIOFILM

The exact mode of action of sex hormones on the growth and development of biofilm is still under research. Indeed, experiments have shown several models to understand the mode of action of sex hormones on biofilm generation; nevertheless, some specific mechanisms are considered as a probable operational mechanism which includes some direct and indirect interaction, as described in Figure 11.4.

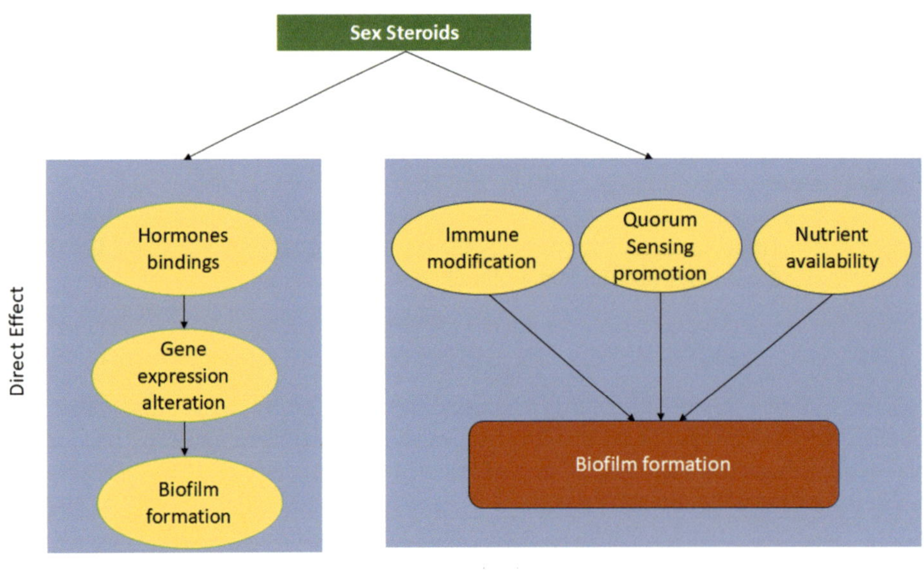

FIGURE 11.4 Representation of the mode of action of sex hormones on biofilm formation.

11.7.1 Direct Interaction

Unlike eukaryotes, bacteria contain protein receptors rather than hormone-specific receptors. The protein receptor may interact with the sex steroids, resulting in binding the regulatory system to transcription factors that change the structural characteristics and DNA-binding affinity and simultaneously transmogrifying the activity of metabolic enzymes. The levels of hormones, along with their binding affinity, influenced bacterial growth. In the case of vaginosis (BV), the bacteria *G. vaginalis* shows higher cauterization and biofilm development during pregnancy, when the progesterone levels remain higher since progesterone may directly stimulate the metabolic pathways and gene expression of the *G. vaginalis*. Again, Cutibacterium acnes, a microbe found mainly in the pilosebaceous unit, exerts its pathogenicity for the development of acne vulgaris, which is directly influenced by androgens, estrogens, progesterone (Elsaie, 2016).

11.7.2 Indirect Interaction

Apart from some direct interaction with the bacterial metabolic system, sex hormones are involved in several subordinary influences upon biofilm formation inside the reproductive system. In this context, quorum sensing, a potential intercommunication system of typical bacteria, produces diverse signalling molecules called autoinducers. Gram-positive and Gram-negative bacteria both potentially form various signals during quorum sensing, facilitating the development of the biofilm. Many studies revealed that sex steroids have an activity over the bacterial quorum-sensing system. The quorum-sensing receptor protein can sometimes interact with sex steroids by which the autoinducers produced by the bacteria may be altered, leading to the modification in the structural conformation and DNA-binding ability of the receptor, and resulting in the altered gene expression. Recent studies showed that testosterone can affect the quorum-sensing system of *Staphylococcus aureus* via affecting the 'agr' autoinducer production, which is a leading cause of distortion of virulence factors such as toxic shock syndrome Toxin -1(TSST-1) and Protein-A (John et al., 2024). *In silico* analysis such as protein docking, molecular docking simulations, and network analysis also confirmed that testosterone and estradiol can interact with proteins associated with biofilm formation. Other studies have shown the effect of sex hormones over autoinducer concentration. Estrogens, specially estradiol (E_1), enhance the concentration of autoinducers, acyl-homoserine lactones (AHL), of *Pseudomonas aeruginosa* whereas they inhibit pyocyanin formation (Al-Zawity et al., 2022). Another study showed that estradiol and estriol can amplify the expression of virulence factors in *Escherichia coli*, a key pathogen in UTIs as the estrogen induces an uncomfortable iron acquisition systems for bacterial biofilm development. Specific genetic encodes like *fimH*, encoding for type 1 fimbriae, which helps to adhere to the urinary tract epithelium at that time (Engelsöy et al., 2021).

Women are more prone to UTIs, for which estrogen is considered as one of the leading causes as it enhances the metabolic activity of the uropathogenic *Escherichia coli*. Androgens, such as testosterone and dihydrotestosterone (DHT), increase the sebum production, furnishing a nutrient-rich environment for *C. acnes* growth and development. Furthermore, androgens have a potential influence on *C. acnes* gene expression, the generation of inflammatory mediators for acne development. Several studies indicate that some bacterial protein receptors may counteract with sex hormones, thereby altering the signaling pathways accordingly to promote bacterial biofilm formation. Other than that, the modification of the cell membrane of the pathogenic bacteria, along with increasing permeability, viscosity and protein activity, and virulence factors due to the lipophilic nature of the sex hormones, facilitate nutrient availability and signal transduction for bacterial colonization.

11.8 EFFECTS OF SEX STEROIDS ON BIOFILM MEDIATED DISEASES

Sex steroids, primarily estrogen, progesterone, and testosterone, play a various roles in the regulation of various physiological processes, like microbial interactions and immune responses. The influence of these sex steroids gets further extended to biofilm-associated diseases (Table 11.2),

TABLE 11.2
The Summarized Presentation of Microbial Diseases Influenced by Sex Steroids

Disease	Example of Pathogens	Hormonal Influence	Impact on Disease	References
Urinary Tract Infections (UTIs)	*E. coli, Klebsiella pneumoniae*, and *Enterococcus faecalis*.	Estrogen strengthens epithelial barriers & maintains *Lactobacillus* balance.	Low estrogen (postmenopause) weakens defenses, allowing more biofilm formation.	**Govindarajan and Kandaswamy (2022)**
Bacterial Vaginosis (BV)	*Gardnerella vaginalis, Atopobium vaginae*	Estrogen promotes *Lactobacillus*, which produces lactic acid to suppress biofilms.	Low estrogen alters vaginal microbiota, promoting biofilm-associated infections.	**Machado et al. (2016)**
Acute Bacterial Prostatitis	*Escherichia coli, Klebsiella pneumoniae*, and *Pseudomonas aeruginosa*	Testosterone promotes inhibition of inflammation	Inflammation in the prostate, abnormal secretion, pain.	**Barone et al. (2023)**
Periodontitis	*Porphyromonas gingivalis, Treponema denticola, Fusobacterium nucleatum*	Estrogen & progesterone alter gum tissue permeability & immune response.	Hormonal fluctuations (pregnancy, menopause) increase biofilm persistence & inflammation.	**Palanisamy (2025)**
Chronic Wounds (Diabetic Ulcers)	*Staphylococcus aureus, Pseudomonas aeruginosa, Enterococcus* spp.	Estrogen promotes angiogenesis & healing, while low estrogen slows repair.	Impaired wound healing due to hormonal imbalances allows biofilm persistence.	**Baba et al. (2017)**
Cystic Fibrosis Lung Infections	*Pseudomonas aeruginosa, Burkholderia cepacia*	Estradiol enhances *Pseudomonas* biofilm growth & virulence factors.	Women with CF experience worse lung infections due to estradiol-driven biofilm changes.	**Kroncke et al. (2016)**
Gastrointestinal Infections	Helicobactor pylori	Estrogen has a protective effect against *H. pylori* colonization.	Low estrogen levels may increase susceptibility to *H. pylori* infection.	**Baba et al. (2017)**

in which microbes adhere to a surface, forming a structural layer of sessile cells encased in a self-produced extracellular matrix (i.e., biofilm formation). This protective biofilm structure creates a defense system for infections. Thus, it works as a shield from antibiotics an host immune responses, becoming a challenge in treatment.

Biofilms are implicated in over 80% of microbial infections in the body, a finding which highlights their clinical significance. Sex steroids can influence this biofilm formation, composition, behavioral changes, and quorum sensing (cell-to-cell communication system) through various mechanisms, significantly affecting the severity and progression of biofilm-associated diseases. For instance, estrogen can enhance the expression of adhesion molecules in certain bacteria, promoting biofilm formation, such as through the enhancement of alginate production in *Pseudomonas aeruginosa*, crucial for biofilm formation. Testosterone, on the other hand, may modulate the host immune response, indirectly affecting biofilm development. It can reduce natural killer cell activity and alter cytokine synthesis, impacting the immune system's ability to combat biofilm-forming pathogens. Similarly, hormonal fluctuations during pregnancy, menopause, or contraceptive usage can alter the vaginal microbiome, increasing susceptibility to infections like bacterial vaginosis. These hormonal effects contribute to the gender disparities observed in the prevalence and severity of several biofilm-associated diseases, such as UTIs, polycystic ovaries, cancer etc.

11.8.1 Urinary Tract Infections

Urinary tract infections (UTI) are one of the most quotidian maladies in developing countries. Daily food habits, a deficiency of proteins, suppressed immune systems, multidrug resistant bacterial pathogenicity, and biofilm formation are among the key causes of such ailments. The risk factors are more frequently found (40–60% once in their lifetime) than their male counterpart (15–20% only) in females (Santos-Marcos et al., 2023). Globally, the number showed 150 million citizens every year, making UTI the second-most biofilm-mediated chronic disease worldwide. Fatal and recurrent UTI are caused mostly by bacteria of the Enterobacteriaceae family, where the biofilm acts as a safeguard, making the bacteria resistant to the multiple drugs. Vaginal estrogen is considered to prevent the occurrence of UTI. Tan-Kim et al., 2023 worked with 5600 hypoestrogenic females, suffering from recurrent UTI. The study found that the application of vaginal estrogen could bring about a decrease in UTI reoccurrence. According to the study, estrogen must have an influence on affecting biofilm formation and the removal of persister bacterial cell causing UTI. Progesterone, on the other hand, does not have any straightforward role, although it may increase the risk for UTI. Testosterone (Wu et al., 2024) has the ability to alter the immune response in both males and females, thereby aggravating for UTI. Hence both the sexes, with hyper-testosterone level, develop chronic cystitis, a bacterial biofilm-mediated disease of persistent bladder inflammation.

11.8.2 Polycystic Ovary Syndrome

Research indicates that an altered gut microbiota, including biofilm formation, may contribute to polycystic ovary syndrome (PCOS)-related issues such as hyperandrogenism, insulin resistance, and metabolic disorders, potentially through mechanisms involving CSFAs, lipopolysaccharides, and the gut–brain axis. Gut microbiota and PCOS are mutually connected to each other due to regulation of production and the secretion of insulin take a vital role in follicle development and sex hormone metabolism.

Firmicutes comprising *Lactobacillus, Clostridium*, and *Ruminococcus* are abundantly present with a diverse variety in the human gut area. *Lactobacillus* has extensively reported in many researches as it has a pivotal role in spreading PCOS. Obese women with an abundance of *Ruminococcaceae*, *Clostridium* and *Lachnospiraceae oribacterium* in the gut are more prone to PCOS. It is observed that *L. plantarum* is associated with isobutyric acid and isovaleric acid production, which has a close relationship with lipid metabolism (He et al., 2022). Again, in the body of postmenopausal women, *L. plantarum* is reported to have a role in reduction of glucose and C-reactive protein, resulting in inflammation in PCOS. Chronic estrogen exposure, or a lack of progesterone due to ovarian dysfunction, may initiate and aggravate this multifactorial disease.

11.8.3 Cancer

Cancer is a common genetic and multifunctional disease, influenced by the host pathogen interaction and environmental factors. Nearly 20% of cancers are linked to the composition and pathogenesis of microorganisms. *Helicobacter pylori*, a key producer of CagA protein, was first reported for its involvement in the development of cancer (Hatakeyama et al., 2017). The biofilm-bound persister cells of *H. pylori* is presumed to be responsible for noncardia gastric cancer. Although the frequency of gastric cancer is higher in males, the exact role of sexual difference cannot be explained. Although estrogen is presumed to play a protective role, experimental results did not reveal any significant relation between sex steroids (testosterone and estradiol) with gastric cancer (Leal et al., 2021). The metabolic manipulation imparted by the biofilm-bound bacteria, causing impaired polyamine metabolism, induces excessive cell growth and tumor formation, leading to cancers of the breast, lung, prostate, and gut (Choi et al., 2023). Sex steroid hormones, such as estrogen and testosterone, play a significant role in the development and progression of breast, prostate, and, potentially, lung and gut cancers. Some cancers are directly hormone-dependent and, hence, a sex specific cancer therapy is practiced (Bakhshi et al., 2024).

Esophageal cancer is developed by the action of *Streptococcus viridans* (Baba et al., 2017). This forms biofilm inside the esophageal region, which may increase the risk of inflammation, caused by dysbiosis. In addition, *Campylobacter concisus*, *Campylobacter rectus*, and *Escherichia coli* have parallel contributions in the development of esophageal cancer. It has recently been found that *Staphylococcus epidermidis* biofilm is associated with the development of breast cancer, which is accomplished by the generation of an inflammatory scenario, leading to the breakage of DNA in breast tissue, and thereby encouraging cellular proliferation, angiogenesis, and genetic instability (Allen-Taylor et al., 2024). Hence, anticancer drugs are designed with biofilm-eradicating properties.

11.8.4 Acute Bacterial Prostatitis

Acute bacterial prostatitis (ABP) refers to the inflammation of the prostate gland, which is primarily caused by ascending urethral infections with Gram-negative bacteria such as *Escherichia coli*, *Klebsiella pneumoniae*, and *Pseudomonas aeruginosa*. The prostate is the most advantageous place to grow and develop biofilm as many secretions of proteins and extra polysaccharides occur which facilitate the acquisition of suitable nutrients to build up multiple layers of biofilm. It is found that testosterone may suppress uropathogenic *Escherichia coli* (UPEC) invasion and colonization, potentially impacting biofilm formation and prostatitis outcomes.

11.8.5 Vaginitis

The vagina is one of the most complex female organs. The terms for the inflammation of the vagina is simply vaginitis, and it is characterized by a polymicrobial relocation from *Lactobacillus* colonies towards the thriving of anaerobic bacteria, including *Gardnerella vaginalis*, *Atopobium vaginae*, *Mobiluncus* spp., and *Prevotella* spp. *G. vaginalis* is considered to be the primary contributor in bacterial vaginosis, introducing the cyclic pathways by adhering to the vaginal epithelium and forming a robust biofilm scaffold. The biofilm of *G. vaginalis* then promotes the co-aggregation and colonization of other bacteria related to vaginitis, leading to the formation of a complex, recalcitrant biofilm (Machado et al., 2016).

Sex steroids, particularly estrogen and progestins, play a significant role in bringing about change in the composition of normal vaginal microflora, and inducing the risk of vaginitis. Although estrogen-containing compounds may help in maintaining an optimal vaginal microenvironment, progesterone shows mixed effects (Ratten et al., 2021).

11.8.6 Cystic Fibrosis

Cystic fibrosis, an acute lung disease, is characterized by inflammation as well as chronic bacterial infection, with bacteria becoming host-adapted pathogens. The multitudinous network of a genetic disorder ascribed to difficulties of cystic fibrosis transmembrane conductance regulator (CFTR), resulting in inflammation of the lungs and severe infection along with a pathogenic attack of *Pseudomonas aeruginosa* in respiratory tracts, and pulmonary exacerbations. *Pseudomonas aeruginosa* and *Staphylococcus aureus* are among the most frequently encountered pathogens in the airways of patients with cystic fibrosis; however, other microorganisms, such as *Burkholderia* spp., *non-tuberculous mycobacteria (NTM)* spp., *Stenotrophomonas maltophilia, Achromobacter* spp., and fungi, including *Aspergillus* spp., have also been identified. Studies based on clinical data are showing that the females' CF is more treacherous than that of the male as the respective respiratory systems are quite different. Females tend to have a thinner respiratory tract, lower lung volume, and the presence of various secretions and particular hormone regulation. On the other side, sexual dysfunction is found to be associated with CF, since 30% of females and 20% of males with CF suffer from sexual dysfunction (Ramel et al., 2024). Indeed, research shows that CFTR expression can be transmogrified as in response to estrogen and progesterone. However, estrogen has adverse consequences on CFTR-independent chloride channel activity, facilitating thicker production of mucus and thereby worsening the condition. Moreover, estrogen also disrupts the interaction between KCNE3 and KCNQ1 (two potential modified genes responsible in CFTR), sequentially resisting the mucociliary secretion in CF airways (Kroncke et al., 2016).

Furthermore, estradiol (E_2) promotes biofilm formation of the *Pseudomonas aeruginosa*, *Pseudomonas citronellolisa*. The biofilm and the EPS produce selective chemicals, such as polysaccharides and alginate, which is involved in pulmonary exacerbations. The administration of progesterone suppressed CFTR expression and alleviated symptoms in this animal model.

11.8.7 Periodontitis

Periodontitis, the most common oral disease, is represented by gum inflammation and tooth decay and attributed to the formation of plaque biofilm on the dental and gingival surfaces (Ray, 2023). Although it is found that a low level of estrogen is associated with increased gingival inflammation, causing a higher incidence of periodontal disease, an increase in ovarian steroid hormones, namely estrogen and progesterone, does intensify the risk of gingival inflammation. Elevated hormone levels promote bacterial proliferation in the oral cavity, leading to a shift in the bacterial flora (Palanisamy, 2025).

Progesterone, surging vascular permeability and favoring gingival secretion, facilitates in the escalation of populations of oral biofilm-forming pathogens, including *Treponema denticola, Tannarella Forsythia*, and *Porphyromonas gingivali*. Periodontitis, a biofilm-dependent inflammatory disease, is linked to microbial dysbiosis, and research suggests that lower bioavailable testosterone levels, particularly in males, may increase the risk of developing periodontitis (Su et al., 2023).

11.9 CONCLUSION

Each micro environment inside the human body is somehow different and complicated from the other and is harbored by a unique group of microbes. These biofilms may benefit the host by fortifying host defenses, enhancing the exchange of nutrients between the microbiota and the host, and interfering with colonization by pathogenic bacteria. After pathogenic infestation and consequent dysbiosis, the functional attributes of these biofilm change. Sex steroid hormones exert different effects on a variety of microbes and play an important role in maintaining the overall homoeostasis in healthy situations.

REFERENCES

Allen-Taylor, D., et al. 'Staphylococcus Epidermidis Biofilm in Inflammatory Breast Cancer and Its Treatment Strategies'. *Biofilm*, vol. 8, December 2024, p. 100220. https://doi.org/10.1016/j.bioflm.2024.100220

Al-Zawity, J., et al. 'Effects of the Sex Steroid Hormone Estradiol on Biofilm Growth of Cystic Fibrosis Pseudomonas Aeruginosa Isolates'. *Frontiers in Cellular and Infection Microbiology*, vol. 12, July 2022, p. 941014. https://doi.org/10.3389/fcimb.2022.941014

Baba, Y., Iwatsuki, M., Yoshida, N., Watanabe, M., & Baba, H. 'Review of the Gut Microbiome and Esophageal Cancer: Pathogenesis and Potential Clinical Implications'. *Annals of Gastroenterological Surgery*, vol. 1, no. 2, 2017, pp. 99–104. https://doi.org/10.1002/ags3.12014

Baker, J. M., et al. 'Estrogen–Gut Microbiome Axis: Physiological and Clinical Implications'. *Maturitas*, vol. 103, September 2017, pp. 45–53. https://doi.org/10.1016/j.maturitas.2017.06.025

Bakhshi, P., et al. 'Sex-Specific Outcomes in Cancer Therapy: The Central Role of Hormones'. *Frontiers in Medical Technology*, vol. 6, February 2024, p. 1320690. https://doi.org/10.3389/fmedt.2024.1320690

Barone, B., Mirto, B. F., Falcone, A., Del Giudice, F., Aveta, A., Napolitano, L., Del Biondo, D., Ferro, M., Busetto, G.M., Manfredi, C., Terracciano, D., Gambardella, R., Pandolfo, S.D., Trama, F., De Luca, C., Martino, R., Capone, F., Giampaglia, G., Sicignano, E., Tataru, O. S., Lucarelli, G., Crocetto, F. The Efficacy of Flogofilm® in the Treatment of Chronic Bacterial Prostatitis as an Adjuvant to Antibiotic

Therapy: A Randomized Prospective Trial. *Journal of Clinical Medicine*, vol. 12, no. 8, 2023 April 9, p. 2784. doi: https://doi.org/10.3390/jcm12082784. PMID: 37109121; PMCID: PMC10142953

Choi, E., et al. 'Biofilm and Cancer: Interactions and Future Directions for Cancer Therapy'. *International Journal of Molecular Sciences*, vol. 24, no. 16, August 2023, p. 12836. https://doi.org/10.3390/ijms241612836

Cohen, I., et al. 'Influence of Immunomodulatory Drugs on the Gut Microbiota'. *Translational Research*, vol. 233, July 2021, pp. 144–161. https://doi.org/10.1016/j.trsl.2021.01.009

Colldén, H., et al. 'The Gut Microbiota Is a Major Regulator of Androgen Metabolism in Intestinal Contents'. *American Journal of Physiology-Endocrinology and Metabolism*, vol. 317, no. 6, December 2019, pp. E1182–E1192. https://doi.org/10.1152/ajpendo.00338.2019

Cornejo Ulloa, P., et al. 'The Effect of Sex Steroid Hormones on the Ecology of in Vitro Oral Biofilms'. *Biofilm*, vol. 6, December 2023, p. 100139. https://doi.org/10.1016/j.bioflm.2023.100139

De Freitas, G. C., et al. 'Effect of Self-performed Mechanical Plaque Control Frequency on Gingival Inflammation Revisited: A Randomized Clinical Trial'. *Journal of Clinical Periodontology*, vol. 43, no. 4, April 2016, pp. 354–358. https://doi.org/10.1111/jcpe.12520

Elsaie, M. L. 'Hormonal Treatment of Acne Vulgaris: An Update'. *Clinical, Cosmetic and Investigational Dermatology*, vol. 9, 2 September 2016, pp. 241–248. https://doi.org/10.2147/CCID.S114830

Engelsöy, U., Svensson, M. A., & Demirel, I. 'Estradiol Alters the Virulence Traits of Uropathogenic Escherichia Coli'. *Frontiers in Microbiology*, vol. 12, 2021, p. 682626. https://doi.org/10.3389/fmicb.2021.682626

Forbes, J. D., et al. 'A Comparative Study of the Gut Microbiota in Immune-Mediated Inflammatory Diseases—Does a Common Dysbiosis Exist?' *Microbiome*, vol. 6, no. 1, December 2018, p. 221. https://doi.org/10.1186/s40168-018-0603-4

Ghosh, S., Saha, I., Dey, A., Lahiri, D., Nag, M., Sarkar, T., … & Ray, R. R. 'Natural Compounds Underpinning the Genetic Regulation of Biofilm Formation: An Overview'. *South African Journal of Botany*, vol. 151, 2022, pp. 92–106.

Golonka, R. M., et al. 'Sex Dimorphic Effects of Bile Acid Metabolism in Liver Cancer in Mice'. *Cellular and Molecular Gastroenterology and Hepatology*, vol. 17, no. 5, 2024, pp. 719–735. https://doi.org/10.1016/j.jcmgh.2024.01.011

Govindarajan, D. K., & Kandaswamy, K. 'Virulence Factors of Uropathogens and their Role in Host Pathogen Interactions'. *Cell Surface (Amsterdam, Netherlands)*, vol. 8, 2022, p. 100075. https://doi.org/10.1016/j.tcsw.2022.100075

Greenhill, C. 'Progestin Production by the Gut Microbiota'. *Nature Reviews Endocrinology*, vol. 20, no. 8, 2024, pp. 446–446. https://doi.org/10.1038/s41574-024-01013-8

Hammouda, Z. K., et al. 'Hormonal Drugs: Influence on Growth, Biofilm Formation, and Adherence of Selected Gut Microbiota'. *Frontiers in Cellular and Infection Microbiology*, vol. 13, March 2023, p. 1147585. https://doi.org/10.3389/fcimb.2023.1147585

Hatakeyama, S., et al. 'Recent Progress and Perspectives on Prostate Cancer Biomarkers'. *International Journal of Clinical Oncology*, vol. 22, no. 2, April 2017, pp. 214–221. https://doi.org/10.1007/s10147-016-1049-y

He, Y., et al. 'Butylated Starch Alleviates Polycystic Ovary Syndrome by Stimulating the Secretion of Peptide Tyrosine-Tyrosine and Regulating Faecal Microbiota'. *Carbohydrate Polymers*, vol. 287, July 2022, p. 119304. https://doi.org/10.1016/j.carbpol.2022.119304

John, M. S., Chinnappan, M., Artami, M., Bhattacharya, M., Keogh, R. A., Kavanaugh, J., Sharma, T., Horswill, A. R., & Harris-Tryon, T. A. *Androgens at the Skin Surface Regulate* S. aureus *Pathogenesis through the Activation of* AGR *Quorum Sensing*, 2024. https://doi.org/10.1101/2024.02.10.579753

Kalaycı-Yüksek, F., Gümüş, D., Güler, V., Uyanık-Öcal, A., & Anğ-Küçüker, M. 'Progesterone and Estradiol alter the Growth, Virulence and Antibiotic Susceptibilities of Staphylococcus Aureus'. *The New Microbiologica*, vol. 46, no. (1), 2023, pp. 43–51.

Kim, Y. S., et al. 'Sex Differences in Gut Microbiota'. *The World Journal of Men's Health*, vol. 38, no. 1, 2020, p. 48. https://doi.org/10.5534/wjmh.190009

Kroncke, B. M., Van Horn, W. D., Smith, J., Kang, C., Welch, R. C., Song, Y., Nannemann, D. P., Taylor, K. C., Sisco, N. J., George, A. L., Meiler, J., Vanoye, C. G., & Sanders, C. R. 'Structural Basis for KCNE3 Modulation of Potassium Recycling in Epithelia'. *Science Advances*, vol. 2, no. 9, 2016, p. e1501228. https://doi.org/10.1126/sciadv.1501228

Leal, Y. A., et al. 'Circulating Levels of Sex Steroid Hormones and Gastric Cancer'. *Archives of Medical Research*, vol. 52, no. 6, August 2021, pp. 660–664. https://doi.org/10.1016/j.arcmed.2021.03.001

Liu, R., et al. 'Dysbiosis of Gut Microbiota Associated with Clinical Parameters in Polycystic Ovary Syndrome'. *Frontiers in Microbiology*, vol. 8, February 2017. https://doi.org/10.3389/fmicb.2017.00324

Machado, D., Castro, J., Palmeira-de-Oliveira, A., Martinez-de-Oliveira, J., & Cerca, N. 'Bacterial Vaginosis Biofilms: Challenges to Current Therapies and Emerging Solutions'. *Frontiers in Microbiology*, vol. 6, 2016, p. 1528. https://doi.org/10.3389/fmicb.2015.01528

McGee, J. S., & Huttenhower, C. 'Of Mice and Men and Women: Sexual Dimorphism of the Gut Microbiome'. *International Journal of Women's Dermatology*, vol. 7, no. 5 Part A, October 2021, pp. 533–538. *PubMed Central*, https://doi.org/10.1016/j.ijwd.2021.10.007

Moadi, L., et al. 'The Effect of Testosterone on the Gut Microbiome in Mice'. *Communications Biology*, vol. 7, no. 1, July 2024, p. 880. https://doi.org/10.1038/s42003-024-06470-5

Nagy, B., et al. 'Key to Life: Physiological Role and Clinical Implications of Progesterone'. *International Journal of Molecular Sciences*, vol. 22, no. 20, October 2021, p. 11039. https://doi.org/10.3390/ijms222011039

Nuriel-Ohayon, M., Neuman, H., Ziv, O., Belogolovski, A., Barsheshet, Y., Bloch, N., Uzan, A., Lahav, R., Peretz, A., Frishman, S., Hod, M., Hadar, E., Louzoun, Y., Avni, O., & Koren, O. 'Progesterone Increases Bifidobacterium Relative Abundance during Late Pregnancy'. *Cell Reports*, vol. 27, no. 3, 2019, pp. 730–736. https://doi.org/10.1016/j.celrep.2019.03.075

Palanisamy, S. 'The Impact of Estrogen on Periodontal Tissue Integrity and Inflammation—A Mini Review'. *Frontiers in Dental Medicine*, vol. 6, February 2025, p. 1455755. https://doi.org/10.3389/fdmed.2025.1455755

Patt, M. W., Conte, L., Blaha, M., & Plotkin, B. J. 'Steroid Hormones as Interkingdom Signaling Molecules: Innate Immune Function and Microbial Colonization Modulation'. *AIMS Molecular Science*, vol. 5, no. 1, 2018, pp. 117–130. https://doi.org/10.3934/molsci.2018.1.117

Pillerová, M., et al. 'On the Role of Sex Steroids in Biological Functions by Classical and Non-Classical Pathways. An Update'. *Frontiers in Neuroendocrinology*, vol. 62, July 2021, p. 100926. https://doi.org/10.1016/j.yfrne.2021.100926

Polikarpova, A. V., et al. 'Immunomodulatory Effects of Progesterone and Selective Ligands of Membrane Progesterone Receptors'. *Steroids*, vol. 145, May 2019, pp. 5–18. https://doi.org/10.1016/j.steroids.2019.02.009

Ramel, S., et al. 'Sexual Dysfunction in Cystic Fibrosis'. *Journal of Cystic Fibrosis*, vol. 23, no. 4, July 2024, pp. 647–652. https://doi.org/10.1016/j.jcf.2024.04.011

Ramírez-Paz-y-Puente, G. A., et al. 'Testosterone and Estradiol Modify the Expression of Adhesins and Biofilm Formation in *Actinobacillus Seminis*'. *FEMS Microbiology Letters*, vol. 370, January 2023, p. fnad048. https://doi.org/10.1093/femsle/fnad048

Ratten, L. K., et al. 'The Effect of Exogenous Sex Steroids on the Vaginal Microbiota: A Systematic Review'. *Frontiers in Cellular and Infection Microbiology*, vol. 11, November 2021, p. 732423. https://doi.org/10.3389/fcimb.2021.732423

Ray, R. R. 'Periodontitis: An Oral Disease with Severe Consequences'. *Applied Biochemistry and Biotechnology*, vol. 195, no. 1, January 2023, pp. 17–32. https://doi.org/10.1007/s12010-022-04127-9

Santos-Marcos, J. A., et al. 'Interaction between Gut Microbiota and Sex Hormones and Their Relation to Sexual Dimorphism in Metabolic Diseases'. *Biology of Sex Differences*, vol. 14, no. 1, February 2023, p. 4. https://doi.org/10.1186/s13293-023-00490-2

Simpson, E. R. 'Sources of Estrogen and their Importance'. *The Journal of Steroid Biochemistry and Molecular Biology*, vol. 86, no. (3–5) 2003, pp. 225–230.

Smith, S. J., et al. 'Examining the Effects of Herbs on Testosterone Concentrations in Men: A Systematic Review'. *Advances in Nutrition*, vol. 12, no. 3, May 2021, pp. 744–765. https://doi.org/10.1093/advances/nmaa134

Su, X., Jin, K., Zhou, X., Zhang, Z., Zhang, C., Li, Y., Yang, M., Huang, X., Xu, S., Wei, Q., Cheng, X., Yang, L., & Qiu, S. 'The Association between Sex Hormones and Periodontitis among American Adults: A Cross-Sectional Study'. *Frontiers in Endocrinology*, vol. 14, 2023, p. 1125819. https://doi.org/10.3389/fendo.2023.1125819

Tan-Kim, J., Shah, N. M., Do, D., & Menefee, S. A. 'Efficacy of Vaginal Estrogen for Recurrent Urinary Tract Infection Prevention in Hypoestrogenic Women'. *American Journal of Obstetrics and Gynecology*, vol. 229, no. 2, 2023, pp. 143.e1–143.e9. https://doi.org/10.1016/j.ajog.2023.05.002

Wasfi, R., Zarkan, A., & Hamed, S. M. 'Emerging Strategies in Combatting and Managing Bacterial Biofilms'. *Frontiers in Cellular and Infection Microbiology*, vol. 13, 2023, p. 1264346.

Wu, R., Pettersson, C., & Demirel, I. 'Testosterone Increases the Virulence Traits of Uropathogenic Escherichia Coli'. *Frontiers in Microbiology*, vol. 15, 2024, p. 1422747. https://doi.org/10.3389/fmicb.2024.1422747

Yoon, K., & Kim, N. 'Roles of Sex Hormones and Gender in the Gut Microbiota'. *Journal of Neurogastroenterology and Motility*, vol. 27, no. 3, July 2021, pp. 314–325. https://doi.org/10.5056/jnm20208

Zhang, C.-Y., et al. 'Gut Microbiota Comparison Between Intestinal Contents and Mucosa in Mice With Repeated Stress-Related Diarrhea Provides Novel Insight'. *Frontiers in Microbiology*, vol. 12, February 2021, p. 626691. https://doi.org/10.3389/fmicb.2021.626691

12 Enteric Hormones and Biofilm

Suhasini Mallick, Anjali Rao, and Rina Rani Ray
Department of Biotechnology, Maulana Abul Kalam Azad University of Technology, Kolkata, India

12.1 INTRODUCTION

Recent research emphasizes the interplay between enteric hormones and biofilms —which are complex communities of microorganisms that adhere to interfaces within the gastrointestinal tract, making them critically dependent on each other. The gastrointestinal tract's primary motive is to nourish the body via the consequent processes of nutritional intake with motility, and the secretion of digestive fluids, followed by digestion and absorption of the nutritional uptake. This is scrupulously synchronized and accomplished by the intrinsic endocrine and neurological framework. The gastrointestinal (GI) tract is known to be the largest endocrine organ in the human body. It produces a decent number of mediators that are required to control the GI tract's functioning, although the nervous system influences many digestive processes. The scientific community has studied several enteric hormones but has extensively pointed that ghrelin, leptin, glucagon-like peptide 1 (GLP-1), and peptide YY (PYY) are among the most significant hormones due to their roles in metabolic processes (De Silva and Bloom, 2012). The hunger hormone, ghrelin, is natively produced in the stomach and acts on the hypothalamus to stimulate appetite. Likewise, leptin is predominantly secreted by the adipose tissue. It is influenced by gut signalling and aids in regulating long-term energy balance by suppressing hunger. Released by the intestines after food intake, the GLP-1 enhances insulin secretion and slows gastric emptying. Lastly, PYY acts as an appetite suppressant, slowing gut motility and reinforcing satiety signals. These hormones help maintain metabolic equilibrium and are integral to GI health (Gissey, Mariolo and Mingrone, 2019).

Apart from serving as a hormonal regulating factor, the GI tract involves a diverse microbial ecosystem. Here, microorganisms form relationships with the host in several ways. The microbial adaptation of the biofilm's formation is one such vital function; ensuring protection and survival of the microbes against environmental stressors, immune responses, and antibiotics (Sharma et al., 2023). Within the GI tract, both gut homeostasis and pathogenic infections deal with their effects, depending on biofilm compositions and interactions with the host cell. However, normal biofilms can support both gut integrity and immune regulation whereas pathogenic ones often cause chronic infections, many inflammatory diseases, and a certain degree of antibiotic resistance. The working capabilities of biofilms to persist on mucosal surfaces and resist eradication orchestrate a significant challenge in treating both minimal and major GI infections (Von Rosenvinge et al., 2012). There have been researchers that show a complex and bidirectional interaction between biofilm formation and enteric hormones where gut-derived hormones are influencing microbial behaviour. This includes biofilm development, and, on the other hand, the microbial communities and biofilms are modulating host hormone secretion (Wang et al., 2023). One perfect example of this is Dipeptidyl peptidase-4 (DPP-4) activity by gut microbiota (GM) which influence the host's metabolic activity. DPP-4 is involved in the proteolysis of key hormones, thus its secretion from the GM can significantly alter the host's protein digestion and, thereby affect overall health (Olivares et al., 2018).

DOI: 10.1201/9781003519041-12

This chapter analyses and shows the connection between enteric hormones and biofilm formation, having most of the focus on their interactions, and the implications for health. It examines the mechanisms through which enteric hormones influence biofilm dynamics, and also how microbial biofilms impact hormonal signalling. Drawing from the current literature, we seek to provide new insights into their roles in both physiological and pathological conditions. Additionally, the chapter also highlights the therapeutic strategies that target these interactions to manage metabolic disorders, gut infections, and other GI diseases.

12.2 BRIEF DESCRIPTION OF ENTERIC HORMONES

The gastrointestinal tract is a highly multifaceted system that plays a fundamental role in digestion, nutrient absorption, metabolic regulation, and overall immune function. Its mechanisms involve enteric hormones, secreted by specialized enteroendocrine cells and leading to the modulation of numerous physiological processes (Mace et al., 2015). Besides controlling digestion and gut motility, the hormones also influence metabolism, energy homeostasis, and appetite regulation (Steinert et al., 2016). There is a dispersion of hormones throughout the intestinal epithelium, releasing them in response to mechanical and chemical stimuli, including food intake, microbial interactions, and the metabolic demands of the host cell.

12.2.1 Synthesis and Release of Major Enteric Hormones

The enteroendocrine cells are found within the mucosal lining of the GI tract. These cells can detect nutrients and other signals in the intestinal lumen and aid in releasing the hormones into the bloodstream or act locally through paracrine and autocrine signalling (Steinert et al., 2016). Some of the important enteric hormones and their sources are provided in Table 12.1.

TABLE 12.1
Some Key Enteric Hormones and Their Sources of Origin

Hormone	Origin
Gastrin	Produced by G cells in the stomach and duodenum. It stimulates gastric acid secretion, helping gastric mucosal growth, and improving motility of the GI tract.
Secretin	Produced by S-cells in the duodenum. It also regulates pH by stimulating bicarbonate secretion from the pancreas to neutralize gastric acid.
Cholecystokinin (CCK)	Secreted by I-cells in the duodenum and jejunum, CCK facilitates digestion by stimulating bile and pancreatic enzyme secretion while also inducing satiety.
Ghrelin	Produced mainly by X/A-like cells in the stomach. It is released during fasting and stimulates hunger by acting on the hypothalamus.
Leptin	Natively secreted by adipose tissue. It is also a gut-derived component that helps regulate satiety and energy balance.
Glucagon-like peptide 1 (GLP-1)	Secreted by L-cells in the small intestine and colon, also known to enhance insulin secretion, delays gastric emptying, and promotes satiety.
Peptide YY (PYY)	Secreted by L-cells, and released postprandially. Plays a role in appetite suppression and slowing intestinal motility.
Motilin	Secreted by M-cells in the small intestine, motilin regulates the migrating motor complex (MMC) and controls gastric and intestinal motility during fasting.

12.2.2 Physiological Functions of Enteric Hormones

Enteric hormones have an important role in numerous regulatory functions within the GI tract and beyond. Their main physiological roles include:

1. **Regulation of Gut Motility**:
   a. Motilin enhances the migrating motor complex (MMC), this enables the proper clearance of undigested material between each meal.
   b. Cholecystokinin (CCK) and Peptide YY (PYY) aid in the reduction of the speed of gastric emptying. In addition, this also helps them to prevent excessive food influx into the intestine.
   c. Ghrelin ensures an increase in gastric motility in preparation for food intake.
2. **Nutrient Absorption and Digestion**:
   a. Gastrin triggers the release of gastric acid in the stomach, which, in turn, activates pepsin, the enzyme required for protein digestion.
   b. CCK pushes the bile and pancreatic enzymes to get released, thereby helping in lipid and protein digestion
   c. Secretin gives rise to the secretion of bicarbonate to improve enzymatic activity by maintaining a neutral pH in the duodenum.
   d. GLP-1 also influences the absorption process through the modulation of gastric emptying and insulin response.
3. **Appetite and Satiety Control**:
   a. Under this point, ghrelin performs as a hunger signal, which further stimulates the food intake through the hypothalamus.
   b. Leptin signals satiety and restrains appetite by acting on hypothalamic neurons.
   c. PYY and CCK are released after meals and ensure the provision of short-term satiety signals which help to reduce further food consumption.
4. **Glucose Homeostasis and Metabolic Regulation**:
   a. GLP-1 amplifies insulin secretion from pancreatic beta cells. But it also inhibits glucagon release, making it the main target for diabetes treatment.
   b. Ghrelin promotes glucose metabolism by modulating insulin sensitivity.
   c. Leptin affects glucose regulation indirectly, by a minor effect on energy expenditure and appetite.
5. **Immune Function and Gut Barrier Integrity**:
   a. Some enteric hormones, such as glucagon-like peptide-2 (GLP-2), play a role in maintaining intestinal barrier function. It also works on reducing gut permeability and inflammation.
   b. Leptin has immunomodulatory functions, leading to immune cell activity within the gut.
   c. Gut hormones interact with the GM, for microbial composition and immune responses (Lai et al., 2024).

12.3 FACTORS AFFECTING ENTERIC HORMONE REGULATION

The major factors that influence enteric hormone regulation are essential in understanding metabolic disorders, obesity, and GI diseases. Several components and host-specific factors lead to the modulation of enteric hormone secretion and function. A few are given below:

12.3.1 Hormonal Interaction

Enteric hormones never show their working process in isolation; they rather interact with multiple endocrine and neurohormonal systems. Hormones like gastrin, secretin and CCK, are the primary enteric hormones that help in digestion and metabolism. Ghrelin stimulates appetite, and leptin, an adipose-derived hormone, suppresses it. In the same manner, insulin and GLP-1 work simultaneously to regulate glucose metabolism. If there is any dysregulation of these interactions, it can lead to metabolic disorders, including diabetes and obesity. The interference between enteric hormones and systemic endocrine pathways throws some light on the complexity of their regulation (Cho and Lim, 2024).

12.3.2 Nutritional Factors

These factors, such as the diet of a person, significantly impacts enteric hormone secretion and function. The macronutrient composition, the frequency of meals, and caloric intake all enhance hormone release. High-protein meals modulate CCK and PYY secretion, promoting the feeling of being satiated. However, high-carbohydrate diets increase insulin and GLP-1 levels. Dietary fibres also increase the short-chain fatty acids (SCFAs) formation, influencing the overall gut hormone release and improving metabolic health. Talking about the nutrient deficiencies, such as those related to essential amino acids, vitamins, and minerals, impairs hormone regulation and gut microbiota composition by affecting enteric hormone balance (Sun et al., 2020).

12.3.3 Psychological and Emotional Factors

The gut–brain axis (GBA) is extremely important in the regulation of enteric hormones. Persons experiencing anxiety and depression have a hormonal imbalance causing problems in the hypothalamic–pituitary–adrenal axis and autonomic nervous system activity. These stress-induced changes in cortisol levels disrupt gut motility and hormone secretion. This further leads to GI disturbances, such as irritable bowel syndrome (IBS). In addition to this, emotional states lead to changes in eating behaviours, with stress majorly linked to increased ghrelin secretion and emotional overeating (Sun et al., 2020).

12.3.4 Circadian Rhythms

The central and outermost biological clocks regulate the circadian rhythms, causing changes in enteric hormone secretion patterns. Certain hormones, such as ghrelin and insulin, show a diurnal rhythm, with the greatest elevation occurring at a selected time of the day. Disturbances in sleep patterns can disrupt these rhythms. Following this disruption there will be metabolic dysregulation and increased susceptibility to obesity and diabetes. Coinciding eating patterns with natural circadian rhythms may enhance metabolic health and optimize enteric hormone function (Martchenko et al., 2020).

12.3.5 Gender Differences

Gender-based physiological differences cause variations in enteric hormone regulation. Hormonal imbalance in menstrual cycle, menopause, and pregnancy leads to gut hormone secretion and appetite control. Oestrogen suppresses ghrelin levels in order to reduce hunger. On the other hand, the hormone progesterone works in the opposite manner. Thus, there are differences in body composition, gut microbiota, and metabolic rates between men and women. This causes distinct hormonal responses in relation to food intake and stress (Santos-Marcos et al., 2023).

12.4 BIOFILMS IN THE GASTROINTESTINAL TRACT

Biofilms provide microorganisms with enhanced protection against the external stimuli of environmental stressors, antimicrobial agents, and host immune responses. Although biofilms significantly contribute to gut homeostasis and symbiosis in certain contexts, they also inflict pathogenic infections, chronic inflammation, and GI disorders (Buret and Allain, 2023; Tiamani et al., 2022; Ye and

Liddle, 2016). The GI tract is often home to several oral multispecies biofilms; however, the diversity of the species can vary depending on the host. They often follow a specific pattern of development in dental plaque: *Streptococcus*, *Actinomyces*, *Veillonella*, *Fusobacterium*, *Porphyromonas*, and *Aggregatibacter* (Perry and Tan, 2023). The most abundant phyla of the human digestive tract are *Bacteroidetes*, *Proteobacteria*, *Actinobacteria*, and *Firmicutes* (Srednicka et al., 2021). Likewise, other studies have reported that, irrespective of the host's underlying condition, the biofilms are relatively more common around the ileal–cecal junction and the ascending colon in comparison to the transverse/descending colon and rectum (Perry and Tan, 2023). Biofilm formation in the GI tract is a process which involves multiple microbial species, but the most commonly found microbes are:

1. **Bacteria**: *E. coli, Enterococcus* spp., *Bacteroides* spp., *Fusobacterium nucleatum, Helicobacter pylori* (which help in gut homeostasis but are sensitive to host-system, thereby also causing infections).
2. **Fungi**: *Candida albicans* (commonly found in GI biofilms and associated with dysbiosis)
3. **Viruses and Phages**: Noroviruses and bacteriophages (these play a role in microbial community modulation, biofilm resilience and infections. Host age-specific species are noticed).
4. **Archaea**: *Methanogenic* archaea contribute to gut microbial metabolism and biofilm stability

The GM comprises several microbes, but bacterial populations tend to be the most common species. Thus, the formation of a bacterial biofilm typically follows a five-stage process (Sharma et al., 2023), which includes:

12.4.1 Initial Adhesion

Microorganisms have developed abilities to interact with different regions of the intestine—mucin layer, epithelial cells, and intestinal crypts which are biofilm-favourable regions. Their initial attachment is facilitated by fimbriae, pili, and adhesion proteins; meanwhile, their excreted surface mucosa enables the microbes to stick. The genera and species of biofilm frequently found in the intestine include *Escherichia coli, Enterococcus faecalis, Bacteroides fragilis* and *Helicobacter pylori.*

12.4.2 Microcolony Formation

Following attachment, the microbes tend to proliferate and form microcolonies that secrete extracellular polymeric substances (EPS). These EPS comprise polysaccharides, proteins, and extracellular DNA (eDNA). They provide mechanical stability and protection to the biofilm along with establishing microbial communication via quorum sensing (QS).

12.4.3 Biofilm Maturation

A biofilm develops into a three-dimensional structure with complex gradients of nutrients and oxygen. Within the biofilm, different species of microbes form either cooperative or competitive associations which increase the variety of microbes present in the biofilm. In particular, anaerobic bacteria such as *Bacteroides* spp. and *Clostridium* spp. live in biofilms characterized by low-oxygen microenvironments.

12.4.4 Dispersal and Colonization

Biofilms that are mature in age release microbial cells periodically into the lumen to colonize other locations within the host's intestines and, in some cases, transfer to other hosts. Biofilm building is catalyzed with ease by changes in the environment, responses from the immune system, or external factors such as antibiotics or dietary elements.

12.5 BIOFILMS FOR GASTROINTESTINAL HEALTH

The GI tract contains biofilms that show dual roles. It acts as both protective and pathogenic factors for gut homeostasis, digestion, and disease pathogenesis. The presence of biofilms can maintain a balanced gut microbiome and intestinal functions. In contrast, pathogenic biofilms are linked to chronic infections, inflammation, and disease progression. The GM that promotes and ensures homeostasis is called eubiosis; meanwhile, the dysregulation of GM is termed as dysbiosis. This has health implications (Srednicka et al., 2021). Beneficial bacteria, such as *Bifidobacterium* and *Lactobacillus*, form protective biofilms that play a vital role in the symbiosis of GM. These biofilms maintain the gut barrier capabilities, preventing pathogen colonization and thus enhancing mucosal defences. Apart from this, it also contributes to the production of SCFAs, serving immune responses, supporting intestinal epithelial health, and promoting anti-inflammatory effects. By stimulating the production of mucins, biofilms strengthen the mucus layer, thereby showing their role in mucosal protection. It shields intestinal epithelial cells from harmful agents, hence reducing the risk of microbial invasion and inflammation. In addition, beneficial biofilms also participate in immune system modulation. They do so to help them train the immune system and maintain immune tolerance to commensal microbes. Such regulations prevent excessive inflammatory responses and contribute to a balanced gut immune environment (Von Rosenvinge et al., 2012).

Leaving aside its protective effects, pathogenic biofilms are involved in various GI diseases. One such example is *Helicobacter pylori*, a bacterial pathogen which forms robust biofilms on the gastric mucosa that cause peptic ulcer disease. *H. pylori*'s biofilm structure shields the bacteria from immune attacks. This leads to resistance to antibiotic therapy, elevating chronic colonization and persistent infection. So, ultimately, this persistence can cause problems such as gastritis and peptic ulcers. In some cases, gastric cancer is also seen. Inflammatory bowel disease (IBD) is mostly connected with altered gut microbiota composition. In this situation, the pathogenic biofilms exacerbate chronic inflammation. Dense biofilm formation by bacteria such as *Escherichia coli* and *Enterococcus faecalis* has been linked to both Crohn's disease and ulcerative colitis.

Nowadays, biofilm formation in colorectal cancer is a major concern. Researchers have found that biofilms enriched with *F. nucleatum* play a major role in colorectal tumorigenesis. It is a growing threat because these biofilms are seen to be influencing proinflammatory signalling pathways leading to tumour progression, and the modulation of the tumour microenvironment. This creates a suitable niche for the organism which supports cancerous growth. Due to the presence of their EPS matrix, biofilms start acting like a reservoir for antibiotic-resistant bacteria. This makes treatment for the infections difficult (Chandra, Srivastava and Kumar, 2019).

12.6 INFLUENCE OF BIOFILMS ON THE IMMUNE SYSTEM

The commensal biofilms help maintain homeostasis and the pathological biofilms bypass immune detection, thereby invoking chronic inflammation. The biofilms' EPS matrix serves as a protective barrier by masking bacteria to the immune system and inhibiting engulfment by immune cells, thereby enabling bacteria under shield to sow the seeds of pathogenicity in the gut. This process helps to perpetuate chronic infection. However, biofilms can actively suppress production of pro-inflammatory cytokines, resulting in an immune system that is more tolerant. This allows further colonization by microbes with ease. With regards to bio-pathogenic microbes, however, such immune avoidance results in sustained infection and inflammatory sequelae. (Sartor, 2018). Biofilm-related disorders of the gut stand out because chronic inflammation is predominant. Pathogenic biofilms can activate epithelial and immune cells by stimulating Toll-like receptors (TLRs), causing long-term inflammation. This leads to the production of key inflammatory mediators, such as IL-6, TNF-α, and IL-1β, which worsen IBD, IBS, and similar disorders. Autoimmune diseases, including rheumatoid arthritis and multiple sclerosis, show the importance of systemic inflammation in disease causation. They are frequently associated with immune activation involving biofilm-producing pathogens due to diabetes. Other than the modulation of immune function, biofilms also affect gut barrier function. Pathogenic biofilms may cause the intestinal barrier to breach by loosening tight junctions between epithelial cells which results in increased intestinal permeability, or leaky gut. This allows for the influx of bacterial toxins, metabolites, and inflammatory agents within the human body's circulation, which is linked with metabolic disorders together with obesity, as well as neuroinflammatory diseases such as Alzheimer's and Parkinson's diseases. Notwithstanding these problems, helpful biofilms contribute to the maintenance of gut homeostasis by interacting with the gut microbiota. These biofilms formed by probiotic bacteria are able to outcompete pathogens, which prevents dysbiosis and lowers the chances of chronic inflammation and infection (Buret and Allain, 2023).

12.7 IMPACT OF ENTERIC HORMONES ON BIOFILM FORMATION

The connection between enteric hormones and biofilm formation showcases a promising area of research, with a significant understanding of GI health, and microbial ecology, leading to overall disease progression. This chapter brings up the idea of enteric hormones influencing biofilm development, and biofilm-forming microorganisms' interaction with enteric hormone secretion and its function.

12.7.1 Influence on Biofilm Development

Novel research indicates that hormones can influence the structure, function, cellular proliferation, adhesion, and biofilm maturation of the host microbial population. The delicate relations between enteric hormones and biofilms tend to have consequences for GI health and disease progression by selectively inhibiting or promoting biofilm development. Let us take ghrelin as an example. Its secretion is done by gastric endocrine cells and is a well-recognized hormone for the control of appetite, gut motility, and immune modulation. Further, it is speculated to have biofilm-suppressive capacity against *H. pylori*. On a different note, some gut bacteria can capture ghrelin and use it as a signalling molecule to facilitate biofilm adhesion and maturation. This unique property of ghrelin suggests that it could bear a dual usage depending on the microbial community. Meanwhile, GLP-1 can affect the modulate microbes by favouring beneficial biofilms and hindering pathogenic biofilms. Moreover, the GLP-1 receptor agonists are suspected of reducing the persistence of biofilms, thereby promoting a good GM and an overall healthier gut environment (Hammouda, Wasfi, and Abdeltawab, 2023).

Peptide YY (PYY) is another gut hormone whose secretion is activated by food intake. It strategically regulates gut motility. However, microbial associations to metabolic PYY is believed to affect the adhesion of bacteria by modifying the mucus secretion and enhancing epithelial barrier function. It may offset the colonization of pathogenic biofilms by offering a shielding effect on the gut environment. This hormone has an action in inter-digestive motility, and motilin is also responsible for stimulating peristaltic waves that cleanse the small intestine during fasting. Fluctuation in motilin levels tends to influence the biofilm stability of commensals, thereby affecting microbial dispersal and gut colonization patterns. Such changes in motilin activity can alter the ratio of commensal and pathogenic biofilms associated with gut health and microbial resilience (Hammouda, Wasfi and Abdeltawab, 2023).

12.7.2 Influence on Biofilm Inhibition

12.7.2.1 Glucagon-Like Peptide-1 (GLP-1)

GLP-1 influences biofilm formation, primarily by impeding microbial adhesion, metabolic activity, and mucus production. The hormone has been linked to alterations in the composition of GM, a reduction in the presence of pathogenic biofilms and the promotion of a balanced microbial environment. In addition, the GLP-1 resists favourable conditions for biofilm persistence by modulating gut physiology and bacterial energy metabolism and enhancing mucus production—leading to a thicker protective layer that prevents bacterial attachment and the subsequent development of biofilms. GLP-1 receptor agonists, commonly prescribed for type 2 diabetes, are clinically relevant since they can reduce the intestinal bacterial biofilms, leading to improved gut barrier function, and reduced inflammation. These findings highlight the potential of GLP-1-based therapies in managing GM imbalances and preventing biofilm-associated GI disorders (De Silva and Bloom, 2012).

12.7.2.2 Peptide YY (PYY) and Motilin

PYY and motilin are key hormones involved in the proper transport of food and microbiota in humans. Their ability to strengthen peristalsis acts as a natural mechanism of biofilm inhibition since increased gut motility obstructs the colonization time, hampering biofilm formation and maturation. This is crucial in preventing small intestinal bacterial overgrowth (SIBO), a liability arising from biofilm-assisted infections. From a clinical perspective, motilin agonists used for the treatment of disorders in gut motility are suggested to be useful in controlling biofilm appendicitis; consequently, improving SIBO and irritable bowel syndrome (IBS). These prospective treatments offer in-depth analysis and remediation of chronic gastrointestinal disorders linked to, or originating from, biofilm (Lesouhaitier et al., 2018).

12.7.2.3 Ghrelin

Ghrelin concentration tends to negatively correlate with infections from biofilms (Nogueira et al., 2022). On the other hand, GM can influence food intake by stimulating ghrelin secretion. Interestingly, studies also showcase that gut-associated ghrelin can disrupt microbial adhesion, thereby halting the adhesion phase of the biofilm formation process. Ghrelin affects the metabolism of GM, leading to alterations in the environment that support biofilm persistence by modifying the way in which a bacterial cell expends energy (Sun et al., 2020). The secretory stimulation of gastric acid by ghrelin negatively favours the survival of pathogenic biofilms, thus restricting their growth. Meanwhile, in chronic gastric infections, lower levels of ghrelin are observed. This underlying consequence implies that there exists an extended control of biofilm infection and that ghrelin plays a role in it (Leeuwendaal, Cryan and Schellekens, 2021).

12.7.3 Impact on Microbial Motility

There are several hormones that modulate the locomotion of microbes, which inversely influences biofilm initiation and dispersion. Taking dopamine and serotonin as an example, they have been linked to altered bacterial movement, putting an impact on biofilm distribution and formation under distinct fluid dynamic conditions (Cambronel et al., 2019).

12.8 IMPACT OF BIOFILMS ON ENTERIC HORMONES

During hormonal secretion, the changes influenced by biofilms can manifest positive effects on gut motility, appetite regulation, and glucose homeostasis. For instance, *E. coli* and *Clostridium difficile* biofilms can effectively disrupt gut motility by influencing motilin- and serotonin-signalling pathways. This can potentially lead to dysmotility syndromes of IBS and SIBO. Thus, dysbiosis originating from biofilm dysregulation can help secrete appetite-regulating hormones along with ghrelin and PYY, further contributing to metabolic disorders such as obesity and anorexia. Some bacteria within biofilms directly ferment dietary fibres to form SCFAs, which could influence GLP-1 secretion, thereby enhancing appetite control. The pathogenic biofilms in the gut might also change GLP-1 secretion, leading to insulin sensitivity and glucose metabolism, ultimately exacerbating conditions like type 2 diabetes (T2D) (Neuman et al., 2015). Disruptions caused in this microbial ecosystem are linked to metabolic disorders such as obesity, T2D, and IBD. In obesity and T2D, GM imbalances affect the secretion of hormones such as PYY, insulin, ghrelin, and GLP-1. These are crucial for appetite regulation and energy homeostasis. If there is an overproduction of ghrelin, it can cause increased food intake; on the other hand, dysregulated GLP-1 might impair insulin sensitivity. Additionally, biofilm-forming bacteria such as *Firmicutes sp.* can upgrade the energy extraction from certain foods, contributing to fat accumulation and rising metabolic dysfunction (LeValley, Tomaro-Duchesneau and Britton, 2020).

Some important biofilm-forming pathogens have been shown to be a direct influence of enteric hormone signalling, paving the way for various disease states. For example, *H. pylori* biofilms might alter the secretion of gastric hormones such as ghrelin and leptin, affecting appetite and metabolism. By forming biofilms on the gastric epithelium, *H. pylori* promote chronic inflammation and alters acid secretion, thereby influencing gut hormone dynamics. Biofilms formed by *F. nucleatum* in the colon have also been linked to colorectal cancer progression, possibly through interactions with hormonal pathways like insulin signalling. These bacteria may suppress beneficial hormonal responses, leading to increased inflammation and tumorigenesis (Boukerb et al., 2021). The metabolites from GM can also be signalling molecules for different cells; that then release the hormones. These, in turn, affect the host's metabolic regulation and create a chain of successive events within. The alterations of GM concerning increased colonic secretions have been associated with IBS and suggest that such modifications can provoke the enteroendocrine cells to exude the abnormal release of hormones. Likewise, species-specific microbe-associated molecular patterns can be recognized by TLRs which, in turn, activate the hormone secretion by enteroendocrine cells. Like serotonin and CCK, secretions can be activated by TLR agonists: lipopolysaccharide and synthetic bacteria lipoproteins (Ye and Liddle, 2016).

Obesity and T2D are metabolic disorders that are becoming a prominent global health concern. Although the precise mechanisms are still unknown, research has demonstrated a robust correlation between human metabolism and gut flora. Some studies examine the effects of gut bacteria, including *Enterococcus faecalis*, on the levels of GLP-1. This incretin hormone helps regulate glucose and hunger. Meanwhile, the bacterial supernatants of the *E. faecalis* strains were found to lower the GLP-1 levels using a human-derived intestinal cell line (NCI-H716). Consequent investigations have also shown that GelE, a metalloprotease released by *E. faecalis*, is the reason for this effect. It was found that GelE directly breaks down GLP-1, weakening its functionality. Furthermore, the study specified that *E. faecalis* supernatants may damage the intestinal epithelial barrier in a GelE-dependent manner. The gut microbiota has a complicated effect on human health that might sometimes go beyond positive or negative results. There are some studies that also illustrate how the degradation of GLP-1 by commensal gut bacteria such as *E. faecalis* could affect metabolic processes, influencing blood sugar homeostasis and increasing the risk of metabolic diseases. The various outcomes deepen our knowledge of host–microbiome interactions and state that the human body's hormone signalling can be directly impacted by microbial secretions (LeValley, Tomaro-Duchesneau and Britton, 2020).

Despite the fact that metabolic disorders and GM are closely intertwined, the exact mechanisms behind these interactions are mostly undiscovered. The L-cells from the GI tract secrete the GLP-1, a vital metabolic regulation by encouraging insulin secretion, increasing insulin sensitivity, increasing β-cell mass, delaying stomach emptying, and decreasing hunger. Thus, making GLP-1 a critical component in glucose metabolism. Due to its significance, scientists looked for microbial strains that can increase GLP-1 secretion as possible treatments for metabolic disorders. They found 45 bacterial strains that could increase GLP-1 secretion by screening more than 1,500 human-derived strains from healthy people. Using 16S rRNA sequencing, all these bacteria were determined to be *Staphylococcus epidermidis*. The 3 kDa peptide Hld (delta-toxin) was found to be present in GLP-1 stimulatory strains but lacking in neutral strains, indicating a possible role for it in raising GLP-1 levels, according to an additional mass spectrometry study. Hld was evaluated in human-derived NCI-H716 cells and jejunal enteroids (designed to have more enteroendocrine cells) to verify this, and the results showed that Hld was enough to increase GLP-1 production on its own. GLP-1 release was stimulated by calcium signalling, thus demonstrating the mechanism underlying this impact. Moreover, in high-fat-fed mice, they were given Hld-producing *S. epidermidis* to evaluate its therapeutic potential. This resulted in notable decreases in obesity and T2D-related indicators. Such a discovery raises the possibility of creating microbial peptide-based treatments for metabolic disorders since it implies that some bacteria from the gut can naturally increase the production of GLP-1. Using the microbiome's innate capacity to control incretin hormones may provide a more physiologically integrated and sustainable therapeutic strategy than the conventional GLP-1 receptor agonists used to treat diabetes. By highlighting the hitherto-undervalued function of commensal gut bacteria in metabolic regulation and paving the way for novel microbiome-based therapies for diabetes and obesity, this study offers vital insights into host–microbiome interactions (Tomaro-Duchesneau et al., 2020).

The main causes of two of the serious global health issues—obesity and diabetes—are decreased energy expenditure, excessive calorie intake, and the ensuing negative energy balance. The rise in diabetes, especially among children, highlights the urgent requirement of safe, rapid, and affordable, treatment options. Emphasizing on their capacity to modify gut microbiota and the effect on metabolic health, this study investigates the possibility of probiotics as a workable therapy approach for diabetic patients. As their mechanisms of action are still not widely understood, live bacteria that colonize the gastrointestinal system and offer a plethora of health benefits remain infrequently used in medicine. In this study, scientists studied how a particular probiotic formulation (VSL#3) impacted metabolic health in several mice models of diabetes and obesity. By modifying the composition of the gut microbiota, VSL#3 administration successfully prevented and reversed insulin resistance and obesity, resulting in positive metabolic results. One of the most significant findings was that VSL#3 promoted the production of GLP-1, a hormone which was important for controlling insulin secretion, hunger, and glucose metabolism. Minimum food consumption and improved glucose tolerance were the results of the rising GLP-1 levels, emphasizing the promise of probiotic-driven hormone regulation as a therapeutic method.

Further research showed that the metabolic advantages of VSL#3 were linked with a rise in butyrate, a SCFA that is necessary for gut health and metabolic control. Researchers also showed that butyrate directly promoted GLP-1 secretion with the help of intestinal L-cell cultures, delivering a molecular explanation for the advantageous benefits of VSL#3. On the basis of these results, probiotics may alter the gut microbiota–SCFA–hormone axis, affecting host metabolism in ways that prevent diabetes and obesity. Probiotics such as VSL#3 provide a long-lasting and non-invasive treatment option for the management of metabolic illnesses by motivating positive bacteria interactions and natural hormone secretion. (Yadav et al., 2013).

Besides putting an impact on the processes outside of the digestive system, gut microorganisms are essential for saving intestinal health. A beneficial human intestine bacterium called *Limosilactobacillus reuteri* has been shown to promote bone health, wound healing, immunological control, gut motility, and social behaviour. These physiological processes are known to be helped by the hormone oxytocin, and prior studies have proved that oxytocin signalling is significant for social behaviour and wound healing mediated by *L. reuteri*. The intestinal epithelium is examined in this work as a potential source of oxytocin, and the mechanism by which *L. reuteri* promotes its release is determined. Researchers discovered that the intestinal epithelium's ability to produce oxytocin is evolutionarily conserved by single-cell RNA sequencing and tissue imaging of the human and animal intestines. The epithelium itself can produce oxytocin, according to another research conducted with human intestine organoids. Crucially, it was shown that *L. reuteri* directly promotes the release of oxytocin in human intestinal tissues and organoids, indicating a new microbial influence on hormone release. The study also found that secretin, a gut hormone made by enteroendocrine cells, is necessary for the release of oxytocin. On a similar note, the enterocytes also produce oxytocin, indicating a two-step signalling system, wherein the *L. reuteri* induces the production of secretin. This integrated approach, in turn, causes the secretion of oxytocin. Notably, not all tested bacteria had this effect, as noted for *E. coli Nissle* and *B. subtilis*, since they were unable to trigger the release of oxytocin. This underscores the special function of *L. reuteri*. However, it is confirmed that oxytocin as an intestinal hormone offers important new information about how particular gut microorganisms influence host health through hormonal interactions. These results positively welcome the microbiome-based treatment approaches that target immunological, neurological, and metabolic disorders. They basically identify it using a direct microbial mechanism for oxytocin release via secretin signalling (Danhof et al., 2023).

It is chastening to understand that periodontal diseases can cause a risk to the overall quality of life. *Fusobacterium nucleatum* is a major reason for periodontal infections that lead to inflammation and tissue damage. Interestingly, the periodontal tissues in rats were studied for the expression of growth hormone secretagogue receptor (GHS-R). Ghrelin is known to interact with GHS-R and influence the response of periodontal cells to inflammation. In addition, the researchers exposed *F. nucleatum* to human gingival fibroblasts (HGFs) added with ghrelin. The study showcased that the GHS-R expression increases in response to periodontitis, suggesting that the *F. nucleatum* also leads to a signalling of the activation of ghrelin. Upon evaluating the wound-healing capacity, ghrelin inhibits the pro-inflammatory cytokines from coming into action, thereby reducing infection-associated inflammation and aiding in the repair of HGFs. This demonstrates the overall connection in this study, which is that bacterial biofilm in periodontitis can upregulate GHS-R expression, and ghrelin may serve as a protective agent, promoting healing and mitigating inflammation. This highlights the potential therapeutic role of enteric hormones towards biofilm-related infections in managing periodontal disease and its inflammatory responses (Nogueira et al., 2022).

12.9 BIDIRECTIONAL RELATIONSHIP BETWEEN ENTERIC HORMONES AND BIOFILMS

This relationship is especially relevant in disease contexts (Figure 12.1). In IBD, altered hormone levels contribute to GM imbalance and the persistence of pathogenic biofilms, exacerbating inflammation and mucosal damage (Duran-Pinedo, Solbiati and Frias-Lopez, 2018). Recognizing how enteric hormones influence biofilm formation presents opportunities for novel therapeutic interventions.

Following that line of discussion, T2D is associated with cardiovascular risk, systemic inflammation, and increased mortality. Like many chronic illnesses, robotic-assisted surgery has both genetic and environmental—including obesity—etiological components. Recent studies suggest that some characteristics of T2D are impacted by gut flora and associated intestinal endocrine function. Clinical evidence that patients with a total colectomy have higher incidence rates of T2D is one such example that supports this claim, indicating how the gut may be important for metabolic control. More is to be learned from studies of germ-free mice which showed these mice are resistant to diet-induced obesity. Obese-associated bacteria, such as *Enterobacter cloacae*, and microbiota from human obese donors have previously been shown to induce colonization. These mice exhibited significant weight gain and poor glucose tolerance. T2D, metabolic diseases, and gut microbiota also implies that we understand the alternation of specific populations of microbes that may trigger T2D and/or obesity (Jensen et al., 2018).

People with so-called 'sweet teeth' tend to be at the epicentre of developing T2D. The free fatty acid (Ffar4) receptor in diabetic human and mouse models is seen to decrease, and is associated with escalated sugar intake. Zhang et al. (2025) found that upon removing the receptor in the mice model, there was a decrease in *Bacteroides vulgatus* gut bacteria, which produces pantothenate. The reduction of this compound manifests an increased sugar preference in the mice. Further, pantothenate

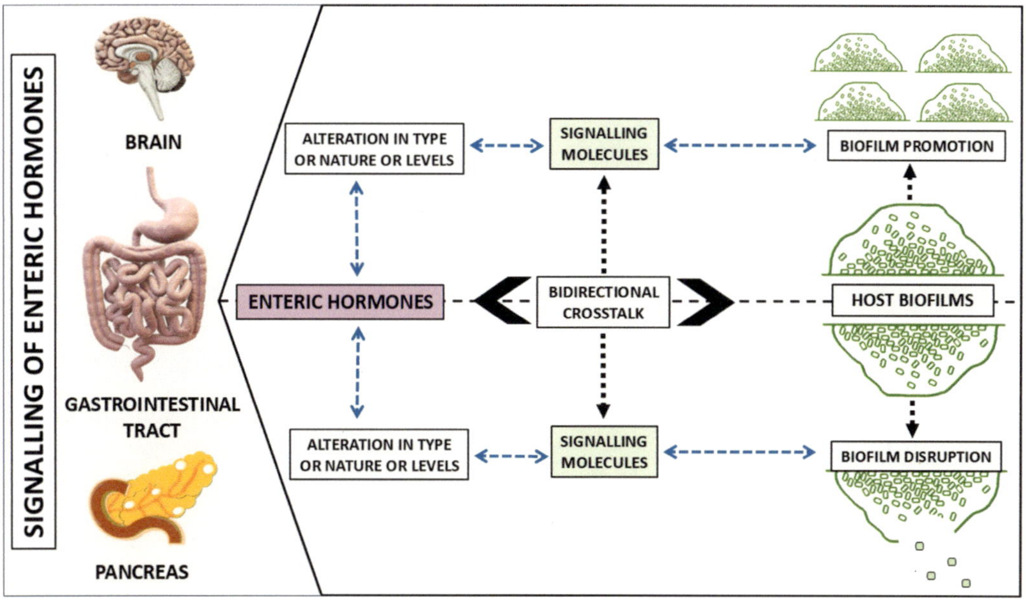

FIGURE 12.1 Schematic representation of the bidirectional crosstalk between enteric hormones and biofilms.

helps in releasing GLP-1. The hormone, in turn, signals the production of FGF21, which is associated with the host's energy regulation. It is thus meant to counteract sweet cravings. This suggests that this receptor–biofilm–hormone influence can be used to target pantothenate production, to navigate diabetes treatment (Zhang et al., 2025).

There is a complex communication between host biofilm systems and the host hormones. Interestingly, Lopes and Sourjik (2018) investigated the chemotactic responses elucidated by *E. coli* in response to different hormones and polyamines. This study shed light on the mechanisms employed by the gut bacteria to influence GI health and metabolism, as well as immune response. The bacteria exhibited a biphasic response to dopamine. It showcased repulsion for concentrations <1mM and attraction at 10mM concentrations of dopamine. With melatonin concentration exceeding 0.1 mM, a strong dose-dependent repellent behaviour was noted and no significant chemotactic behaviour was recorded for serotonin. This illustrates that *E. coli* has the ability to physiologically detect and respond to the different concentrations of host hormones in the GI tract, which poses wider consequences in the health of the host. It also illustrates how bacterial chemotaxis may be applied as a marker for the bacterial niche and as a signal of the hormonal well-being of the host (Lopes and Sourjik, 2018). This reveals a cross-talk between enteric hormones and biofilms, which are mutually interdependent to a significant degree.

The link between biofilm formation and enteric hormones is clinically important concerning diseases in the gastrointestinal tract and metabolic disorders. Dysfunction of both systems can worsen obesity, T2D, IBD, and various gastrointestinal infections. These persistent biofilm-associated infections, in conjunction with increasing antibiotic resistance, complicate treatment approaches that require novel solutions for managing these intricate relationships. With appropriate biofilm-targeting therapies, integrating hormone regulation approaches may provide better treatment options for metabolic and gastrointestinal diseases.

Here, things examined are the clinical consequences of the malfunctioning of the enteric hormones and biofilm production, its ramifications on metabolic disorders, the problem of antibiotic treatment failure, and possible ways of dealing with these multifaceted problems (Giordano and Giannoudis, 2024). The relationship is reciprocal and is augmented by the paradigms of:

12.9.1 Hormones and Biofilm Composition Related to Diet

Diet has a direct impact from the gut microbiota and also the hormonal imbalance and biofilm formation. Different diets can directly interfere during biofilm formations. Diets rich in fats encourage dysbiosis of the gut, and biofilm formation increases the gut inflammation and permeability (Acharya et al., 2023). On the other hand, the use of prebiotics and probiotics helps in increasing the beneficial bacteria and modulation of hormonal signalling like the increase of GLP-1 and serotonin production while reducing biofilm-caused infections (Akimbekov et al., 2022). Fruits and vegetables polyphenols possess anti-biofilm activity and altering the hormone secretion influences the insulin and estrogen balance (Rodríguez-Daza et al., 2021).

12.9.2 Differences between the Genders' Interaction with Hormones and the Gut

In women, ovaries and estrogen lead to hormonal changes that affect microbial composition and biofilm activity, eventually modifying disease risk and treatment response. *Lactobacillus* species elevate in women's microbiota due to the presence of estrogen, allowing for protective effects against damaging biofilm formation. This leads to improved gut health and lowered inflammation (Holingue et al., 2020). Also, males are more likely to have *Bacteroides* and *Prevotella* species, which are known to increase inflammation and predispose individuals to biofilm-related infections. These distinctions in the composition of microbiota between women and men determines the risk of developing different gut diseases and could be a consideration for future targeted treatments based on hormonal profiles (Jašarević, Morrison and Bale, 2016).

12.9.3 Oxidative Stress from Hormone–Gut Interfaces

The interaction of oxidative stress at the level of biofilm formation in the gut with hormonal factors determines inflammatory responses and the disease progression. One of the primary constituents of oxidative stress includes the metabolism of estrogen, which proceeds through a line of steps that produce reactive oxygen species (ROS) capable of disturbing the homeostasis of the gut microbial community. The sustained inflammation initiated as a result of gut dysbiosis amplifies oxidative stress and biofilm-associated antibiotic resistance, worsening the disease severity (Rogers et al., 2016).

12.9.4 Integration in the Gut–Brain Axis

The GBA serves as a two-way communication channel between the gastrointestinal system and the central nervous system, driven through neural, endocrine, immune, and metabolic systems. Such an extensive system in its entirety is necessary for neurological functions of cognition and emotion. These aspects of emotional and cognitive regulation are profoundly affected by the GM; as they modulate the action of critical neurotransmitters, including dopamine, serotonin, and gamma-aminobutyric acid (GABA). Excessive organism disruption, particularly in conjunction with biofilm formation, can interrupt these neurotransmitter circuits, resulting in mood disorders like depression and anxiety. (Sun, Li and Nie, 2020). Probiotic and prebiotic therapies, dietary changes, and even stress relief all qualify as therapeutic options for mitigating biofilm-associated cognitive deficits. (Rogers et al., 2016).

Many reasons, like hunger and energy metabolism, determine the amount of movement an animal performs. Specialized sensors located in the gut communicate with the brain to modulate movements based on metabolic signals. The gut microbiota profoundly impacts locomotor activity through GLP-1 signalling, as this study that explored the interaction of gut bacteria, gut hormones, and brain transmission link suggests. Mice given a cocktail of antibiotics had augmented GLP-1 levels but reduced locomotor activity. Once the mice were administered blockers that obstructed the gut–brain signaling axis or the GLP-1 receptor, they resumed normal levels of activity, signifying that GLP-1 is likely responsible for reduced activity levels.

Further investigation exposes that certain antibiotic aided in elevating the GLP-1 levels more, which lowered the movement even further. It is noteworthy that with the inclusion of some useful bacteria—*L. reuteri* and *Bacteroides thetaiotaomicron*—to the germ-free mice, the GLP-1 levels lowered, and the mice became more active. Hence, these bacterial species are likely significant for modulating locomotor behaviour and suppressing GLP-1 expression (Lai et al., 2024). Considering everything, this tells us that gut microbes alter movement by changing GLP-1 levels in the brain through a vagal nerve-dependent pathway. These findings highlight the complex modifications within the gut–brain axis and influence the importance of intestinal microorganisms for the function and behaviour of the brain.

12.10 INTERVENTIONS IN ENTERIC HORMONE–BIOFILM RESEARCH

The convergence of metagenomics, microfluidics, and advanced imaging techniques is revolutionizing our understanding of the dual roles of enteric hormones–biofilm behaviour in health and diseases. Advanced imaging techniques used for detailed high-resolution visualization of biofilm–hormone dynamics include confocal laser scanning microscopy, super-resolution microscopy, fluorescence, *in situ* hybridization, two-photon microscopy, multiphoton fluorescence imaging, Raman spectroscopy, and more (Valverde-Pozo et al., 2024). Future research integrating multi-omics approaches (such as transcriptomics, proteomics, metagenomics, and metabolomics) will provide a more comprehensive picture of the molecular pathways driving biofilm formation and enteric hormone signalling. The use of novel human organoid–biofilm model systems is required for accessing the comprehensive coupling of the two systems (Engevik et al., 2020). Further, GM can be used as a competent biosensor for investigating underlying health ailments; thus using them clinically can offer a fresh perspective on the hormone–biofilm relationship; especially in understanding gender-biased hormonal settings. Likewise, engineered hormone therapy for GI microbiota and microfluidic models that simulate gut environments in real time can help us derive feedback on the dynamic nature of the host–biofilm scenario (Valiei et al., 2023). Additionally, advancements in AI-driven predictive models will help stimulate gut microbial responses to hormone therapies, guiding the development of personalized treatments for biofilm-associated infections and metabolic disorders (Li et al., 2022).

Several critical research gaps persist, limiting a comprehensive understanding of their combined influence of enteric hormones and biofilms on GI health. One major area of uncertainty lies in the molecular mechanisms of interaction between enteric hormones and biofilm formation. Most research in this field has been conducted using animal models or in vitro studies, which fail to fully capture the complexity of the human gut microbiota and its interaction with host hormones (Clarke et al., 2014). Conducting clinical trials that explore the effects of hormonal therapies, probiotics, or biofilm-disrupting drugs on gut health and metabolic disorders is crucial for translating these findings into effective treatments. The presence of comorbid conditions related to microbiome imbalance or hormonal deficiencies can significantly affect biofilm formation and antibiotic resistance. A compromised gut microbiome can weaken protective mucus layers and impair immune responses, exacerbating intestinal inflammation and discomfort. The loss of microbial diversity may lead to chronic inflammation and autoimmunity by failing to properly train the immune system, increasing susceptibility to immune-related disorders and hormonal imbalances.

On the flip side, hormonal deficiencies, imbalances, malfunctioning, and decline can also significantly impact GM composition, further exacerbating health conditions. Hormonal imbalances have also been linked to a higher risk of osteoporosis, cardiovascular diseases, and cognitive decline, all of which can be influenced by gut microbial changes. Such imbalances contribute to increased complications in disease management, that are not well reported (de Vos et al., 2022). Hormones other than enteric can have a significant command over the functioning of the biofilms. Such ascendancy can dictate biofilm-induced enteric hormonal fluctuations compromising both the microbial and the hormonal homeostasis of the host. Offshooting a new challenge in managing biofilm-associated infections is their inherent antibiotic resistance, which may create unprecedented implications on the overall hormonal balance (Wallace et al., 2020). This necessitates the strategic exploration of alternative therapeutic strategies. Many enteric hormones–biofilm relationships are yet to be explored and there lies significant research gaps that need to be filled.

12.11 CONCLUSION

In this chapter, we have outlined the designation, physiological functioning, and constraints for the release of enteric hormones, and further laid the context of the impact of biofilms on them. Research on the intriguing dynamic interaction between the enteric hormones and the host biofilms; irrespective of their origin is still ongoing. It highlights the critical role commensal microorganisms play in influencing hormone metabolism, with these hormones, in turn, influencing the microbiota itself, even though the precise signalling mechanisms that define these interactions are still not fully understood. Researching how biofilms and enteric hormones influence health and illness is an insightful opportunity, especially with the advancement of enteric hormone–gut–microbiota research. With substantial evidence of the hormone–biofilms interplay, there is a compelling need for research to understand the molecular elements influencing the formation and persistence of biofilms, the hormonally compromised influence on host microbiota and the cross-talk occurring among the host's polymicrobial biofilms. Further, since there are exceptional conditions in both enteric hormones and biofilms concerning non-communicable diseases and ailments, it can give a deeper understanding and may help in predictive prognosis at the early onset of such conditions. In the end, this comprehensive system can maintain homeostatic balance by regulating physiological functions such as hunger regulation, metabolic activity, and immunological activity. The control of these interactions suggests new, promising therapy targets and opens up a new way to preserve host health and treat infections but maintain beneficial biofilm species.

REFERENCES

Acharya, K. D., Graham, M., Raman, H., Parakoyi, A. E. R., Corcoran, A., Belete, M., Ramaswamy, B., Koul, S., Sachar, I., Derendorf, K., Wilmer, J. B., Gottipati, S., & Tetel, M. J. (2023). Estradiol-mediated protection against high-fat diet induced anxiety and obesity is associated with changes in the gut microbiota in female mice. *Scientific Reports*, *13*(1). https://doi.org/10.1038/s41598-023-31783-6

Akimbekov, N. S., Digel, I., Yerezhepov, A. Y., Shardarbek, R. S., Wu, X., & Zha, J. (2022). Nutritional factors influencing microbiota-mediated colonization resistance of the oral cavity: A literature review. *Frontiers in Nutrition*, *9*. https://doi.org/10.3389/fnut.2022.1029324

Boukerb, A. M., Cambronel, M., Rodrigues, S., Mesguida, O., Knowlton, R., Feuilloley, M. G. J., Zommiti, M., & Connil, N. (2021). Inter-kingdom signaling of stress hormones: Sensing, transport and modulation of bacterial physiology. *Frontiers in Microbiology*, *12*. https://doi.org/10.3389/fmicb.2021.690942

Buret, A. G., & Allain, T. (2023). Gut microbiota biofilms: From regulatory mechanisms to therapeutic targets. *The Journal of Experimental Medicine*, *220*(3). https://doi.org/10.1084/jem.20221743

Cambronel, M., Tortuel, D., Biaggini, K., Maillot, O., Taupin, L., Réhel, K., Rincé, I., Muller, C., Hardouin, J., Feuilloley, M., Rodrigues, S., & Connil, N. (2019). Epinephrine affects motility, and increases adhesion, biofilm and virulence of Pseudomonas aeruginosa H103. *Scientific Reports*, *9*(1). https://doi.org/10.1038/s41598-019-56666-7

Chandra, N., Srivastava, A., & Kumar, S. (2019). Bacterial biofilms in human gastrointestinal tract: An intricate balance between health and inflammatory bowel diseases. *World Journal of Pharmacology*, *8*(3), 26–40. https://doi.org/10.5497/wjp.v8.i3.26

Cho, H., & Lim, J. (2024). The emerging role of gut hormones. *Molecules and Cells*, 100126. https://doi.org/10.1016/j.mocell.2024.100126

Clarke, G., Stilling, R. M., Kennedy, P. J., Stanton, C., Cryan, J. F., & Dinan, T. G. (2014). Minireview: Gut microbiota: The neglected endocrine organ. *Molecular Endocrinology*, *28*(8), 1221–1238. https://doi.org/10.1210/me.2014-1108

Danhof, H. A., Lee, J., Thapa, A., Britton, R. A., & Di Rienzi, S. C. (2023). Microbial stimulation of oxytocin release from the intestinal epithelium via secretin signaling. *Gut Microbes*, *15*(2). https://doi.org/10.1080/19490976.2023.2256043

De Silva, A., & Bloom, S. R. (2012). Gut hormones and appetite control: A focus on PYY and GLP-1 as therapeutic targets in obesity. *Gut and Liver*, *6*(1), 10–20. https://doi.org/10.5009/gnl.2012.6.1.10

de Vos, W. M., Tilg, H., Van Hul, M., & Cani, P. D. (2022). Gut microbiome and health: mechanistic insights. *Gut*, *71*(5), 1020–1032. https://doi.org/10.1136/gutjnl-2021-326789

Duran-Pinedo, A. E., Solbiati, J., & Frias-Lopez, J. (2018). The effect of the stress hormone cortisol on the metatranscriptome of the oral microbiome. *NPJ Biofilms and Microbiomes*, *4*(1). https://doi.org/10.1038/s41522-018-0068-z

Engevik, M. A., Danhof, H. A., Chang-Graham, A. L., Spinler, J. K., Engevik, K. A., Herrmann, B., Endres, B. T., Garey, K. W., Hyser, J. M., Britton, R. A., & Versalovic, J. (2020). Human intestinal enteroids as a model of *Clostridioides difficile*-induced enteritis. *American Journal of Physiology-Gastrointestinal and Liver Physiology*, *318*(5), G870–G888. https://doi.org/10.1152/ajpgi.00045.2020

Giordano, V., & Giannoudis, P. V. (2024). Biofilm formation, antibiotic resistance, and infection (BARI): The triangle of death. *Journal of Clinical Medicine*, *13*(19), 5779. https://doi.org/10.3390/jcm13195779

Gissey, L. C., Mariolo, J. C., & Mingrone, G. (2019). Roles of gut hormones in the regulation of food intake and body weight. In *Endocrinology* (pp. 75–88). https://doi.org/10.1007/978-3-319-46933-1_13

Hammouda, Z. K., Wasfi, R., & Abdeltawab, N. F. (2023). Hormonal drugs: Influence on growth, biofilm formation, and adherence of selected gut microbiota. *Frontiers in Cellular and Infection Microbiology*, *13*. https://doi.org/10.3389/fcimb.2023.1147585

Holingue, C., Budavari, A. C., Rodriguez, K. M., Zisman, C. R., Windheim, G., & Fallin, M. D. (2020). Sex differences in the gut–brain axis: Implications for mental health. *Current Psychiatry Reports*, *22*(12). https://doi.org/10.1007/s11920-020-01202-y

Jašarević, E., Morrison, K. E., & Bale, T. L. (2016). Sex differences in the gut microbiome–brain axis across the lifespan. *Philosophical Transactions of the Royal Society B Biological Sciences*, *371*(1688), 20150122. https://doi.org/10.1098/rstb.2015.0122

Jensen, A. B., Sørensen, T. I., Pedersen, O., Jess, T., Brunak, S., & Allin, K. H. (2018). Increase in clinically recorded type 2 diabetes after colectomy. *eLife*, *7*. https://doi.org/10.7554/elife.37420

Lai, T., Tsai, Y., Liou, C., Fan, C., Hou, Y., Yao, T., Chuang, H., & Wu, W. (2024). The gut microbiota modulate locomotion via vagus-dependent glucagon-like peptide-1 signaling. *NPJ Biofilms and Microbiomes*, *10*(1). https://doi.org/10.1038/s41522-024-00477-w

Leeuwendaal, N. K., Cryan, J. F., & Schellekens, H. (2021). Gut peptides and the microbiome: focus on ghrelin. *Current Opinion in Endocrinology Diabetes and Obesity*, *28*(2), 243–252. https://doi.org/10.1097/med.0000000000000616

Lesouhaitier, O., Clamens, T., Rosay, T., Desriac, F., Louis, M., Rodrigues, S., Gannesen, A., Plakunov, V. K., Bouffartigues, E., Tahrioui, A., Bazire, A., Dufour, A., Cornelis, P., Chevalier, S., & Feuilloley, M. G. (2018). Host peptidic hormones affecting bacterial biofilm formation and virulence. *Journal of Innate Immunity*, *11*(3), 227–241. https://doi.org/10.1159/000493926

LeValley, S. L., Tomaro-Duchesneau, C., & Britton, R. A. (2020). Degradation of the incretin hormone glucagon-like peptide-1 (GLP-1) by enterococcus faecalis metalloprotease GelE. *mSphere*, *5*(1). https://doi.org/10.1128/msphere.00585-19

Li, P., Luo, H., Ji, B., & Nielsen, J. (2022). Machine learning for data integration in human gut microbiome. *Microbial Cell Factories*, *21*(1). https://doi.org/10.1186/s12934-022-01973-4

Lopes, J. G., & Sourjik, V. (2018). Chemotaxis of Escherichia coli to major hormones and polyamines present in human gut. *The ISME Journal*, *12*(11), 2736–2747. https://doi.org/10.1038/s41396-018-0227-5

Mace, O. J., Tehan, B., & Marshall, F. (2015). Pharmacology and physiology of gastrointestinal enteroendocrine cells. *Pharmacology Research & Perspectives*, *3*(4). https://doi.org/10.1002/prp2.155

Martchenko, A., Martchenko, S. E., Biancolin, A. D., & Brubaker, P. L. (2020). Circadian rhythms and the gastrointestinal tract: Relationship to metabolism and gut hormones. *Endocrinology*, *161*(12). https://doi.org/10.1210/endocr/bqaa167

Neuman, H., Debelius, J. W., Knight, R., & Koren, O. (2015). Microbial endocrinology: the interplay between the microbiota and the endocrine system. *FEMS Microbiology Reviews*, *39*(4), 509–521. https://doi.org/10.1093/femsre/fuu010

Nogueira, A. V. B., Nokhbehsaim, M., Damanaki, A., Eick, S., Beisel-Memmert, S., Kirschneck, C., Schröder, A., Cirelli, T., Leguizamón, N. D. P., Cirelli, J. A., & Deschner, J. (2022). Effect of bacterial infection on ghrelin receptor regulation in periodontal cells and tissues. *International Journal of Molecular Sciences*, *23*(6), 3039. https://doi.org/10.3390/ijms23063039

Olivares, M., Schüppel, V., Hassan, A. M., Beaumont, M., Neyrinck, A. M., Bindels, L. B., Benítez-Páez, A., Sanz, Y., Haller, D., Holzer, P., & Delzenne, N. M. (2018). The potential role of the dipeptidyl Peptidase-4-Like activity from the gut microbiota on the host health. *Frontiers in Microbiology*, *9*. https://doi.org/10.3389/fmicb.2018.01900

Perry, E. K., & Tan, M. (2023). Bacterial biofilms in the human body: prevalence and impacts on health and disease. *Frontiers in Cellular and Infection Microbiology*, *13*. https://doi.org/10.3389/fcimb.2023.1237164

Rodríguez-Daza, M. C., Pulido-Mateos, E. C., Lupien-Meilleur, J., Guyonnet, D., Desjardins, Y., & Roy, D. (2021). Polyphenol-mediated gut microbiota modulation: Toward prebiotics and further. *Frontiers in Nutrition*, *8*. https://doi.org/10.3389/fnut.2021.689456

Rogers, G. B., Keating, D. J., Young, R. L., Wong, M., Licinio, J., & Wesselingh, S. (2016). From gut dysbiosis to altered brain function and mental illness: Mechanisms and pathways. *Molecular Psychiatry*, *21*(6), 738–748. https://doi.org/10.1038/mp.2016.50

Santos-Marcos, J. A., Mora-Ortiz, M., Tena-Sempere, M., Lopez-Miranda, J., & Camargo, A. (2023). Interaction between gut microbiota and sex hormones and their relation to sexual dimorphism in metabolic diseases. *Biology of Sex Differences*, *14*(1). https://doi.org/10.1186/s13293-023-00490-2

Sartor, B. (2018). Intestinal bacterial biofilms modulate mucosal immune responses. *Journal of Immunological Sciences*, *2*(2), 13–18. https://doi.org/10.29245/2578-3009/2018/2.1122

Sharma, S., Mohler, J., Mahajan, S. D., Schwartz, S. A., Bruggemann, L., & Aalinkeel, R. (2023). Microbial biofilm: A review on formation, infection, antibiotic resistance, control measures, and innovative treatment. *Microorganisms*, *11*(6), 1614. https://doi.org/10.3390/microorganisms11061614

Srednicka, P., Juszczuk-Kubiak, E., Wójcicki, M., Akimowicz, M., & Roszko, M. Ł. (2021). Probiotics as a biological detoxification tool of food chemical contamination: A review. *Food and Chemical Toxicology*, *153*, 112306. https://doi.org/10.1016/j.fct.2021.112306

Steinert, R. E., Feinle-Bisset, C., Asarian, L., Horowitz, M., Beglinger, C., & Geary, N. (2016). Ghrelin, CCK, GLP-1, and PYY(3–36): Secretory controls and physiological roles in eating and glycemia in health, obesity, and after RYGB. *Physiological Reviews*, *97*(1), 411–463. https://doi.org/10.1152/physrev.00031.2014

Sun, L., Li, J., & Nie, Y. (2020). Gut hormones in microbiota-gut-brain cross-talk. *Chinese Medical Journal*, *133*(7), 826–833. https://doi.org/10.1097/cm9.0000000000000706

Tiamani, K., Luo, S., Schulz, S., Xue, J., Costa, R., Mirzaei, M. K., & Deng, L. (2022). The role of virome in the gastrointestinal tract and beyond. *FEMS Microbiology Reviews*, *46*(6). https://doi.org/10.1093/femsre/fuac027

Tomaro-Duchesneau, C., LeValley, S. L., Roeth, D., Sun, L., Horrigan, F. T., Kalkum, M., Hyser, J. M., & Britton, R. A. (2020). Discovery of a bacterial peptide as a modulator of GLP-1 and metabolic disease. *Scientific Reports*, *10*(1). https://doi.org/10.1038/s41598-020-61112-0

Valiei, A., Aminian-Dehkordi, J., & Mofrad, M. R. K. (2023). Gut-on-a-chip models for dissecting the gut microbiology and physiology. *APL Bioengineering*, *7*(1), 011502. https://doi.org/10.1063/5.0126541

Valverde-Pozo, J., Paredes, J. M., García-Rubiño, M. E., Girón, M. D., Salto, R., Alvarez-Pez, J. M., & Talavera, E. M. (2024). Advanced imaging methodology in bacterial biofilms with a fluorescent enzymatic sensor for pepN activity. *Biosensors*, *14*(9), 424. https://doi.org/10.3390/bios14090424

Von Rosenvinge, E. C., O'May, G. A., Macfarlane, S., Macfarlane, G. T., & Shirtliff, M. E. (2012). Microbial biofilms and gastrointestinal diseases. *Pathogens and Disease*, *67*(1), 25–38. https://doi.org/10.1111/2049-632x.12020

Wallace, M. J., Fishbein, S. R. S., & Dantas, G. (2020). Antimicrobial resistance in enteric bacteria: current state and next-generation solutions. *Gut Microbes*, *12*(1), 1799654. https://doi.org/10.1080/19490976.2020.1799654

Wang, Y., Xu, S., He, Q., Sun, K., Wang, X., Zhang, X., Li, Y., & Zeng, J. (2023). Crosstalk between microbial biofilms in the gastrointestinal tract and chronic mucosa diseases. *Frontiers in Microbiology*, *14*. https://doi.org/10.3389/fmicb.2023.1151552

Yadav, H., Lee, J., Lloyd, J., Walter, P., & Rane, S. G. (2013). Beneficial metabolic effects of a probiotic via butyrate-induced GLP-1 hormone secretion. *Journal of Biological Chemistry*, *288*(35), 25088–25097. https://doi.org/10.1074/jbc.m113.452516

Ye, L., & Liddle, R. A. (2016). Gastrointestinal hormones and the gut connectome. *Current Opinion in Endocrinology Diabetes and Obesity*, *24*(1), 9–14. https://doi.org/10.1097/med.0000000000000299

Zhang, T., Wang, W., Li, J., Ye, X., Wang, Z., Cui, S., Shen, S., Liang, X., Chen, Y. Q., & Zhu, S. (2025). Free fatty acid receptor 4 modulates dietary sugar preference via the gut microbiota. *Nature Microbiology*. https://doi.org/10.1038/s41564-024-01902-8

13 Catecholamines and Biofilm

Rina Rani Ray
Department of Biotechnology, Maulana Abul Kalam Azad University of Technology, Kolkata, India

13.1 INTRODUCTION

Biofilm is a microbial aggregation, which is found to be formed on various biotic surfaces such as dental enamel and cardiac tissue, resulting in various recurrent diseases. On the other hand, biofilm formed on abiotic surfaces such as prosthetics and surgical implants, stents, contact lenses and catheters leads to the failure of devices, persistent infection causing regular inflammation at the implant area and chronic infection, which is not susceptible to cure by antibiotics. Since the internal milieu of the body greatly influences the microbial community and biofilm formation, extensive research is going on to check the effect of physical and mental stress on biofilm formation. Hormone, the chemical coordinator of all physiological functioning, must be one of the significant regulators of microbial growth and biofilm formation. Stress brings about notable alterations in the serum level of many hormones, the most noteworthy of which is catecholamines. Hence it is expected that various hormones coming under the nomenclature of catecholamines have a significant effect on biofilm formation or eradication.

This chapter deals with the effect of various catecholamine hormones on biofilm, formed by various microbes.

13.2 CATECHOLAMINES

Catecholamines are a group of hormones which are released from adrenal medulla. These are basically neurotransmitters released to maintain homoeostasis after having some stress. Structurally, they are monoamine neurotransmitters. This group of hormones include epinephrine (E), norepinephrine (NE), and dopamine (Figure 13.1). For example, dobutamine is a synthetic catecholamine which is used in the treatment of heart diseases. These hormones share a structural similarity, having a catechol molecule with a side-chain amine. A catechol or pyrocatechol or 1,2-dihydroxybenzene is a natural benzene derivative with two hydroxyl groups with the molecular formula $C_6H_4(OH)_2$. They prepare the body under stress conditions by inducing a fight, flight and fright response.

The biosynthesis of catecholamines involve a multi-step process, deriving from an L-tyrosine precursor. It is converted to dopa and dopamine with the help of tyrosine hydroxylase and dopa decarboxylase, respectively. Dopamine is now converted to norepinephrine by dopamine decarboxylase and then to epinephrine with the help of phenyl-ethanolamines N-methyl transferase (PNMT). These are neurotransmitters produced by the nerve cells and the medulla part of the adrenal glands. The major catecholamine hormones, namely adrenalin and noradrenaline, are released from adrenal medulla under various stresses, to help the body to avoid various adverse situations (Figure 13.2).

Under nervous tension and traumatic conditions, these hormones cause the body to adopt the strategy to fright (to realize the severity of the situation), fight (to combat with the situation) or flight (to escape from the situation).

Under such situations, the blood level of catecholamine rises, leading to an increase in blood pressure, heart rate and breathing rate, which leads to an increase in blood flow to vital visceral organs to prepare the body to accept the stress. They increase the sugar metabolism to release more

DOI: 10.1201/9781003519041-13

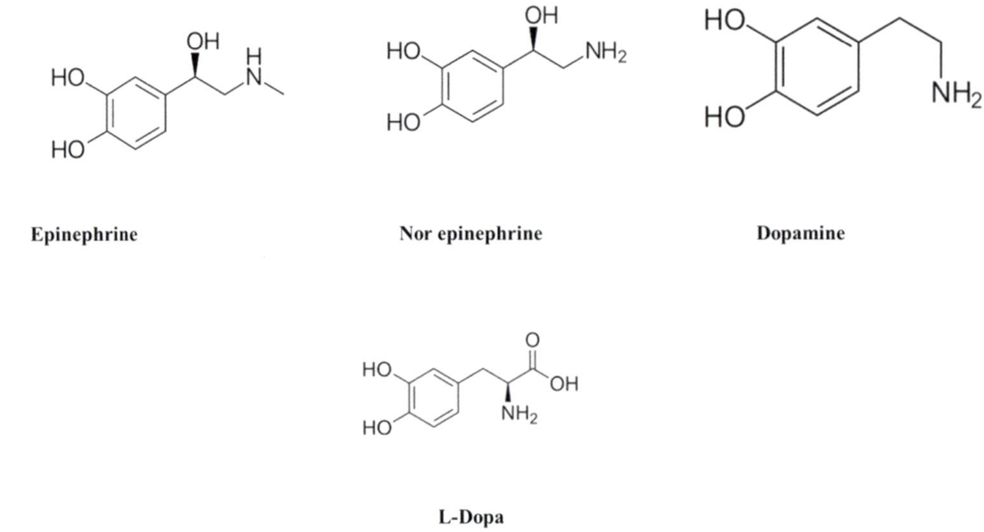

FIGURE 13.1 Structure of catecholamine group of hormones.

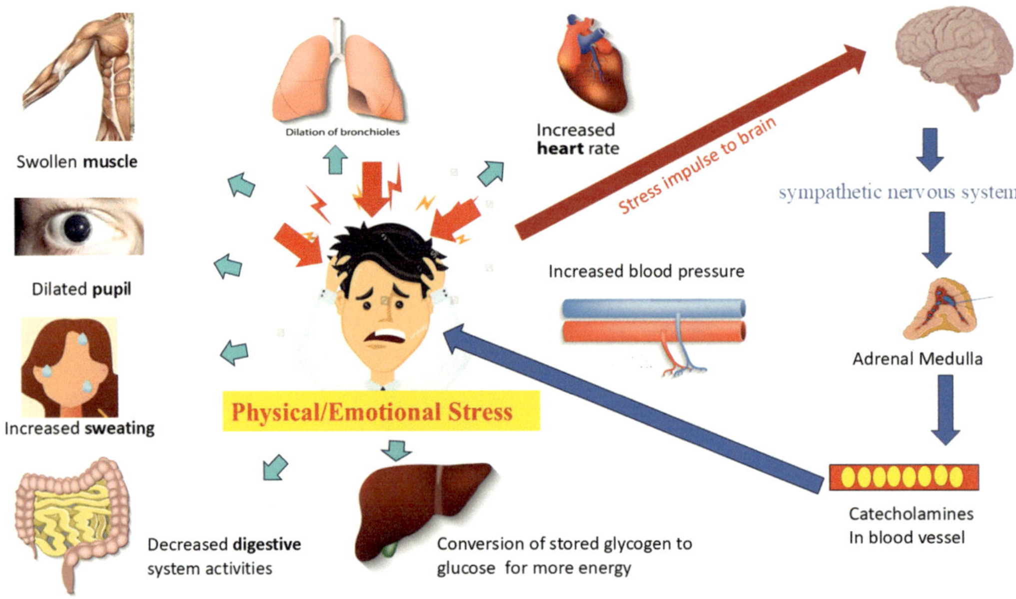

FIGURE 13.2 Stress management by catecholamines.

energy and break down stored fat in adipose tissue to mobilize free fatty acids. Such metabolic changes lead to thermogenesis, and the generated heat is dissipated by increased sweating. Catecholamines increase muscle strength and mental alertness, thereby making the body prepared to combat the stressed condition and maintain a steady state both physically and mentally. Accordingly, these hormones are known as stress hormones.

13.3 THE MODE OF ACTION OF CATECHOLAMINES

Catecholamines bind with specific receptors present on the plasma membrane of cells. Of these hormones, epinephrine (E) and norepinephrine (NE) bind to α and β adrenergic receptors, which are actually the subtypes of G-protein coupled receptors. Dopa binds to another G-coupled receptor, the dopaminergic receptors, having five subtypes. Interestingly, antagonists of adrenergic and dopaminergic receptors found in eukaryotes are also able to inhibit the bacterial catecholamine-induced effects (Sharaff & Freestone, 2011).

Catecholamines can also help in establishing communication between bacteria and their hosts, by binding the specific adrenergic receptors, a phenomenon called inter-kingdom signalling. Catecholamines can increase the growth and virulence of some pathogenic bacteria in vitro after binding with specific receptor proteins as QseC in *E coli*.

Exposure to epinephrine and norepinephrine can increase the virulence of enterohemorrhagic *Escherichia coli*. Epinephrine binds and signals through the QseC receptor.

Additionally, there are some more catecholamine receptors found in bacteria, like histidine sensor kinases QseC and QseE (Hughes et al., 2009). A highly active bacterial catecholamine receptor antagonist, LED209, has identified that can neutralize the effects of catecholamines. It is also found in *Salmonella enterica* (Moreira & Sperandio, 2012) as adrenergic sensor kinase.

Conversely, dopamine, which is a form of catecholamine hormone, can function as both a promoter and a disruptor of biofilms. While it has the ability to act as a signaling molecule that encourages biofilm development in certain bacteria, it can also break down the biofilm structure by inducing polymerization. Dopamine may enhance bacterial pathogenicity by boosting protein metabolism and energy release through the activation of genes linked to quorum-sensing pathways. It can interact with ferrous ions to expedite polymerization in biofilms. Research indicated that both adrenergic and dopaminergic antagonists exhibited a high degree of specificity in their modes of action, which did not involve the obstruction of catecholamine-mediated iron acquisition. The utilization of radiolabeled N suggested that adrenergic antagonists might inhibit catecholamine uptake.

The current findings indicate that a specific pathogen's responsiveness to a certain hormone is influenced by the anatomical area of the host where the disease manifests, as well as the neuroanatomical specificity associated with the hormone's production; both factors have a coincidental anatomical relationship. Consequently, this report proposes that pathogens predominantly affecting the gastrointestinal tract have developed response mechanisms to neuroendocrine hormones like norepinephrine and dopamine, but not to epinephrine, which are present in the enteric nervous system (Freestone et al., 2007).

13.4 ROLE OF CATECHOLAMINES ON MICROBES

A two-way interaction between microbes and human neuroendocrine factors is suggested by the proponent of the concept of "microbial endocrinology" (Lyte & Ernst, 1992). Since under stressful conditions, a body becomes more susceptible to various microbial attack, it can be presumed that the stress hormones or catecholamines play some role in the microbial growth and the increase in virulence (Vinasco et al., 2019) production up to 20–40 times (Boukerb et al., 2021).

13.4.1 Effect of Catecholamines on the Growth and Biofilm of Gram-Positive Bacteria

Studies have demonstrated that exposure to catecholamines can increase the formation of biofilms in various Gram-positive bacteria, most of them being potent pathogens. Many of them attack the live tissue and also, if they come into contact with abiotic surfaces, damage prosthetics and implants. Research suggests that both epinephrine (or adrenaline) and nor epinephrine (or nor adrenalin), two stress hormones, can generally enhance the growth of some Gram-positive bacteria, particularly by stimulating biofilm formation. However, the exact effect can vary depending on the specific bacterial species and the concentration of epinephrine present.

Cutibacterium acnes, an opportunistic pathogen, forms biofilms on skin, infecting skin glands and causing inflammation and skin diseases such as acne and pustulosis. It is also concerned with various biofilm-mediated diseases like endophthalmitis, endocarditis, septic arthritis, osteomyelitis, chronic prostatitis, sarcoidosis, synovitis, hyperostosis, and osteitis syndrome. It is also known for its ability to form biofilms on medical devices, frequently causing prosthesis infections. It is found that both epinephrine and norepinephrine promote biofilm production in *C acnes* (Gannesen et al., 2021). The transmembrane protein, histidine kinase (KdpD), is thought to act as the sensor of epinephrine in *C. acnes*.

Another acne-causing microorganisms *Propionibacterium acnes*, an aerotolerant anaerobic bacterium, is found to exist on all external and internal surfaces, including skin, oral and gastrointestinal epithelia and conjunctiva. It was found that the cells became dispersed from already formed biofilm on the atherosclerotic carotid artery with the release of lipase and protease after an interaction between the transferrin-bound iron and norepinephrine (NE).

This sudden plaque rupture under the action of NE released under stress may lead to cardiac tissue damage, heart attacks and strokes, which may prove fatal (Lanter & Davies, 2015).

Skin bacteria like *Micrococcus luteus* can be affected by epinephrine in terms of changes in biofilm composition. *M. luteus* can form biofilms on abiotic surfaces, including plastic, and microfluidic channels. Biofilms can damage the underlying material and pose a health risk to humans. While the exact mechanism has not yet been made completely clear, it's believed that bacteria might possess receptors that interact with epinephrine, potentially triggering signaling pathways that lead to altered gene expression and biofilm development.

Staphylococcus aureus, being capable of forming multilayered biofilm, augments the antibiotic resistance, leading to the life-threatening infections of various organs. Both norepinephrine and epinephrine can significantly promote the formation of *Staphylococcus aureus* biofilms by facilitating iron acquisition for the bacteria (Flannagan et al., 2022).

Catecholamines make the body adapted to stress conditions by encouraging energy expenditure. To bring back the expended energy, the E/NE-exposed cells require more iron.

Hence, under persistent and severe stress conditions, the regulation of the hypothalamic–pituitary–adrenal (HPA) axis is hampered, leading to the stimulation of hepcidin hormone production from the liver. This plays a pivotal role in iron absorption and distribution and maintains iron homeostasis in body. In iron-limited conditions, *Staphylococcus aureus* biofilm formation is found to be encouraged (Johnson et al., 2008).

Among the cytotoxic alpha-toxin release regulating loci of *Staphylococcus aureus*, namely, agr, sarA and sae, sae and agr are found to be essential and hence for biofilm formation, whereas sarA appears to have a less-significant role. The *ica* operon is required for biofilm formation in iron-restricted growth conditions (Johnson et al., 2008).

Blood infections caused by *Staphylococcus aureus* typically occur due to biofilm development on central venous catheters (CVC) and haemodialysis catheters, which are significant healthcare-related infections.

Another bacterium, namely *Staphylococcus epidermidis*, is a natural inhabitant of skin. As an opportunistic pathogen, if it can enter the human body via any invasive device. It causes virulence and proves itself to be the most common medical device contaminant. The bacterium now moves from the human skin and enters the bloodstream after forming a colony on the medical device. The findings of Neal et al. (2001) clearly confirmed that the inotropic catecholamines significantly promoted the growth of *Staphylococcus epidermidis*.

The capability of catecholamine inotropic agents to promote the growth of bacteria and the formation of biofilms might contribute to the onset of colonization of intravascular catheters and subsequent catheter-associated infections. When *S. pneumoniae* is exposed to catecholamines, its growth is enhanced, potentially facilitating greater bacterial colonization in the respiratory tract and thereby raising the risk of infection.

Studies have indicated that epinephrine can significantly enhance biofilm formation in various Gram-positive bacteria, even if it doesn't directly increase their growth rate. The impact of epinephrine on different Gram-positive bacteria can vary depending on the species. For example, some research shows that skin bacteria like *Micrococcus luteus* can form biofilms on surfaces, including skin, plastic, and microfluidic channels. Biofilms can damage the underlying material and pose a health risk to humans.

While the exact mechanism is not completely elucidated, it's believed that bacteria might possess receptors that interact with epinephrine, potentially triggering signalling pathways that lead to altered gene expression and biofilm development. Epinephrine has some effects at gene expression levels and shows a multifaceted effect on biofilm formation in *Micrococcus luteus* (Gannesen et al., 2021).

Scientists reported the inhibitory effect of norepinephrine on biofilm formation by *Kytococcus schroeteri*, a human skin commensal (Danilova et al., 2021) as the biofilms formed on glass in the presence of NE were thinner and flatter than those in the controls, although no significant effect on metabolic activity of the cells were found. Hence, they opined that the initial stages of the switching from planktonic to sessile biofilm lifestyle and matrix synthesis were probably affected by NE.

The gastrointestinal disease-causing Gram-positive bacteria, namely *Campylobacter jejuni*, *Listeria monocytogenes*, *Aeromonas hydrophila*, and *Streptococcus pneumoniae*, cause diarrhoea, nausea, vomiting etc. The growth profiles of these bacteria are found to be enhanced by E and NE (Niu et al., 2023).

Streptococcus pneumoniae, which causes both non-invasive (such as pneumonia) and invasive diseases (such as meningitis) in the presence of catecholamines, experiences enhanced growth and increased colonization (Gonzales et al., 2010). NE stimulates growth of *S. pneumoniae* by allowing access to the iron within host transferrin, an iron-binding protein through inotrope-delivery of transferrin-iron and internalization of the inotrope. NE also markedly increased *S. pneumoniae*-biofilm formation. Gene profiling reveals that the expression of *pspA* and *pspC* genes for host colonization is triggered in the presence of NE, leading to biofilm formation and the onset of pneumococcal pneumonia (Sandrini et al., 2014).

Enterococcus faecalis, another Gram-positive gut resident, is often used in dairy industry as starter cultures and also as probiotics. Cambronel et al. (2020) worked with different strains of *E. faecalis*, including pathogenic and probiotic strains. They observed the biofilm-inducing effect of epinephrine on all such strains, precisely on increasing the volume of biofilm. However, NE was found to have less effect, while both E and NE could increase the adhesion of the bacterial cell on gut epithelial cells and skin cells of the eukaryotic host.

The VicK protein was assumed to be the probable adrenergic sensor for E and NE in *E. faecalis*, whereas QseC acts as the adrenergic receptor in *E. coli*.

13.4.2 Effect of Catecholamines on the Growth and Biofilm of Gram-Negative Bacteria

Among various Gram-negative bacteria, *Pseudomonas aeruginosa* is regarded to be one of the most important plant and animal pathogen. It can attack various human tissues, especially the lungs, causing cystic fibrosis and triggering lung damage. Keeping itself protected by biofilm, it often acquires resistance to multiple antibiotic drugs. Dickson et al. (2014) identified an in vitro relationship between catecholamines and the growth of several bacterial species, notably including key members of the lung microbiome such as *Pseudomonas aeruginosa*. Additionally, Cambronel et al. (2019) demonstrated that epinephrine enhanced the adhesion and biofilm formation of *P. aeruginosa*, as well as its virulence in an in vivo model using *Galleria mellonella* larvae. In the case of *P. aeruginosa* strain PA14, the las quorum-sensing pathway may play a role in the bacterium's response to norepinephrine, which subsequently regulates its virulence and motility (Hegde et al., 2009).

Furthermore, a separate investigation on the *P. aeruginosa* strain PAO1 revealed that norepinephrine could inhibit siderophore production while promoting the expression of toxA, a critical virulence factor associated with exotoxin A. New research has found that in *P. aeruginosa* PAO1, dopamine serves as an exogenous quorum sensing (QS)-signalling molecule that enhances the bacterial virulence and pathogenicity through the manifold increase of biofilm formation with associated quorum-sensing molecules. Transcriptomic analysis showed that dopamine promoted the expression of virulence genes *psl, alg, lasA, rhlABC, rml*, and *phz* in *P. aeruginosa* PAO1 (Xiang et al., 2024).

It is found that the addition of various catecholamines (NE, E, dopamine and dopa) to cultures can dramatically increase the growth of various Gram-negative bacteria, namely *E. coli, Salmonella enterica, V. cholerae, H. pylori, Yersinia enterocolitica*, and *P. aeruginosa*, but this is dependent on the type of hormone employed and species of bacteria used (Lyte & Ernst, 1992).

N was found to be more efficient in enhancing biofilm formation by *Salmonella enteritidis* strains at 12°C accomplished by three main genes, namely csgD, fimA, and adrA (Hiller et al., 2019).

Enterotoxigenic E. coli (ETEC), *Salmonella* sp., and *Vibrio* spp. are major enteropathogenic bacteria, often showing virulence and were found to be stimulated in presence of catecholamines (Niu et al., 2023).

Catecholamines have been shown to increase the virulence of bacteria and activate genes associated with virulence. In strains of enterohemorrhagic *E. coli* (EHEC), the key virulence genes include Shiga toxins (Stx), the flagella regulon, and the LEE genes, which are responsible for encoding the type III secretion system, the adhesin known as intimin, and its receptor, Tir. It was found that E- and NE-exposed strains of *E. coli* O157:H7 have activated type III secretion and increased LEE1 transcription.

It was also opined that E and NE act as chemo attractants, enhancing the motility of various bacteria, including *S. enterica* serovar Typhimurium, *E. coli* O157:H7, *Campylobacter jejuni* NCTC 11168, and other pathogenic strains such as *Pseudomonas fluorescens* MFN1032 *Pseudomonas aeruginosa* PAO1/H103, *P. aeruginosa* PA14, and *Vibrio* sp (Rushing & Santoro, 2022).

It is found that asymptomatic strains of ETEC of gut start showing pathogenic symptoms if the host experiences stress and releases catecholamines, although the mode of action is still unclear.

The growth of *E coli* was markedly enhanced in the presence of both NE and E, although the growth was more pronounced in presence of NE. It was found that the parasympathetic stress response of the host may contribute to the pathogenesis of some bacteria, such as *E. coli, Klebsiella pneumoniae*, and *Pseudomonas aeruginosa*. But the growth and biofilm formation in other tested enteropathogenic bacteria, namely *Enterobacter sp*, *Salmonella choleraesuis subsp. cholleraesuis* serotype *typhi*, and *Shigella boydii*, was not affected in presence of catecholamines. *Bacteroides fragilis*, a natural inhabitant of the colon, is also unaffected by catecholamines (Belay et al., 2003).

Among various oral pathogens, the growth of *Porphyromonas gingivalis*, a natural component of the oral microbiota, was not significantly affected by any catecholamine hormone. In contrast, the growth of *Prevotella intermedia* and *Eikenella corrodens* was inhibited by norepinephrine and nearly all stress hormones examined. Different hormones exert varying effects on the growth of bacteria associated with periodontitis in vitro. It seems that bacterial viability is more significantly impacted than bacterial metabolic activity. Additionally, the growth of *Fusobacterium nucleatum* and *Tannerella forsythia* is stimulated by several stress hormones, which may exacerbate pathological conditions (Jentsch et al., 2013).

This demonstrates that the enhancement of bacterial growth by catecholamines is not a generalized phenomenon (Figure 13.3). Rather, the enhancement of bacterial growth by catecholamines is specific to certain species (Belay et al., 2003). The catecholamines are found to enhance H_2S production within the oral environment by different anaerobic periodontal pathogens, including *Fusobacterium nucleatum*, *Porphyromonas endodontalis*, and *Prevotella intermedia* and thereby possibly increasing halitosis (malodor) in patients under stress conditions (Calil et al., 2014).

Vibeio harveyi, an important marine pathogen for fish and aquatic organisms, causes various infective diseases with the symptoms of gastro-enteritis, skin and tissue necrosis, tail rot disease in

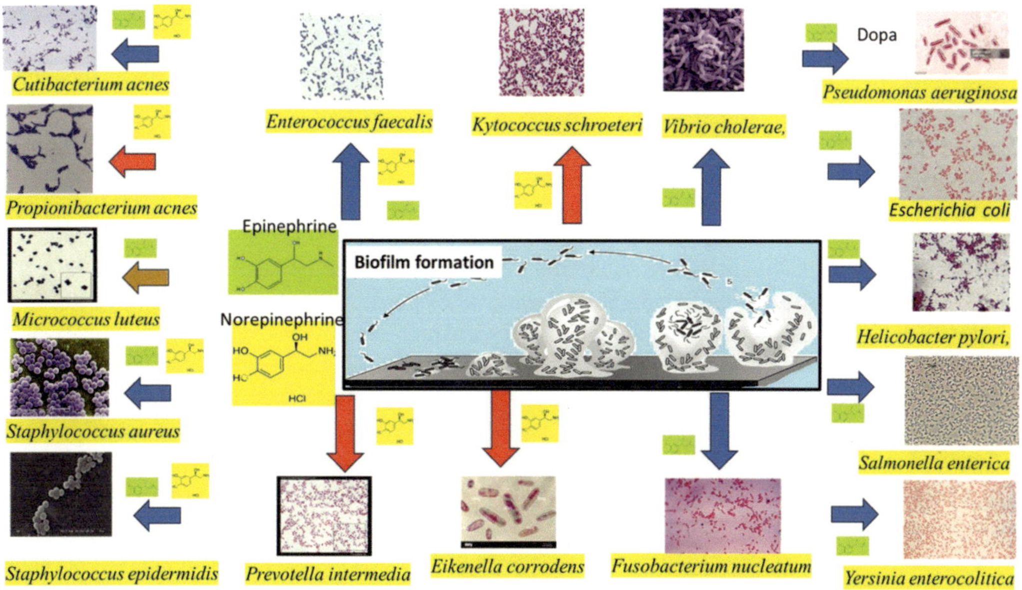

FIGURE 13.3 Effect of epinephrine and norepinephrine on the biofilm formation of various bacteria.

fish, and blindness. Yang et al. (2014) reported that supplementation of NE and dopamine in a growth medium could enhance the growth of *V. harveyi* along with biofilm production marked by the enhanced production of exopolysaccharides, siderophore, and swimming motility with boosted flagellar movement and growth in the presence of a very low amount of iron. Exposure to catecholamines like NE and dopamine make the bacterium more virulent towards brine shrimp larvae (Yang et al., 2014). Hence, the pretreatment of *V. harveyi* with catecholamines will lead to the significant decrease in the survival of brine shrimp larvae.

Since these effects can be reduced or nullified in the presence of antagonists of bacterial or eukaryotic catecholamine receptors, it is crystal clear that whenever the pathogenic bacterium comes into contact with a higher concentration of catecholamine, released as a result of stress, the expression of virulence factors is stimulated to form a new biofilm colony (Yang et al., 2014).

The stress-associated hormones E and NE of the human host act as signal molecules for the production of virulence factors in *V. cholerae*, and promote bacterial growth by sequestration of iron from the host (Halang et al., 2015). It was found that both E- and NE-enhanced *Helicobacter pylori* growth in a dose-dependent strain-independent fashion, with NE being more effective than E (Doherty et al., 2009).

Brachyspira pilosicoli is a Gram-negative, anaerobic spirochete that is associated with hosts and inhabits the intestinal tract of both animals and humans. It is responsible for conditions such as focal colitis and chronic diarrhoea, which may present with symptoms such as abdominal pain, flatulence, gastrointestinal bleeding, rectal discharge, watery diarrhea, and pseudoappendicitis. Research has indicated that NE can influence its virulence factors; specifically, elevated levels of norepinephrine enhance the attraction of *B. pilosicoli* to mucin and promote a strong adherence to intestinal cells, facilitating biofilm development.

13.5 EFFECT OF CATECHOLAMINES ON *CANDIDA* BIOFILM

Research suggests that catecholamines, stress hormones like norepinephrine and epinephrine, may potentially influence the formation and virulence of *Candida* biofilms, although the exact mechanisms are still being investigated, with some studies indicating that elevated catecholamine levels could potentially promote *Candida* biofilm development and resistance to antifungal treatments; however, further research is needed to fully understand this relationship. Catecholamines are released during stress situations, and their presence can impact various bodily functions including immune response.

Research suggests that catecholamines, stress hormones such as N and E, may influence the formation and virulence of *Candida* biofilms, though the exact mechanisms are still being investigated. Some studies suggest that elevated catecholamine levels may promote *Candida* biofilm development and resistance to antifungal treatments; however, more research is needed to fully understand this relationship. Catecholamines are released under stressful events, and their presence can affect a variety of body activities, including the immunological response.

Moreover, it is anticipated that catechol's anti-virulence mechanism against *C. albicans* may involve the activation of the quorum-sensing molecule farnesol. This molecule's accumulation naturally inhibits hyphal elongation and the essential process of biofilm formation by enhancing the expression of TUP1, a negative regulator of hyphal growth. The results from the serial passage experiment further confirmed the prolonged non-fungicidal antihyphal effectiveness of catechol.

This study has effectively highlighted an additional pharmacological aspect of catechol in combating the virulence of *C. albicans* (ATCC 10231). Moreover, it is anticipated that catechol's anti-virulence mechanism against *Candida albicans* may involve the activation of the quorum-sensing molecule farnesol, which inhibits hyphal growth and its necessary precursor—biofilm formation—by enhancing the expression of TUP1 (a negative regulator of hyphal development). The serial passage experiments further confirmed the prolonged non-fungicidal—anti-hyphal effectiveness of catechol over an extended duration. The current research has effectively shown an additional pharmacological aspect of catechol in combating the virulence of *C. albicans* (ATCC 10231).

According to current research, NE, at its physiological concentration, does not appear to significantly affect the growth or formation of *Candida* biofilms, since the high blood level of NE does not show the stimulation of growth, dimorphism, stress tolerance and virulence in *C albicans*. However, at sub-minimal concentrations, NE displayed remarkable anti-virulence activity, due to the inactivation of hydrolase, the virulence enzyme of *Candida* (Jothi et al., 2021). On the contrary, very high levels of catecholamines support cell adhesion, increased hyphal development and production of EPS components and thereby promote biofilm formation in *Candida*. This may be the underlying cause of the development of *Candida* infections on catheters in patients suffering from chronic stress with consequent high catecholamine levels.

On the other hand, E effectively enhances the biofilm formation by *Candida*, although the exact mechanisms are still yet to be known, However, the positive effect is dependent on the target species of *Candida* and the concentration E used.

E could possibly encourage the creation and growth of *Candida* biofilms, although the precise mechanisms are still under research. Some studies indicate that physiological levels of E can enhance biofilm biomass and matrix elements in specific microbial species, including various *Candida* strains; however, the impact may differ based on the particular *Candida* species and E concentration applied.

Studies have indicated that epinephrine, at physiological concentrations, can stimulate biofilm formation in certain bacteria and may have a similar effect on some *Candida* species, potentially by altering gene expression related to biofilm development. Epinephrine may influence the composition of the biofilm matrix, increasing the production of extracellular polymeric substances (EPS) which contribute to the biofilm's structural integrity.

Some research suggests that epinephrine could activate stress response pathways in *Candida*, potentially leading to increased biofilm formation as a survival strategy under stressful conditions.

This discovery might be important for individuals suffering from high stress or conditions linked to elevated epinephrine, as it could raise their vulnerability to *Candida* infections, especially those related to biofilms. Some research indicates that epinephrine enhances *Candida* biofilm formation, while other studies report no notable impact, suggesting possible species-specific variations in reaction. The level of epinephrine is vital, since elevated doses can lead to varying effects when compared to normal physiological levels. Further studies are needed to completely grasp the mechanisms that underlie the interaction between epinephrine and the formation of *Candida* biofilms, especially the particular signalling pathways that are engaged.

13.6 CONCLUSION

From various experimental observations, it is now clear that the catecholamines have receptors on both eukaryotic and bacterial cells, which made the communication between these cells easier. There is a general tendency of these catecholamines to encourage biofilm formation in various microbes through the release of virulent factors. Excessive or continued exposure of these hormones cause damage to prosthetics and implants by encouraging biofilm growth on them. Although there is no such antibiofilm activities of E and NE, dopamine, due to its configurational similarity to quorum-sensing molecules, can take part in ferrous-assisted *in situ* polymerization that results in the disintegration of the biofilm matrix and biofilm destruction.

REFERENCES

Belay, T., Aviles, H., Vance, M., Fountain, K., & Sonnenfeld, G. (2003). Catecholamines and in vitro growth of pathogenic bacteria: enhancement of growth varies greatly among bacterial species. *Life Sciences*, *73*(12), 1527–1535.

Boukerb, A. M., Cambronel, M., Rodrigues, S., Mesguida, O., Knowlton, R., Feuilloley, M. G. J., Zommiti, M., Connil, N. (2021). Inter-kingdom signaling of stress hormones: Sensing, transport and modulation of bacterial physiology. *Frontiers in Microbiology*, *12*, 690942.

Calil, C. M., Oliveira, G. M., Cogo, K., Pereira, A. C., Marcondes, F. K., Groppo, F. C. (2014). Effects of stress hormones on the production of volatile sulfur compounds by periodontopathogenic bacteria. *Brazilian Oral Research*, *28*(01), 01–08.

Cambronel, M., Nilly, F., Mesguida, O., Boukerb, A. M., Racine, P. J., Baccouri, O., Borrel, V., Martel, J., Fécamp, F., Knowlton, R., Zimmermann, K., Domann, E., Rodrigues, S., Feuilloley, M., & Connil, N. (2020). Influence of catecholamines (epinephrine/norepinephrine) on biofilm formation and adhesion in pathogenic and probiotic strains of enterococcus faecalis. *Frontiers in Microbiology*, *11*, 1501.

Cambronel, M., Tortuel, D., Biaggini, K., Maillot, O., Taupin, L., Réhel, K., Rincé, I., Muller, C., Hardouin, J., Feuilloley, M., Rodrigues, S., & Connil, N. (2019). Epinephrine affects motility, and increases adhesion, biofilm and virulence of *Pseudomonas aeruginosa* H103. *Scientific Reports*, *9*(1), 20203.

Danilova, N. D., Geraskina, O. V., Diuvenji, E. V., Feofanov, A. V., Plakunov, V. K., & Gannesen, A. V. (2021). Inhibitory effect of norepinephrine on biofilm growth of the human skin commensal *Kytococcus schroeteri* H01. *Microbiology*, *90*, 666–669.

Dickson, R. P., Martinez, F. J., & Huffnagle, G. B. (2014). The role of the microbiome in exacerbations of chronic lung diseases. *Lancet (London, England)*, *384*(9944), 691–702. https://doi.org/10.1016/S0140-6736(14)61136-3

Doherty, N. C., Tobias, A., Watson, S., & Atherton, J. C. (2009). The effect of the human gut-signalling hormone, norepinephrine, on the growth of the gastric pathogen *helicobacter pylori*. *Helicobacter*, *14*(3), 223–230.

Flannagan, R. S., Brozyna, J. R., Kumar, B., Adolf, L. A., Power, J. J., Heilbronner, S., & Heinrichs, D. E. (2022). In vivo growth of *Staphylococcus lugdunensis* is facilitated by the concerted function of heme and non-heme iron acquisition mechanisms. *Journal of Biological Chemistry*, *298*(5), 101823.

Freestone, P. P., Haigh, R. D., & Lyte, M. (2007). Blockade of catecholamine-induced growth by adrenergic and dopaminergic receptor antagonists in *Escherichia coli* O157: H7, *Salmonella enterica* and *Yersinia enterocolitica*. *BMC Microbiology*, *7*, 1–13.

Gannesen, A. V., Schelkunov, M. I., Geras' Kina, O. V., Makarova, N. E., Sukhacheva, M. V., Danilova,N. D., Ovcharova, M. A., Mart'yanov, S. V., Pankratov, T. A., Muzychenko, D. S., Zhurina, M. V., Feofanov, A. V., Botchkova, E. A., & Plakunov, V. K. (2021). Epinephrine affects gene expression levels and has a complex effect on biofilm formation in *Micrococcus luteus* strain C01 isolated from human skin. *Biofilm*, *3*, 100058.

Gonzales, X. F., Falconi, F. Q., Rojas, G. C., & Vidal, Y. L. (2010). *Streptococcus pneumoniae* treatment with catecholamines affects growth based on strain isolation site with a lack of effect on biofilm formation. *International Journal of Infectious Diseases*, *14*, e398.

Halang, P., Toulouse, C., Geißel, B., Michel, B., Flauger, B., Müller, M., Voegele, R. T., & Steuber, J. (2015). Response of *Vibrio cholerae* to the catecholamine hormones epinephrine and norepinephrine. *Journal of Bacteriology*, *197*(24), 3769–3778.

Hegde, M., Wood, T. K., & Jayaraman, A. (2009). The neuroendocrine hormone norepinephrine increases *Pseudomonas aeruginosa* PA14 virulence through the las quorum-sensing pathway. *Applied Microbiology and Biotechnology*, *84*, 763–776.

Hiller, C. C., Lucca, V., Carvalho, D., Borsoi, A., Borges, K. A., Furian, T. Q., & do Nascimento, V. P. (2019). Influence of catecholamines on biofilm formation by *Salmonella Enteritidis*. *Microbial Pathogenesis*, *130*, 54–58. https://doi.org/10.1016/j.micpath.2019.02.032

Hughes, D. T., Clarke, M. B., Yamamoto, K., Rasko, D. A., & Sperandio, V. (2009). The QseC adrenergic signaling cascade in enterohemorrhagic *E. coli* (EHEC). *PLoS Pathogens*, *5*(8), e1000553.

Jentsch, H. F. R., März, D., & Krüger, M. (2013). The effects of stress hormones on growth of selected periodontitis related bacteria. *Anaerobe*, *24*, 49–54.

Johnson, M., Cockayne, A., & Morrissey, J. A. (2008). Iron-regulated biofilm formation in *Staphylococcus aureus* Newman Requires ica and the Secreted Protein Emp. *Infection and Immunity*, *76*(4), 1756–1765. https://doi.org/10.1128/IAI.01635-07

Jothi, R., Sangavi, R., Kumar, P., Pandian, S. K., & Gowrishankar, S. (2021). Catechol thwarts virulent dimorphism in *Candida albicans* and potentiates the antifungal efficacy of azoles and polyenes. *Scientific Reports*, *11*(1), 21049. https://doi.org/10.1038/s41598-021-00485-2

Lanter, B. B., & Davies, D. G. (2015). Propionibacterium acnes recovered from atherosclerotic human carotid arteries undergoes biofilm dispersion and releases lipolytic and proteolytic enzymes in response to norepinephrine challenge in vitro. *Infection and Immunity*, 83, 3960–3971

Lyte, M., & Ernst, S. (1992). Catecholamine induced growth of Gram negative bacteria. *Life Sciences*, *50*(3), 203–212.

Moreira, C. G., & Sperandio, V. (2012). Interplay between the QseC and QseE bacterial adrenergic sensor kinases in *Salmonella enterica serovar Typhimurium* pathogenesis. *Infection and Immunity*, *80*(12), 4344–4353. https://doi.org/10.1128/IAI.00803-12

Neal, C. P., Freestone, P. P., Maggs, A. F., Haigh, R. D., Williams, P. H., & Lyte, M. (2001). Catecholamine inotropes as growth factors for *Staphylococcus epidermidis* and other coagulase-negative staphylococci. *FEMS Microbiology Letters*, *194*(2), 163–169. https://doi.org/10.1111/j.1574-6968.2001.tb09463.x

Niu, L., Gao, M., Wen, S., Wang, F., Shangguan, H., Guo, Z., Zhang, R., & Ge, J. (2023). Effects of catecholamine stress hormones norepinephrine and epinephrine on growth, antimicrobial susceptibility, biofilm formation, and gene expressions of enterotoxigenic *Escherichia coli*. *International Journal of Molecular Sciences*, *24*(21), 15646.

Rushing, J., & Santoro, N. (2022). Current opinion in endocrine and metabolic research perimenopause: Utility of testing. *Current Opinion in Endocrine and Metabolic Research*, *27*, 100402.

Sandrini, S., Alghofaili, F., Freestone, P., & Yesilkaya, H. (2014). Host stress hormone norepinephrine stimulates pneumococcal growth, biofilm formation and virulence gene expression. *BMC Microbiology*, *14*, 1–12.

Sharaff, F., & Freestone, P. (2011). Microbial endocrinology. *Central European Journal of Biology*, *6*, 685–694.

Vinasco, K., Mitchell, H. M., Kaakoush, N. O., Castaño-Rodríguez, N. (2019). Microbial carcinogenesis: Lactic acid bacteria in gastric cancer. *Biochimica et Biophysica Acta (BBA) – Reviews on Cancer*, *1872*(2), 188309.

Xiang, S. L., Xu, K. Z., Yin, L. J., Rao, Y., Wang, B., & Jia, A. Q. (2024). Dopamine, an exogenous quorum sensing signaling molecule or a modulating factor in *Pseudomonas aeruginosa*? *Biofilm*, *8*, 100208. https://doi.org/10.1016/j.bioflm.2024.100208

Yang, Q., Anh, N. D., Bossier, P., & Defoirdt, T. (2014). Norepinephrine and dopamine increase motility, biofilm formation, and virulence of *Vibrio harveyi*. *Frontiers in microbiology*, *5*, 584. https://doi.org/10.3389/fmicb.2014.00584

14 Effect of Melatonin and Local Hormones on Biofilm

Rina Rani Ray
Department of Biotechnology, Maulana Abul Kalam Azad University of Technology, Kolkata, India

14.1 INTRODUCTION

With the increasing global interest in gaining knowledge into the effects of hormones on microbes, a wide range of studies are being embarked upon. Since the beginning of the twentieth century, scientists have begun to discover the role of hormones on microbes residing inside the body as natural flora or potent pathogens. Since, the homoeostasis of the internal milieu is maintained by the coordinated function of endocrine glands and hormones, they must have some effect on microbial lifestyle. Among various endocrine glands, the pineal gland was called the "Seat of the Soul" as it can obtain information directly from the environment about the light–dark cycle and the photoperiod and makes the body aware through a hormone, called melatonin. Melatonin synchronizes the circadian rhythms through maintaining a proper sleep cycle. It is known that sleep is immensely influenced by gut microbiota, the dysbiosis of which brings fragmentation of normal sleep. Hence it can be presumed that melatonin has some direct effect on microbial virulence and quorum sensing. Melatonin, while primarily known as a hormone produced by the pineal gland, is also synthesized locally in various tissues and organs. It acts as a local modulator of hormone activity and influences processes like steroidogenesis and immune function (Cipolla-Neto & Amaral, 2018) On the other hand, local hormones are produced by nerve and gland cells and bind to either neighbouring cells or the same type of cell as that which produced them. These local hormones, namely histamine, prostaglandin, erythropoietin, etc., perform specific physiological functions and have some indirect effects on the normal microflora of the body, although there has been relatively research into such relationships.

The present chapter discusses the effect of melatonin and some local hormones on the formation and eradication of microbial biofilm and the enhancement of virulence in the pathogens.

14.2 SOURCES OF MELATONIN

In addition to the pineal gland, melatonin is also synthesized in various extra-pineal tissues, such as the brain, retina, gut, liver, and skin, acting locally and potentially contributing to cell protection and homeostasis.

14.3 CHEMICAL CHARACTERISTICS OF MELATONIN

Melatonin ($C_{13}H_{16}N_2O_2$, N-Acetyl-5-methoxytryptamine) is a biogenic acetamide in which one of the hydrogens attached to the nitrogen atom is replaced by a 2-(5-methoxy-1H-indol-3-yl) ethyl group. It is synthesized in mammalian pinealocytes from the amino acid tryptophan, which is hydroxylated to 5-hydroxytryptophan and then decarboxylated to serotonin. In turn, serotonin is acetylated to *N*-acetylserotonin and is then methylated to form 5-methoxytryptamine to produce melatonin.

DOI: 10.1201/9781003519041-14

14.4 MODE OF ACTION OF MELATONIN

In humans, the primary site for melatonin production is the pineal gland, while minor contributions come from the retina, lymphocytes, bone marrow, gastrointestinal tract, and thymus. Melatonin, which functions as a neurohormone, is released into the bloodstream where it can penetrate all body tissues. The synthesis and release of melatonin are inhibited by light and stimulated by darkness. Melatonin primarily exerts its effects through various mechanisms, with melatonin receptor 1 (MT1) and melatonin receptor 2 (MT2) being the two membrane receptors found in many mammalian organs. There exists a nuclear receptor belonging to the RORα/RZR family for this hormone.

The membrane receptors are mostly located in the central nervous system, although some are to be found in the peripheral nervous system and in the brain. Melatonin, after binding with MT1 and MT2 receptors attached on the cell surface, activates them and acts through G-protein. The activation of MT1 and MT2 receptors leads to the inactivation of adenylate cyclase, and, in turn, to the inhibition of the cyclic AMP (cAMP) production and the triggering of multiple signal pathways via a cascade effect. It also inhibits the formation of cGMP. In addition melatonin is reported to bind to various cytoplasmic and mitochondrial binding proteins, including quinone reductase-2, calmodulin, and calreticulin (Li & Sun, 2022). Melatonin also shows pleiotropic effects, depending on both receptor-mediated and receptor-independent pathways (Pechanova et al., 2014).

14.5 FUNCTIONAL ATTRIBUTES OF MELATONIN

Melatonin plays a crucial role in regulating sleep cycles and circadian rhythms. Furthermore, melatonin functions as an antioxidant, effectively neutralizing excess free radicals produced in the body due to its anti-excitatory and anti-inflammatory properties. Although, at first, it was known for its ability to control the photoperiodism and circadian rhythms, gradually other functions of this hormone are come to known, like modulation of sleep, mood, the regulation of metabolic events, and activities as antioxidant (Figure 14.1)

Melatonin is associated with numerous other functions, including oncostatic effects, hypnotic properties, immune regulation, reproductive processes, the timing of puberty, mood disorders, and transplantation. A lack of adequate melatonin production or synthesis has been linked to the

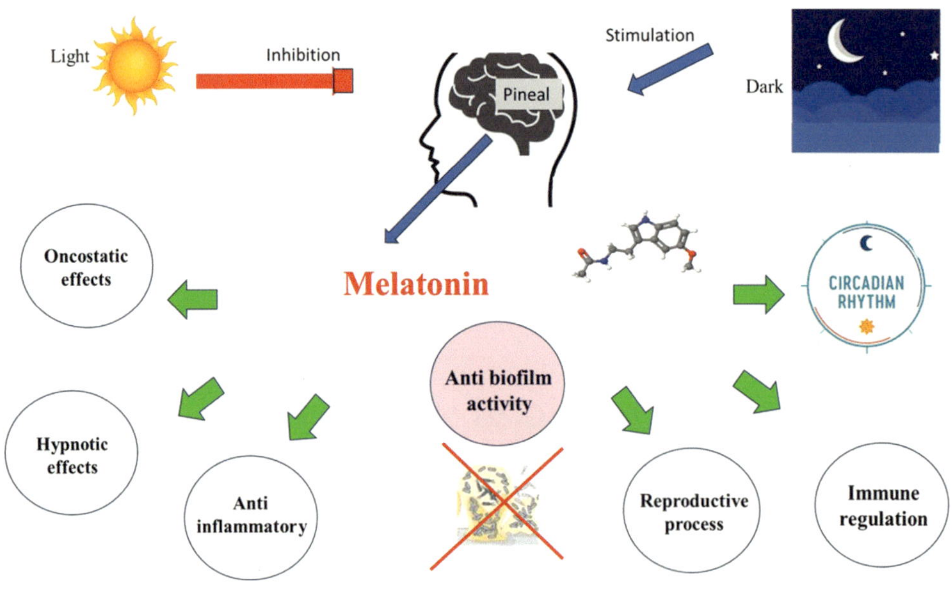

FIGURE 14.1 Functional attributes of melatonin hormone.

emergence of various disorders, such as breast cancer and neurodegenerative diseases. Melatonin shows promising potential as an analgesic in conditions characterized by pain. Over the last century, there has been increasing interest in utilizing melatonin for the treatment of diverse ailments, including inflammatory diseases, gastrointestinal issues, cancer, mood disorders, and more. Several melatonin agonists have been developed, and are commonly employed in the treatment of these conditions. Apart from having a role as a hormone, melatonin acts as a central nervous system depressant, an anticonvulsant, a radical scavenger, an immunological adjuvant, as metabolites of human and mouse and as a geroprotector and may have a role in expanding the life span of animals (Hardeland, 2017).

Interestingly melatonin is reported for its efficacy to inhibit the growth of some pathogenic microbes by affecting some vital functions and directly inhibiting biofilm formation. On that note, attempts are made to use melatonin as an alternative therapeutic measure to control a number of bacterial and viral diseases.

14.6 ANTIMICROBIAL ACTIVITIES OF MELATONIN

Several reports are available regarding the antimicrobial properties of melatonin, including its antibacterial and antifungal effects (Chen et al., 2018), and it was opined that such activity of melatonin is possibly attributable to the indole present in its structure.

Melatonin not only exerts antibacterial activities against Gram-negative and Gram-positive bacteria, but also affects the members of other bacterial groups, like *Mycobacterium tuberculosis*. It shows protective actions against bacterial attack at different levels of infection (He et al., 2021). The antimicrobial effect of this hormone is found to be exerted via the modulation of various transcription factors (Li & Sun, 2022).

A number of antibiotic-resistant bacteria could be killed by various dosages of melatonin, like *P. aeruginosa* (carbapenem-resistant and ATCC 27853 strains), *A. baumannii; S. aureus* (MRSA and ATCC 29123 strains), *Mycobacterium tuberculosis* (MDR strains TBRI 40, TBRI 204 and H37Rv), and *M. bovis* (Tekbas et al., 2008). Using its antibacterial effects, melatonin has been reported to protect the plants such as *Arabidopsis* and *Nicotiana tabacum* against the attack of *Pseudomonas syringae pv. tomato DC3000* (Pst DC3000) (Zhao et al., 2015).

Melatonin exhibited bactericidal activity against *Porphyromonas gingivalis*, *Aggregatibacter actinimycetemcomitans, and Fusobacterium nucleatum*, even at low concentrations (Ganganna et al., 2021).

14.7 ANTIBIOFILM ACTIVITIES OF MELATONIN

14.7.1 Inhibition of Fungal Biofilm

Yang et al. (2014) observed the anti-biofilm effect of melatonin on *Candida parapsilosis*, an opportunistic pathogen, which can cause a serious infection if it somehow manages to get inside the body through a wound, or along with some medical implants. Although the exact mechanism is not known, melatonin was found to be more effective in planktonic forms than in sessile forms. However, it is found that the fungicidal activity of the conventional antifungals against *Candida* sp. increase if they are applied along with melatonin with a significant fall in minimum inhibitory concentration (MIC) and minimum biofilm eradication concentration (MBEC) values (Kilinçel et al., 2019).

A similar observation was stated by Zhang et al. (2017), who found that the application of melatonin, along with the antifungal properties, will make it more effective in controlling *Phytophthora infestans* (*P. infestans*), the causal agent of potato late blight.

Melatonin can be used as a photosensitizer in photodynamic treatment (PDT) to reduce fungal infections, particularly those caused by *Candida albicans*, through producing singlet oxygen and efficiently inhibiting biofilm development (Duterte et al., 2024).

14.7.2 Inhibition of Bacterial Biofilm

Xanthomonas oryzae pv. *oryzae* (Xoo), a bacterial pant pathogen, is the causal organism of rice blight. Melatonin treatment (200 μg/mL) not only hampers the normal cell division and cell growth of the pathogen, but also reduces the motility and inhibited biofilm formation. The antibacterial and antibiofilm activities of melatonin on Xoo were accomplished by the down-regulation of genes involved in metal binding, catalytic, and metabolic activities (Chen et al., 2018).

The biofilm formation and motility in another strain, *Xanthomonas oryzae* pv. *oryzicola* (Xoc), the pathogen causing rice leaf streak, can also be inhibited by melatonin through effective reduction of allied gene expression (Chen et al., 2019). The overall pathogenicity of Xoc has been reduced by the reduction of colonization, decreased bacteria growth, and reduced biofilm production with lower swimming motility.

Melatonin not only showed antimicrobial effects against planktonic cells of *Porphyromonas gingivalis*, but also inhibited biofilm formation, reduced the already matured biofilms, and lessened biofilm viability by killing the sessile cells of this oral pathogen (Zhou et al., 2016).

Further, the hormone could reduce the biofilm of *Staphylococcus aureus* ATCC 29213, the causal organism of a wound (Romić et al., 2016). In addition, it was successfully used to treat bovine mastitis, a cattle disease, caused commonly by *Staphylococcus aureus*, *Streptococcus agalactiae* and *Escherichia coli*, which is presumed to be accomplished by the anti-inflammatory and anti-oxidative activities of this hormone and its metabolic derivatives (Li, and Sun, 2022).

Melatonin induces differential gene expression related to biofilm formation and generation of fimbria and thereby enhances colony formation by *Klebsiella aerogenes*, a nosocomial pathogen (Graniczkowska et al., 2022).

Three multidrug-resistant pathogenic bacteria, namely *Escherichia coli, Pseudomonas aeruginosa* and *Klebsiella pneumoniae*, are generally reported from the urinary tract of cancer patients. The biofilm-mediated antibiotic resistance made their treatment difficult, which emphasizes the need of a natural biofilm eradicator. The experimental observations of Shamsi et al. (2025) indicates that melatonin, being an antioxidant hormone, could remove the profuse biofilm produced by all these three uro-pathogens, of which the extent of *E coli* biofilm was much high than those of the rest two (Shamsi et al., 2025).

Hence, melatonin exhibits antibiofilm activity by interfering with biofilm development and potentially disrupting bacterial metabolic processes, offering a potential alternative therapy for infections caused by Gram-negative pathogens. For cancer patients, this can be recommended as a complementary treatment for biofilm removal.

Recent studies highlight a possible gut microbiota-modulating role of melatonin, with possible implications for the treatment of these pathologies (Iesanu et al., 2022). Melatonin-treated mice exhibited improved bacterial clearance, which clearly reveals its bactericidal activities (Schaefer et al., 2025).

Melatonin can also regulate the secretion of some hormones as a local modulator, known as local hormones.

14.8 LOCAL HORMONES

Local hormones are a large group of signalling molecules, which, instead of circulating through blood vessels, act on a cell located at the vicinity of its origin. Instead of well-defined endocrine gland they are secreted from any cell, or nerve endings. Unlike the regular hormones, these hormones are stimulated and deactivated rapidly. Major local hormones include: histamine, serotonin, prostaglandins, erythropoietin, and eicosanoids (Figure 14.2).

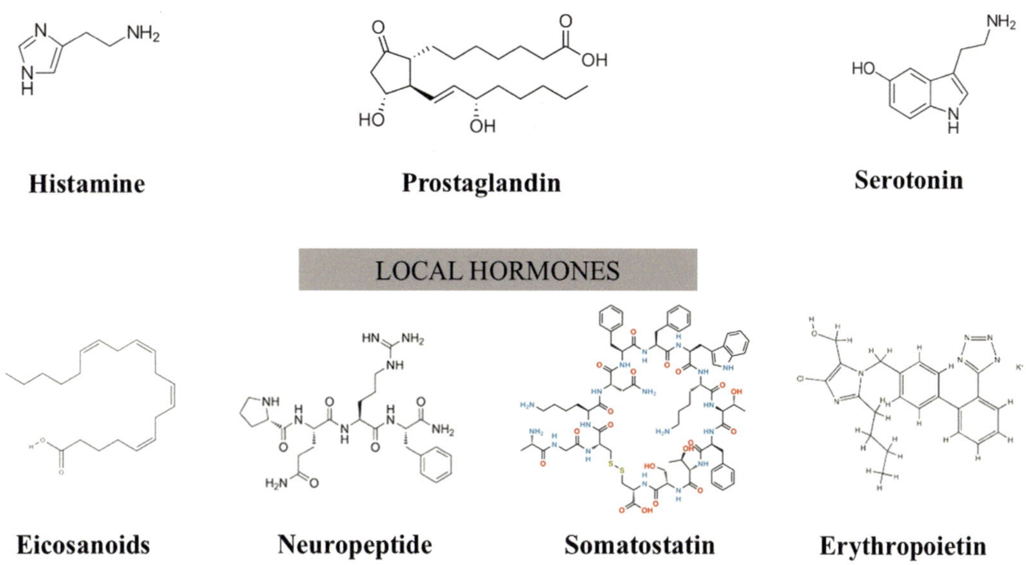

FIGURE 14.2 Chemical configurations of local hormones.

14.8.1 Histamine

14.8.1.1 Source and Chemistry

Histamine is a chemical messenger primarily synthesized and stored in mast cells and in basophils of blood released only the influence of allergen or after the injury. It is a biogenic amine produced from L-histidine by the action of the enzyme L-histidine decarboxylase, using pyridoxal-5′-phosphate as a cofactor. Histamine is found to be involved in the stimulation of local immune responses and also in the regulation of various physiological functions in the gut. It also acts as a neurotransmitter.

14.8.1.2 Effect on Microbial Growth and Biofilm Formation

Histamine, a biogenic amine, is synthesized by some specific microbial strains present as gut microbiota, such as *E. coli, Lactobacillus vaginalis, Pseudomonas fluorescens, Morganella morganii, Citrobacter freundii, Proteus vulgaris, P. mirabilis, Enterobacter aerogenes*, and *Fusobacterium ulcerans* (Barcik et al., 2019), to stimulate intestinal epithelial histamine receptors for improving the health of the gut (Engevik et al., 2024).

Histamine enhances biofilm formation in *Pseudomonas aeruginosa*, making it more resistant to various antibiotics (Dib, 2025). *Lentilactobacillus parabuchneri*, a biofilm producer, was found to accumulate histamine in biofilm to cause various diseases. Finally, histamine was found to play a role in biofilm formation, possibly inducing bacterial resistance, virulence, and inter-species communication.

14.8.2 Prostaglandin

14.8.2.1 Source and Chemistry

Prostaglandins, initially found in semen and named after the prostate gland, are a group of lipids with hormone-like activities, formed at the sites of infection or wound. These are derived from arachidonic acid and are a group of twenty-carbon unsaturated fatty acids.

14.8.2.2 Effect on Microbial Growth and Biofilm Formation

The major types of prostaglandins are PGE2, PGI2 (prostacyclin), PGD2, and PGF2α. All of these are involved in processes such as inflammation, fever generation, and reproduction.

Of these, prostaglandin E2 (PGE2) enhances bacterial biofilm formation and encourages bacterial persistence and virulence. PGE2 not only promotes the development of Candida biofilm (Liu et al., 2016), but also encourages the formation of a polymicrobial biofilm formed by *Candida albicans* and *Staphylococcus aureus* (Krause et al., 2015).

14.8.3 Erythropoietin (EPO)

14.8.3.1 Source and Chemistry

This local hormone is primarily secreted from the kidneys, and promotes the production of red blood capsules (RBCs) in response to low oxygen levels. It is a glycoprotein hormone composed of 165 amino acids containing a huge amount of carbohydrate.

14.8.3.2 Effect on Microbial Growth and Biofilm Formation

Erythropoietin (EPO) has been demonstrated to have both protective and potentially harmful effects on bacterial infections and inflammation, influencing bacterial biofilm formation and resolution. While EPO can help with infection resolution and antibiotic action, it can also hinder pathogen clearance and, in certain situations, may even aggravate inflammation. Experimental observations indicate that EPO increased the antibiotic actions in infections initiated by pathogens like *Escherichia coli* and *Staphylococcus aureus* (Liang et al., 2021).

EPO can help in the growth and proliferation of certain pathogens, like *Escherichia coli* and *Staphylococcus aureus* while bringing about growth inhibition of other bacteria like *Pseudomonas aeruginosa* (Mittal et al., 2019). In *Salmonella*-infected mice, EPO treatment caused an increased risk of the death of the victim as a result of impaired pathogen clearance due to the reduced formation of anti-microbial effector molecules (e.g. TNF-α and NO) (Nairz et al., 2011). EPO treatment was found to induce dysbiosis of the gut microbiota (Sarrabayrouse et al., 2025).

14.8.4 Serotonin

14.8.4.1 Source and Chemistry

This local hormone is synthesized from tryptophan, mostly in the gastrointestinal tract and also, to a lesser extent, in the brain (90: 10). Serotonin (α- 5- hydroxy-tryptamine) is a monoamine hormone.

14.8.4.2 Effect on Microbial Growth and Biofilm Formation

It is found that serotonin, being a neurotransmitter, can interact with bacterial proteins while signalling between members of the gut microbiota and host (Everett et al., 2022). It can act as a signalling molecule that influences bacterial biofilm formation and virulence factor production, particularly in *Pseudomonas aeruginosa*, through the activation of bacterial quorum-sensing pathways (Knecht et al., 2016). Serotonin also improves stress resistance, aggregation, and promotes biofilm formation in the probiotic *Enterococcus faecium* (Scardaci et al., 2021).

Serotonin enhances biofilm development both in vitro and in vivo in mice (Knecht et al., 2016) and the biofilm-promoting role of serotonin was confirmed by the reduction of the biomass and the viability of already formed biofilm *Cryptococcus gattii* by a selective serotonin reuptake inhibitor (SSRI) (Viveiro et al., 2024).

On the other hand, Kumar et al. (2020) found that serotonin downregulates the expression of virulence-causing genes both in enterohemorrhagic *E. coli* (EHEC) and in its murine counterpart, *Citrobacter rodentium* (Kumar et al., 2020). On a similar tone, Jugder et al. (2022) reported that in the mammalian host, the variation of dietary tryptophan by commensal microbe changes serotonin levels. They found that at high bacterial density of *V. cholerae*, the quorum-sensing regulator HapR, which is responsible for the conversion of tryptophan to serotonin, decreases virulence by decreasing tryptophan uptake and thereby increasing the relative amount of serotonin. This affects intestinal inflammation and extends host survival.

14.8.5 Somatostatin

14.8.5.1 Source and Chemistry

Somatostatin, a cyclic a neuropeptide known as the growth-inhibiting hormone, is produced from multiple sites, like the hypothalamus, the gastrointestinal (GI) tract, and the pancreas.

14.8.5.2 Effect on Microbial Growth and Biofilm Formation

Somatostatin was found to have a promoting effect on the formation of bacterial biofilm, but, in some cases, it displays as an inhibitory action. Somatostatin showed a dose-dependent inhibition of *H. pylori* growth and can modulate it by disrupting biofilm matrix and inhibiting its proliferation (Lesouhaitier et al., 2019).

However, other studies have indicated that somatostatin can promote biofilm growth in other bacterial species. For instance, one study found that somatostatin promoted the growth of *Pseudomonas aeruginosa* biofilms. This suggests that the effect of somatostatin on biofilm formation may be species-specific

14.8.6 Eicosanoids

14.8.6.1 Source and Chemistry

Eicosanoids are lipid-based signalling molecules like prostaglandins and leukotrienes that regulate spermatogenesis in mammals.

14.8.6.2 Effect on Microbial Growth and Biofilm Formation

It plays a role in innate immune responses and can influence biofilm formation and host–pathogen interactions, particularly in fungal and bacterial infections. Production of these eicosanoids could affect pathogen-clearance and the infection dynamics of polymicrobial biofilm (Fourie et al., 2017).

Eicosanoids and other derivatives of arachidonic acid increase biofilm formation in various Candida spp., *C. glabrata*, *C. parapsilosis*, and *C. tropicalis* (Mishra et al., 2014). Eicosanoids like PGE_2 encourages the bacterial growth and biofilm formation of *S. aureus* by increasing the attachment of bacterium to the human fibronectin (Wang et al., 2017).

14.8.7 Neuropeptides

Neuropeptides are short chains of amino acids, produced by neurons that can operate as local hormones or paracrine-signalling molecules, impacting neighbouring cells without entering the bloodstream and modulating a variety of body activities. Among these are Neuropeptide Y (NPY), Substance P, enkephalin, and Pituitary Adenylate Cyclase-Activating Polypeptide (PCAP).

14.8.7.1 Effect on Microbial Growth and Biofilm Formation

Some neuropeptides exhibit antimicrobial activity, contributing to the body's defense against pathogens. These neuropeptides, like PACAP and Substance P, can directly kill bacteria and fungi, or modulate immune responses to combat infections.

Neuropeptides exert somewhat anti-infective efficacy indirectly by increasing the innate and adaptive immunity of the host. They are playing, to some extent, a dual role of displaying antimicrobial activity against bacteria, fungi, and protozoa but also enhancing virulence and pathogenicity.

TABLE 14.1
Source, Chemistry, Functions and Action on Biofilm of Local Hormones

Name of Local Hormone	Source	Chemistry	Functional Attributes	Biofilm Formation
Histamine	Mast cell, Basophil	biogenic amine synthesized from L-histidine	Localized immune response, allergic reaction, Vasodilatation, Gastric juice secretion	Promoted
Prostaglandin	Different tissues all over the body, Including prostate gland	Arachidonic acid derived unsaturated 20-carbon carboxylic acids,	Inflammation, pain mediator, uterine contractions during labour, blood clotting	Promoted
Serotonin	Platelet, Brain, Intestinal wall	5-hydroxy derivative of tryptamine	Smooth muscle contraction, blood clotting, intestinal secretion, regulates sleep, mood, appetite.	Promoted in *Enterococcus faecium*, Declined in other
Eicosanoid	Ubiquitous	20-carbon chain with various functional groups	Inflammation, pain mediator, uterine contractions during labour, blood clotting	Promoted
Neuropeptide	Cell body of neuron	Short peptides	Neurotransmission, neuromodulation.	Inhibited
Somatostatin	Hypothalamus, Pancreas D cell, G I tract.	cyclic tetra-decapeptide	Inhibit the release of insulin, glucagon, growth hormone and thyroid-stimulating hormone	Promoted in *P aeruginosa*, Disrupted in *H pylori*
Erythropoietin	Kidneys	glycoprotein	Production of red blood cells	Mixed

Note: The gastrointestinal local hormones are purposefully excluded, since they are discussed elaborately in separate chapter.

These neuropeptides not only regulate important phases of microbial pathogenesis such as adhesion, biofilm formation, invasion, intracellular lifestyle, dissemination, etc., including their anti-infective properties, but also exhibit detrimental effects (Augustyniak et al., 2021).

NPs block microbial attachment to host epithelial and endothelial cells, resulting in the inhibition of biofilm formation both in vivo and in vitro (Augustyniak et al., 2012).

In conclusion, it can be asserted that different local hormones show different type of action towards biofilm (Table 14.1).

14.9 CONCLUSION

Melatonin, which is synthesized by pineal and several additional sources, acts as a local modulator, in common with other local hormones synthesized by various nerve terminals and gland cells.

Both melatonin and local hormones are showing some direct or indirect effect on microbial growth and biofilm formation. Melatonin, show an overall antibiofilm activities, while most of the local hormones discussed exhibit a biofilm-promoting trend. However, the neuropeptides perform a dual role, often showing some antimicrobial activities.

REFERENCES

Augustyniak, D., Kramarska, E., Mackiewicz, P., Orczyk-Pawiłowicz, M., & Lundy, F. T. (2021). Mammalian neuropeptides as modulators of microbial infections: Their dual role in defense versus virulence and pathogenesis. *International Journal of Molecular Sciences*, *22*(7), 3658.

Augustyniak, D., Nowak, J., & Lundy, F. T. (2012). Direct and indirect antimicrobial activities of neuropeptides and their therapeutic potential. *Current Protein & Peptide Science*, *13*(8), 723–738. https://doi.org/10.2174/138920312804871139

Barcik, W., Pugin, B., Brescó, M. S., Westermann, P., Rinaldi, A., Groeger, D., Van Elst, D., Sokolowska, M., Krawczyk, K., Frei, R., Ferstl, R., Wawrzyniak, M., Altunbulakli, C., Akdis, C. A., & O'Mahony, L. (2019). Bacterial secretion of histamine within the gut influences immune responses within the lung. *Allergy*, *74*(5), 899–909. https://doi.org/10.1111/all.13709

Chen, X., Sun, C., Laborda, P., He, Y., Zhao, Y., Li, Z., et al. (2019). Melatonin treatments reduce the pathogenicity and inhibit the growth of *Xanthomonas oryzae* pv. *oryzicola*. *Plant Pathology*, *68*, 288–296. https://doi.org/10.1111/ppa.12954

Chen, X., Sun, C., Laborda, P., Zhao, Y., Palmer, I., Fu, Z. Q., Qiu, J., & Liu, F. (2018). Melatonin treatment inhibits the growth of *Xanthomonas oryzae* pv. *oryzae*. *Frontiers in Microbiology*, *9*, 2280. https://doi.org/10.3389/fmicb.2018.02280

Cipolla-Neto, J., & Amaral, F. G. D. (2018). Melatonin as a hormone: New physiological and clinical insights. *Endocrine Reviews*, *39*(6), 990–1028.

Dib, K. (2025). Histamine promotes *Pseudomonas aeruginosa* biofilm formation and renders *Pseudomonas aeruginosa* biofilms more resistant to gentamicin and azithromycin. *Medical Principles and Practice: International Journal of the Kuwait University, Health Science Centre*, 1–12. Advance online publication. https://doi.org/10.1159/000544111

Duterte, M. M. D., Morales, N. P., Pitiphat, W., Puthongking, P., & Damrongrungruang, T. (2024). Effects of photodynamic therapy using bisdemethoxycurcumin combined with melatonin or acetyl-melatonin on C. Albicans. *Scientific Reports*, *14*(1), 23082.

Engevik, K., Edens, R., Engevik, A., Engevik, M., & Horvath, T. (2024). Production of histamine by diverse gut bacteria can activate intestinal HRH1 and promote calcium mobilization. *Physiology*, *39*(S1), 2142.

Everett, B. A., Tran, P., & Prindle, A. (2022). Toward manipulating serotonin signaling via the microbiota–gut–brain axis. *Current Opinion in Biotechnology*, *78*, 102826.

Fourie, R., Ells, R., Kemp, G., Sebolai, O. M., Albertyn, J., & Pohl, C. H. (2017). *Pseudomonas aeruginosa* produces aspirin insensitive eicosanoids and contributes to the eicosanoid profile of polymicrobial biofilms with *Candida albicans*. *Prostaglandins, Leukotrienes and Essential Fatty Acids*, *117*, 36–46.

Ganganna, A., Rudariah, C. B., Rao, R., & Prakash, V. M. (2021). Antibacterial activity of melatonin against prime periodontal pathogens: An: In vitro: Study. *Journal of International Oral Health*, *13*(2), 164–168.

Graniczkowska, K. B., Shaffer, C. L., & Cassone, V. M. (2022). Transcriptional effects of melatonin on the gut commensal bacterium *Klebsiella aerogenes*. *Genomics*, *114*(2), 110321. https://doi.org/10.1016/j.ygeno.2022.110321

Hardeland, R. (2017). Melatonin as a geroprotector: healthy aging vs. extension of lifespan. In *Anti-aging drugs: From basic research to clinical practice*. Edited by Alexander M. Vaiserman. RSC.

He, F., Wu, X., Zhang, Q., Li, Y., Ye, Y., Li, P., Chen, S., Peng, Y., Hardeland, R., & Xia, Y. (2021). Bacteriostatic potential of melatonin: Therapeutic standing and mechanistic insights. *Frontiers in Immunology*, *12*, 683879. https://doi.org/10.3389/fimmu.2021.683879

Iesanu, M. I., Zahiu, C. D. M., Dogaru, I. A., Chitimus, D. M., Pircalabioru, G. G., Voiculescu, S. E., Isac, S., Galos, F., Pavel, B., O'Mahony, S. M., & Zagrean, A. M. (2022). Melatonin-microbiome two-sided interaction in dysbiosis-associated conditions. *Antioxidants (Basel, Switzerland)*, *11*(11), 2244. https://doi.org/10.3390/antiox11112244

Jugder, B. E., Batista, J. H., Gibson, J. A., Cunningham, P. M., Asara, J. M., & Watnick, P. I. (2022). *Vibrio cholerae* high cell density quorum sensing activates the host intestinal innate immune response. *Cell Reports*, *40*(12), 111368 https://doi.org/10.1016/j.celrep.2022.111368

Kilinçel, Ö., Çaliskan, E., Sahin, I., Öztürk, C. E., Kiliç, N., & Öksüz, S. (2019). The effect of melatonin on antifungal susceptibility in planktonic and biofilm forms of Candida strains isolated from clinical samples. *Medical Mycology*, *57*(1), 45–51. https://doi.org/10.1093/mmy/myx157

Knecht, L. D., O'Connor, G., Mittal, R., Liu, X. Z., Daftarian, P., Deo, S. K., & Daunert, S. (2016). Serotonin activates bacterial quorum sensing and enhances the virulence of pseudomonas aeruginosa in the host. *EBioMedicine*, *9*, 161–169. https://doi.org/10.1016/j.ebiom.2016.05.037

Krause, J., Geginat, G., & Tammer, I. (2015). Prostaglandin E2 from Candida albicans stimulates the growth of Staphylococcus aureus in mixed biofilms. *PloS one*, *10*(8), e0135404.

Kumar, A., Russell, R. M., Pifer, R., Menezes-Garcia, Z., Cuesta, S., Narayanan, S., MacMillan, J. B., Sperandio, V., & Sperandio, V. (2020). The serotonin neurotransmitter modulates virulence of enteric pathogens. *Cell Host & Microbe*, *28*(1), 41–53.

Lesouhaitier, O., Clamens, T., Rosay, T., Desriac, F., Louis, M., Rodrigues, S., Gannesen, A., Plakunov, V. K., Bouffartigues, E., Tahrioui, A., Bazire, A., Dufour, A., Cornelis, P., Chevalier, S., & Feuilloley, M. G. J.

(2019). Host peptidic hormones affecting bacterial biofilm formation and virulence. *Journal of Innate Immunity*, *11*(3), 227–241. https://doi.org/10.1159/000493926

Li, H., & Sun, P. (2022). Insight of melatonin: The potential of melatonin to treat bacteria-induced mastitis. *Antioxidants (Basel, Switzerland)*, *11*(6), 1107. https://doi.org/10.3390/antiox11061107

Liang, F., Guan, H., Li, W., Zhang, X., Liu, T., Liu, Y., Mei, J., Jiang, C., Zhang, F., Luo, B., & Zhang, Z. (2021). Erythropoietin promotes infection resolution and lowers antibiotic requirements in *E. coli*- and *S. aureus*-initiated infections. *Frontiers in Immunology*, *12*, 658715. https://doi.org/10.3389/fimmu.2021.658715

Liu, X., Wang, D., Yu, C., Li, T., Liu, J., & Sun, S. (2016). Potential antifungal targets against a Candida biofilm based on an enzyme in the arachidonic acid cascade—A review. *Frontiers in Microbiology*, *7*, 1925

Mishra, N. N., Ali, S., & Shukla, P. K. (2014). Arachidonic acid affects biofilm formation and PGE2 level in *candida albicans* and non-*albicans* species in presence of subinhibitory concentration of fluconazole and terbinafine. *Brazilian Journal of Infectious Diseases*, *18*, 287–293.

Mittal, A., Singh, V., Chowdhary, S., Moideen, A., Kumar, D., Maniar, K., Bhattacharyya, R., & Banerjee, D. (2019). The effect of recombinant human erythropoietin on bacterial growth: A dual-edged sword. *Kidney Diseases*, *5*(2), 81–90.

Nairz, M., Schroll, A., Moschen, A. R., Sonnweber, T., Theurl, M., Theurl, I., Taub, N., Jamnig, C., Neurauter, D., Huber, L. A., Tilg, H., Moser, P. L., & Weiss, G. (2011). Erythropoietin contrastingly affects bacterial infection and experimental colitis by inhibiting nuclear factor-κB-inducible immune pathways. *Immunity*, *34*(1), 61–74.

Pechanova, O., Paulis, L., & Simko, F. (2014). Peripheral and central effects of melatonin on blood pressure regulation. *International Journal of Molecular Sciences*, 15, 17920–17937. https://doi.org/10.3390/ijms151017920

Romić, M. D., Klarić, M. Š., Lovrić, J., Pepić, I., Cetina-Čižmek, B., Filipović-Grčić, J., & Hafner, A. (2016). Melatonin-loaded chitosan/Pluronic® F127 microspheres as in situ forming hydrogel: An innovative antimicrobial wound dressing. *European Journal of Pharmaceutics and Biopharmaceutics*, *107*, 67–79

Sarrabayrouse, G., Joulain, C., Bessoles, S., Chiron, A. S., Abina, A. M., & Hacein-Bey-Abina, S. (2025). Erythropoietin supplementation induces dysbiosis of the gut microbiota and impacts mucosal immunity in a non-diseased mouse model. *Frontiers in Immunology*, *15*, 1465410.

Scardaci, R., Manfredi, M., Barberis, E., Scutera, S., Marengo, E., & Pessione, E. (2021). Serotonin exposure improves stress resistance, aggregation, and biofilm formation in the probiotic Enterococcus faecium NCIMB10415. *Microbiology Research*, *12*(3), 606–625.

Schaefer, K. R., Rogers, S., Faber, Z. et al. (2025). Efficacy of melatonin treatment in a cystic fibrosis mouse model of airway infection. *Scientific Reports*, *15*, 1849. https://doi.org/10.1038/s41598-025-85948-6

Shamsi, S. S., Ahmed, K. M., Al-Griw, H. H., Abobaker, A. M., Abdulnabi, F. A., Alhashemi, M. A., Alshebani, H. O., & Hassan, N. F. (2025). Combating biofilm: Melatonin's role in mitigating multidrug-resistant uropathogens in cancer. *Egyptian Journal of Medical Microbiology*, *34*(3), 121–129. https://doi.org/10.21608/ejmm.2025.327626.1358

Tekbas, O. F., Ogur, R., Korkmaz, A., Kilic, A., & Reiter, R. J. (2008). Melatonin as an antibiotic: New insights into the actions of this ubiquitous molecule. *Journal of Pineal Research*, 44, 222–226. https://doi.org/10.1111/j.1600-079X.2007.00516.x

Viveiro, L. R. D. G., Rehem, A. R., Santos, E. L. D. S., Carmo, P. H. F. D., Junqueira, J. C., & Scorzoni, L. (2024). In vitro effects of selective serotonin reuptake inhibitors on *Cryptococcus gattii* capsule and biofilm. *Pathogens and Disease*, *82*, ftae001.

Wang, Y., Ren, B., Zhou, X., Liu, S., Zhou, Y., Li, B., Jiang, Y., Li, M., Feng, M., & Cheng, L. (2017). Growth and adherence of *Staphylococcus aureus* were enhanced through the PGE2 produced by the activated COX-2/PGE2 pathway of infected oral epithelial cells. *PLoS One*, *12*, e0177166.

Yang, H. P., Tsang, P. C. S., & Tsang, W. K. P. (2014). Melatonin inhibits biofilm formation in *Candida parapsilosis*. *Journal de Mycologie Medicale*, *24*(4), 360.

Zhang, S., Zheng, X., Reiter, R. J., Feng, S., Wang, Y., Liu, S., Jin, L., Li, Z., Datla, R., & Ren, M. (2017). Melatonin attenuates potato late blight by disrupting cell growth, stress tolerance, fungicide susceptibility and homeostasis of gene expression in *phytophthora infestans*. *Frontiers in Plant Science*, *8*, 1993. https://doi.org/10.3389/fpls.2017.01993

Zhao, H., Xu, L., Su, T., Jiang, Y., Hu, L., & Ma, F. (2015). Melatonin regulates carbohydrate metabolism and defenses against *Pseudomonas syringae* pv. tomato DC3000 infection in *Arabidopsis thaliana*. *Journal of Pineal Research*, *59*(1), 109–119. https://doi.org/10.1111/jpi.12245

Zhou, W., Zhang, X., Zhu, C. L., He, Z. Y., Liang, J. P., & Song, Z. C. (2016). Melatonin receptor agonists as the "perioceutics" agents for periodontal disease through modulation of *porphyromonas gingivalis* virulence and inflammatory response. *PloS One*, *11*(11), e0166442. https://doi.org/10.1371/journal.pone.0166442

15 Hormonal Drugs and Biofilm

Bandita Dutta, Arina Guha, Samriddhi Kumari, and Rina Rani Ray
Department of Biotechnology, Maulana Abul Kalam Azad University of Technology, Kolkata, India

15.1 INTRODUCTION

Hormonal therapy is the name given to a technique of using hormones or special drugs to help control body functions, fix hormone problems, and treat hormone-related diseases. It is used for many critical health conditions (Zhao et al., 2025), including hormone replacement therapy (HRT), fertility treatments, managing cancer, metabolism problems, and care for transgender people. The use of hormones as medicine is generally beneficial, although their use for long periods of time does have risks, which may include heart issues, hormone imbalances, and higher chances of getting cancer (Brogna et al., 2025). Hormonal drugs are special types of medicines that can copy, replace, or slow down our body's hormonal regulations. The use and control of these drugs are very crucial as they help treat special body diseases, illnesses connected to birth and pregnancy, cancers that can grow faster because of the actions of hormones. These health problems are connected to body metabolism, as they keep the body's different biological tasks in a controlled mechanism (Zhao et al., 2025). Through their attachment to specific hormone receptors, they are able to activate or inhibit the hormone-dependent signaling pathways and thus control metabolism, growth, immune functions, and reproductive cycles. In the present day, the most popular hormonal drugs are insulin for diabetes, levothyroxine for thyroid diseases, estrogen and progesterone for birth control, and corticosteroids for inflammation (Brogna et al., 2025). The development of artificial hormone formulations and specific hormone replacement therapy has enhanced their effectiveness with fewer complications and improved patient adherence. Since hormones have a significant role on the microbiome and the structural and functional attributes of biofilm formed in the body, HRT may be pivotal role in terms of both the onset and the management of various biofilm-mediated diseases. Hence, patients may be treated with either natural or synthetic hormones to maintain the proper homoeostatic state of the body by restoring the normal metabolic scenario and resisting biofilm-borne diseases. This chapter deals with the role and application of various hormones for the management of such diseases.

15.2 HORMONAL DRUGS

Hormonal drugs are used in several medical fields (Table 15.1) to control different diseases. Endocrine therapy is required in the case of both hypothyroidism and diabetes, whereas hormone replacement therapies like levothyroxine and insulin are used to maintain metabolic stability (Mei et al., 2025). In reproductive medicine, ovulation, menstruation, and pregnancy are regulated by hormonal drugs and oral contraceptives and gonadotropin treatments are used in family planning and fertility management. In the case of hormone-sensitive cancers, for example breast and prostate cancer, estrogen and androgen blockers are used to slow the growth of tumors (Brogna et al., 2025). GLP-1 RAs are also in use for obesity and metabolic disorders, in order to expand the usages of these drugs in disease management (Zhao et al., 2025). However, such therapeutic pills are not

DOI: 10.1201/9781003519041-15

TABLE 15.1
Different Hormonal Drugs and Their Significant Uses

Category	Subcategory	Examples	Uses
Glucocorticoids	-	Prednisolone, Dexamethasone	Anti-inflammatory, immunosuppressive
Mineralocorticoids	-	Fludrocortisone	Regulates sodium and water balance
Sex Hormones	Estrogens	Estradiol, Ethinyl estradio	Hormone replacement, contraception
	Progestins	Medroxyprogesterone, Norethindrone	Contraception, menstrual disorders
	Androgens	Testosterone Methylitestosterone	Hypogonadism muscle growth
Thyroid & Antithyroid Drugs	Thyroid Hormones	Levothyroxine (T4), Liothyronine (T3)	Hypothyroidism treatment
	Antithyroid Drugs	Methimazole Propyithiouracil	Hyperthyroidism treatment
Pituitary Hormones	Anterior Pituitary	Growth hormone (GH), ACTH	Growth disorders, adrenal stimulation
	Posterior Pituitary	Oxytocin, Vasopressin (Desmopressin)	Labor induction, diabetes insipidus
Pancreatic Hormones	Insulin	Regular insulin, NPH insulin, Insulin glargine	Diabetes management
	Glucagon	Glucagon	Hypoglycemia treatment
Parathyroid & Bone Drugs	Parathyroid Hormone Analogs	Teriparatide	Osteoporosis treatment
	Calcitonin	Calcitonin	Bone metabolism regulation
Gonadotropins	-	FSH, LH	Fertility treatments

without their dangers and thus need to be taken with caution. Steroid and sex hormones, for example, have been known to have cardiovascular side effects and metabolic disturbances and their use over long periods can cause cancer promotion (Che et al., 2025). Furthermore, hormonal alternation can impact the microbiome, leading to the growth of bacterial biofilms that may increase the risk of infections (Mei et al., 2025). Studies on bioidentical hormones and precision medicine are ongoing to develop better ways of using hormones with less adverse effects in both acute and long-term patient management. Future endocrine pharmacology research will also emphasize the development of new, safer, and more specific hormonal therapies for patients.

15.3 CLASSIFICATION OF HORMONAL DRUGS

Hormonal preparations are among the most significant categories of medications in healthcare. Most hormones are normally secreted in extremely small doses; therefore, replacement hormones are frequently engineered to be given in quantities equivalent to those naturally secreted in the body.

Hormonal drugs may be classified in different categories (Figure 15.1) based on their mode of action.

15.3.1 Hormone Replacement Medications

Hormone replacement therapy (HRT) is a medical treatment that replaces or supplements the hormones that your body is lacking through the help of synthetic or bioidentical hormones. It is used across many medical fields and is not limited to reproductive health. It also includes metabolic, endocrine, and oncological treatments.

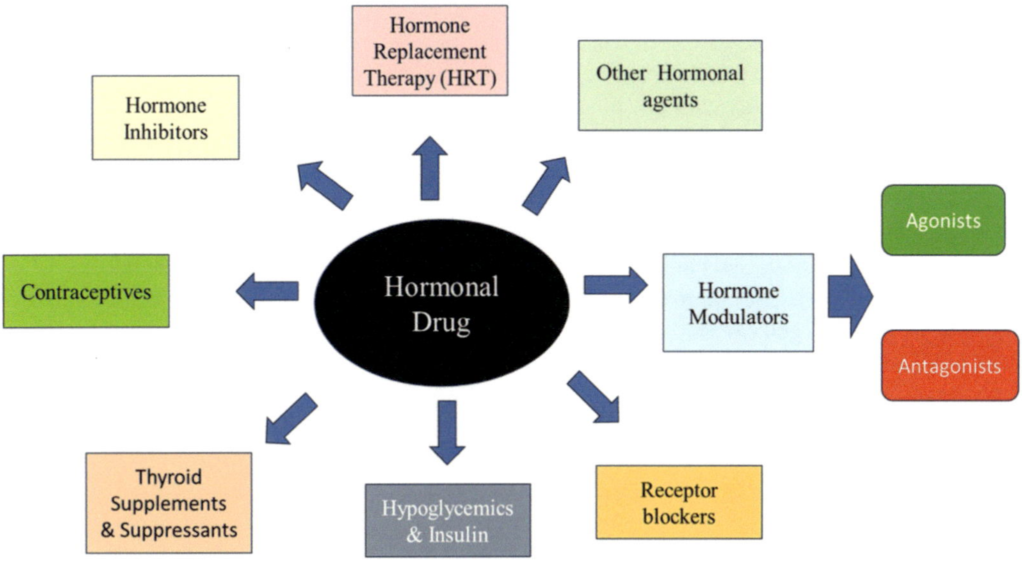

FIGURE 15.1 Various categories of hormonal drugs.

15.3.1.1 Steroids

Steroid drugs are used in hormone replacement therapy (HRT). These include both corticosteroids and anabolic steroids, which regulate metabolism, immune response, inflammation and reproductive functions. Adrenal insufficiency, inflammatory disorders, and anabolic hormone replacement are all classic uses of steroid therapy (Ornetti et al., 2025). Steroids work by binding to nuclear receptors and thereby altering gene transcription. Glucocorticoids such as hydrocortisone decrease inflammation and suppress the immune system while anabolic steroids such as testosterone increase both muscle mass and bone density (Saretzki, 2024). Adrenal insufficiency is a condition in which the adrenal glands are not producing sufficient corticosteroids and this requires the patient to receive steroid hormone replacement for life. A clinical trial of hydrocortisone replacement therapy found that it increased survival rates and decreased the incidence of adrenal crisis in patients with adrenal insufficiency (Marcu et al., 2024). Steroids are also used in the management of rheumatoid arthritis, asthma, and other autoimmune diseases because of their calming effect. A randomized controlled trial showed that low-dose glucocorticoid treatment was effective in reducing inflammation with minimal side effects in patients with rheumatoid arthritis (Ornetti et al., 2025). Anabolic steroids like testosterone are used in the treatment of hypogonadism, sarcopenia, and cachexia to improve muscle and bone strength. A placebo-controlled study established that testosterone therapy led to important improvements in muscle strength and bone mass in hypogonadal men. Glucocorticoids are used in immunosuppressive therapy to prevent organ rejection in transplant patients. A longitudinal study of prednisone therapy for kidney transplant patients showed that it decreased the rate of graft rejection (Ghasemi et al., 2024; Madacki-Todorović et al., 2024).

15.3.1.2 Estrogen and Progesterone Pills

Estrogen and progesterone are key hormones. They control female body functions such as menstrual cycles and pregnancy. They also help keep bones strong, hearts healthy, and minds sharp. Many women take these hormones, known as HRT (Hulubă et al., 2025). HRT helps women going through menopause, using birth control, or dealing with certain health issues. Estrogen sticks to certain cells, which are called estrogen receivers (ERα and ERβ). This helps cells multiply and control the body's metabolism. Progesterone works with progesterone receivers (PRs). This helps control the lining of the womb and supports pregnancy. Having both hormones helps to maintain a good hormonal balance. This prevents too much growth of the womb lining. When a woman goes through menopause, she makes less estrogen. This can cause problems like hot flushes, weak bones, and mood swings, all of which can be helped by taking hormones. One study showed that taking estrogen with progesterone helps with hot flushes and keeps bones strong in women after menopause (Ruan and Mueck, 2024). Birth control pills use fake estrogen and progesterone to stop ovulation, thereby helping prevent pregnancy. Researchers showed that these pills are very good at preventing pregnancy and also help with irregular periods. Women with a health issue called polycystic ovary syndrome (PCOS) can experience hormonal issues and irregular ovulation. Hormone therapy helps to regularize their periods and controls their androgens. A study found that hormone therapy helps reduce androgen levels and improves ovulation in women with PCOS (Onyeso et al., 2025). After menopause, less estrogen can lead to weak bones (osteoporosis). Taking estrogen helps keep bones strong. A long-term study showed that women taking estrogen had fewer bone breaks related to weak bones. Estrogen also helps keep hearts healthy by reducing risks for heart disease and stroke. A combined study showed that women taking HRT after menopause had healthier cholesterol levels and improved heart function.

15.3.1.3 The Androgen Class of Drugs

Testosterone and its derivatives are essential hormones that control not only male reproductive function but also muscle mass, bone density, and metabolism. These are used in hormone replacement therapy (HRT) to treat hypogonadism, muscle-wasting disorders, and gender-affirming hormone therapy. Androgen drugs boost muscle growth, produce more red blood cells, and increase libido by connecting to androgen receptors in our body in specific tissues which helps in gene expression. There are different ways to administer testosterone, including injections, skin patches, gels, or even pills. This helps your body keep a stable testosterone level (Hahamyan and Basaria, 2024). Some people's bodies simply do not produce enough testosterone. This problem is called hypogonadism. It causes tiredness, the loss of muscle, a lack of desire, and bone loss. The main treatment for this problem is to administer more testosterone. In one particular trial, this treatment helped men with low testosterone get stronger, grow more dense bone, and enjoy better sexual function, proving its efficacy (Qin et al., 2024).

15.3.2 Contraceptives

Birth control medicines are either hormonal or non-hormonal, and they're made to guide reproductive processes. Hormonal birth control often contains made-up estrogens, progestins, or both, which copy the actio of body sex hormones to affect the reproductive system. These medicines come from steroid hormones and have a specific type of structure common to all steroid molecules.

Estrogen birth controls such as ethinyl estradiol, have a change at C-17, which improves how well the body absorbs it and how stable it is. Progestin medicines, including levonorgestrel, norethisterone, and drospirenone, have changes at C-17 and C-13, like adding methyl or ethyl, which betters how selective they are and how long they work.

Non-hormonal birth controls include copper IUDs and spermicides ('sperm-killers'). Copper-based IUDs exudes ionic copper, which changes cervical and uterine fluids, interfering with sperm health. Spermicides, like nonoxynol-9, are materials that ruin sperm covers. Advances in medicine delivery systems, like long-acting shots, implants, and skin patches, have improved the effectiveness of birth controls, how easy they are to take, and how safe they are (Stanczyk et al., 2024).

15.3.3 Thyroid Supplements and Suppressants

15.3.3.1 Thyroid Drugs

Thyroid hormone medications are synthetic thyroid hormones that mimic the structure and function of the endogenous thyroxine (T4) and triiodothyronine (T3). These medications are based on a distinctive diphenyl ether backbone which is central to their function. They are small, iodinated tyrosine residues compounds.

Levothyroxine (T4) and liothyronine (T3) are the most commonly used synthetic thyroid hormones. The reason for this is that liothyronine is structurally more active because it has three iodine atoms whereas levothyroxine has four. These hormones are stabilized by ether bonds and then deiodinate in peripheral tissues to form the more active T3 form. Formulation improvements have led to long-acting thyroid hormone preparations with altered hydrophilicity, receptor affinity, and metabolic stability. To achieve a more balanced thyroid hormone profile, there are combination medications with T4 and T3 in various formulations. These structural modifications assist in improved absorption, bioavailability, and durability to acquire the best possible therapeutic outcomes (Mondal et al., 2016).

15.3.3.2 Thyroxin or Levothyroxine Pills

Thyroxine, or T4, is an important hormone from the thyroid gland. It's necessary for managing energy use, supporting growth, and keeping our hearts strong. Levothyroxine, a man-made form of thyroxine, is commonly used in a treatment called hormone replacement therapy. This helps people with conditions such as an underactive thyroid, those who've had their thyroid gland removed, and those with metabolic issues. Levothyroxine works just like natural T4 because it changes into another thyroid hormone, triiodothyronine or T3, in our body's tissues. This helps control how quickly we use energy, keep us warm, and create proteins. It attaches to thyroid hormone spots inside our cells, called receptors. This affects how genes work and the balance of energy in our body (Pu et al., 2025). An underactive thyroid, or hypothyroidism, happens when the thyroid gland doesn't create enough thyroid hormones. This can make people feel tired, gain weight, and have trouble thinking. Levothyroxine is the preferred treatment for restoring thyroid hormone levels to normal. One study with strict controls showed that levothyroxine treatment greatly improved how the body uses energy and how well the brain works in people with hypothyroidism (Alam et al., 2025). Those who have had a thyroidectomy (the surgical removal of the thyroid gland) must take levothyroxine for the rest of their lives to replace their missing hormones. A longitudinal study showed that LT4 therapy was effective in preventing hypothyroidism symptoms and maintaining metabolic stability in patients following thyroidectomy. Subclinical hypothyroidism (SCH) is characterized by mild elevation of TSH with normal T4 values; treatment is generally indicated in symptomatic patients. A meta-analysis of randomized controlled trials (RCTs) revealed that levothyroxine therapy was more effective than placebo in improving lipid profiles and cardiovascular risk in patients with SCH (Pu et al., 2025). Thyroid hormones are vital for fetal brain development and maternal metabolic functions. Levothyroxine is prescribed to women with hypothyroidism during pregnancy to ensure healthy fetal development. A clinical trial found that levothyroxine treatment was associated with lower rates of preterm birth and improved neurodevelopmental outcomes in their children (Goodarzi-Khoigani and Aminorroaya, 2025). New evidence shows that levothyroxine therapy can control metabolic dysfunction-associated steatotic liver disease (MASLD) through enhancing lipid metabolism and insulin sensitivity. Thyroid hormone receptor activation was found to enhance liver function and lipid profiles in a clinical study of MASLD patients, suggesting levothyroxine as a possible metabolic therapy (Ramya et al., 2025; Pu et al., 2025).

15.3.3.3 Antithyroid Agents

Antithyroid medications is a process where tiny molecules collectively try to suppress functionality of thyroid hormones. The thionamide group, including methimazole, propylthiouracil (PTU), and carbimazole, are mainly associated in this medication process. These antithyroid agents contain a thiourea or thioamide group, which helps them to integrate with thyroid hormones producing enzymes.

Every thionamide medication is associated with a molecular framework containing a complex ring of atoms with a sulfur group wedged in. This sulfur plays a significant role in the mechanism of drugs. Methimazole and carbimazole are derived from a thiazole, a five-atom architecture, while propylthiouracil (PTU) resembles a pyrimidine. These differences in the framework of each drug make them function differently in the body, including in terms of how it gets around, how it's broken down, and how long it sticks around. For example, carbimazole is a prodrug, which evolves into methimazole in your body to better medication.

There are other antithyroid drugs, like potassium iodide (KI) and lithium carbonate, which are able to inhibit thyroid hormone production. Radioactive iodine (I-131) relies on its radioactive decay for the inhibition of hormone production. With the advancement in drug design antithyroid medications are targeted, specific, long-lasting and more effective with fewer side effects (Suryandari et al., 2024).

15.3.3.4 Parathyroid Hormone and Analogs

Parathyroid hormone analogs and calcimimetics are associated to synthesize parathyroid drugs and related medicines. They mainly control the stability of calcium and phosphate levels in the body. These drugs are usually small molecules or peptides modelled to imitate or control the natural parathyroid hormone (or PTH). They also mimic activity of a similar hormone protein called PTHrP.

PTH analog is one of the example of teriparatide and abaloparatide, which are peptide pieces directly derived from human PTH. They contain N-terminal amino acids from PTH, which are crucial for activating body receptors. These analogs maintain a spiral shape and undergo specific modifications that mold them, enhancing binding with receptors and prolonging their duration. The drug abaloparatide has been modified in its spiral structure, making it connect better with PTH receptors.

On the other hand, calcimimetics such as cinacalcet and etelcalcetide are smaller drug molecules. They focus on the receptor that senses calcium in human bodies. They are made around something known as a benzenesulfonamide core, which enhance their binding and modulation abilities. Etelcalcetide is a peptide-based calcimimetic that uses disulfide bonds in order to boost its structure and make enduring effects. Smart design strategies have enabled producing stable versions of these drugs and specialized release systems that work to optimize their positive effects while minimizing differences in mode of actions

15.3.4 Hormone Inhibitors

15.3.4.1 5-Alpha-Reductase Inhibitors (5αRIS)

These small molecule medicines hinder the action of 5-alpha-reductase (S5αR), which converts testosterone to dihydrotestosterone (DHT). Steroidal or non-steroidal inhibitors are designed for imitating natural steroid substrates or attach to the enzyme's active site. Finasteride and dutasteride, commonly referred to as 5αRIs, are synthetic 4-azasteroid derivatives with structural similarities to testosterone. They have a modified azasteroid nucleus, which allows them to bind competitively and permanently to the enzyme, preventing testosterone from being converted to DHT. Finasteride comprises a single nitrogen alteration in the steroidal ring, whereas dutasteride includes extra fluorine substitutions, making it more effective and capable of blocking several S5αR isoforms (Type I and II).

Natural 5α-reductase inhibitors (5αRIs) isolated from herbal extracts usually bear specific structures that are common in polyphenols, flavonoids, or terpenoids. Compounds such as epigallocatechin gallate from green tea, β-sitosterol from saw palmetto, and curcumin from turmeric function as enzyme inhibitors but are not acting as synthetic inhibitors. These natural compounds commonly interfere with enzymes by directly binding the active site of the enzyme or by inducing oxidative stress, which then affects enzyme function. Unlike synthetic inhibitors, these have a different structure but the common target: to reduce the activity of an enzyme by changing its interaction with the substrate (Arya et al., 2024; Dai et al., 2025; Ru et al., 2024).

15.3.4.2 Adrenal Corticosteroid Inhibitors

Substances that are adrenal corticosteroid inhibitors, whether synthetic or natural, prevent the adrenal cortex from producing corticosteroids or from acting on them. These inhibitors prevent the conversion of cholesterol to active corticosteroids, such as glucocorticoids and mineralocorticoids, by affecting the steroidogenesis pathway enzymes. In the structure, the inhibitors are non-steroidal, which act by enzyme inhibition, or steroid-based, which act as corticosteroids. Steroid inhibitors of the adrenal corticosteroids are divided into two categories depending on how they function. Steroidal inhibitors such as metyrapone work by competing for 11β-hydroxylase (CYP11B1), an enzyme that is involved in the final steps of cortisol production, using steroids that are similar to the endogenous corticosteroid precursors. The non-steroidal inhibitors, e.g. ketoconazole, have imidazole rings that interact with cytochrome P450 enzymes to affect several steps in the production of adrenal steroids. Alkylation and fluorination are examples of such alterations which have been used to enhance their binding specificity and metabolic stability, thereby enhancing their efficacy and biological half-life (Mukhtar and Haris, 2024).

15.3.4.3 Prolactin Inhibitors Class of Drug

Prolactin inhibitors are a class of dopamine agonists that control prolactin secretion through the stimulation of dopamine D2 receptors in the anterior pituitary. These drugs are small molecules that were designed to mimic dopamine, a natural neurotransmitter that prevents prolactin secretion.

Based on structure, prolactin inhibitors can be divided into ergot and non-ergot dopamine agonists. Ergot derivatives, bromocriptine and cabergoline, have a tetracyclic ergoline core that increases the affinity to and the duration of binding to dopamine receptors. Substitutions with halogens and alkyl groups enhance the lipophilicity and the metabolic stability, which, in turn, increases the half-life.

Non-ergot dopamine agonists, including quinagolide, are benzazepine or tetrahydroisoquinoline derivatives that activate selectively the D2 receptor. Long-acting, and highly specific, prolactin inhibitors have been developed to improve drug formulation with fewer doses and side effects and with potent receptor binding affinity (Lv et al., 2025).

15.3.5 Hypoglycemics & Insulin

Insulin is a hormone that our body manufactures in the beta cells of pancreas. It helps our cells use sugar for energy and the regulation of lipids. When someone has diabetes, their body either cannot use insulin in the correct way or have low secretions of insulin. Doctors provide insulin medicines through oral pills or intravenous injections to these patients. Insulin medications function by attaching to target cell insulin receptors, which sets off a series of signals that encourage the absorption of glucose, the production of glycogen, and the metabolism of fats. Doctors can pick from different kinds of insulin: Rapid, short, intermediate, and long-acting insulin. Each type works for a different span of time. This helps keep sugar levels stable throughout the day. Type 1 diabetes (T1D) happens when our body destroys cells that make insulin—an autoimmune disorder by beta cell destruction. So, people with T1D need exogenous insulin shots. In a clinical test, rapid-working insulin did a better job of keeping mealtimes sugar levels in control and lowered hypoglycemia in T1D patients. People with type 2 diabetes (T2D) often start off with diabetes pills, but may need insulin shots if their bodies don't make enough anymore. A longtime study showed that for T2D patients, starting insulin shots early leads to better sugar control and fewer disease troubles compared to simply using pills. Science has come up with a machine that works like a pancreas. It automatically controls sugar levels. An experiment showed people using this half-robot pancreas had better sugar control and less nighttime low-sugar level dangers when compared with regular insulin shots. Finally, there's diabetes of pregnancy (GDM), which requires careful sugar control to keep mother and baby safe. Insulin shots are shown to lessen the chance of having a big-sized baby and dangerous low-sugar levels (hypoglycemia) in newborns compared to just lifestyle changes (Raposo et al., 2025; Saputra et al., 2025).

15.3.6 Receptor Blockers

15.3.6.1 Melanocortin Receptor Agonists Class of Drug

The action of endogenous melatonin; alpha, beta melanocyte-stimulating hormone (α, β-MSH) and adrenocorticotropic hormone (ACTH). These drugs target melanocortin receptors (MC1R-MC5R), which are G-protein-coupled receptors (GPCRs) that are involved in normal physiology.

The peptide-based MCR agonists, including bremelanotide and afamelanotide, are structural analogs of α-MSH, in which the molecule has been chemically altered through disulfide bonding and N terminal acetylation to increase receptor affinity and metabolic stability. These alterations lead to better efficacy, specificity, and temporal action of the compounds as opposed to the native melanocortins.

Non-peptide small molecule melanocortin agonists like setmelanotide are designed to selectivity activate specific MC4R with better oral bioavailability. These compounds are usually heterocyclic ring systems such as indoles, triazoles, or benzimidazoles, which increase receptor affinity and pharmacokinetic profile. This has been possible due to the advancement in molecular modelling and peptide engineering to produce long-acting, selective, and very stable melanocortin agonists to enhance their therapeutic uses (Todorovic et al. 2024).

15.3.7 Antagonists

15.3.7.1 Growth Hormone Antagonists

Protein-based biologics popularly referred to as growth hormone (GH) antagonist drugs are developed to prevent endogenous GH from interacting with its receptor, the growth hormone receptor (GHR). These drugs are altered variants of the native growth hormone and are usually produced by site-directed mutagenesis to alter the hormone's receptor-binding characteristics while maintaining its overall architecture. Pegvisomant, a genetically modified GH analog with some amino acid substitutions that disable receptor activation, is a well-known GH antagonist. However, although structurally it resembles the 191-amino acid polypeptide chain of natural GH, it has mutations in crucial binding sites that prevent the GH receptor from dimerizing correctly. Pegvisomant is also subjected to pegylation, which is the process of coupling PEG chains to the molecule to enhance the stability and solubility, and the circulation half-life. Monoclonal antibodies and substances that target the GH–GHR interaction are used to develop other GH antagonists. These substances are different from the peptide-based antagonists in that they are often based on small molecule scaffolds to prevent the receptor binding either intracellularly or extracellularly. The development of GH antagonists through GH antagonists through protein engineering and molecular modelling is crucial in making the antagonists more specific and less immunogenic (Basu et al., 2021; Szepesi et al., 2024).

15.3.7.2 Sex Hormones and Antagonists

Sex hormone drugs are man-made or naturally occurring versions of substances like androgens, estrogens, and progestogens. They're made to act like or control the hormones inside us. Most of these drugs are steroids, and they all have a design with four rings in common.

These drugs come in various forms, like tablets, injections, patches, and implants. Changing their structure helps absorption and stability, and makes sure the drug gets to the right place in the body. Advances in nanotechnology and enclosing the drug to slowly release it have also helped by reducing how often you have to take the drug while keeping it effective.

15.3.7.3 Androgens and Antiandrogens

Simply put, androgens are a type of hormone. They're made up of steroids and share a four-loop structure. This unique formation is why all steroid hormones look alike. Naturally found androgens, like testosterone and DHT, have a special OH group at C-17 and a ketone at C-3. These are key for binding to, and interacting with, the androgen receptor (AR). People have created synthetic androgens too, like methyltestosterone, oxandrolone, and fluoxymesterone. The creation of these androgens involves changes to improve how well they're absorbed and how long they last in the body. The changes include tinkering with the C-17α position. This protects the molecule from liver metabolism and makes it safe to take orally. Some androgens also have elements like fluorine or chlorine added to make them bond better to the receptor and increase their effect.

Certain compounds, called antiandrogens, can block signals onto androgen receptors, or androgen formation. These can be in either steroidal or non-steroidal forms. Steroidal antiandrogens, like cyproterone acetate and spironolactone, possess a core structure with four rings, like steroids. Although modification of ketones or lactones affect their interactions with receptors which leads to competition between testosterone and DHT to bind with androgen receptors.

There are non-steroidal antiandrogens like flutamide, bicalutamide, and enzalutamide with a core structure of anilide or benzothiazole, help them to link with the androgen receptor without copying the steroid backbone. Commonly, they include alternatives like fluorine or chlorine which enables their structure stable and extend their 'half-life'. New efforts in drug design have brought about selective androgen receptor degraders (SARDs). These enhance the androgen receptors to break down, and deliver, more long-lasting effects for conditions affected by androgens (Gupta et al., 2024; Javan and Youseftabar-Miri, 2025).

15.3.7.4 Estrogens and Antagonists

Estrogen drugs are a type of hormone that's built like estrane, a complex core. This core is like a special grid. It has an aromatic A-ring, which is a specific circle-like grouping, and other groups at two points named C-3 and C-17. Three main of these hormones made in our bodies are estradiol (E2), estrone (E1), and estriol (E3). Of these, estradiol is the strongest. Sometimes, we make estrogens in labs. These are called synthetic estrogens. Ethinyl estradiol and diethylstilbestrol (DES) are such examples. These are changed to be easier to use by our bodies and stable for longer. It's like packing a lunchbox so it won't spoil till lunchtime! For example, ethinyl estradiol has an ethynyl addition. This tricky change prevents its fast processing in the liver, making it stick around longer. Some estrogens are tied together, like estrone sulfate and equilin. They are made to mix better with water and control their release. This is like having a slow-drip water bottle for the day instead of a regular one.

Anti-estrogens, also called estrogen blockers, are split into two groups. The first are SERMs, including tamoxifen and raloxifene, which have a special structure that lets them connect to estrogen receptors in certain tissues. The second group are SERDs. Of these, fulvestrant is an example. This is a compound that breaks down estrogen receptors and forms lasting changes in them. This increases their ability to connect with the receptors, but also destabilizes them. Drug development has lately led to new SERDs that aren't steroids. They're meant to be more easily taken orally and to more accurately target therapy (Hall, 2025; Kiyama and Wada-Kiyama, 2024).

15.3.7.5 Gonadotropin Releasing Hormones and Antagonists

GnRH (gonadotropin-releasing hormone) medicines are basically hormones made in a lab. They are designed to work just like the real GnRH that your body makes, which is a tiny protein containing ten linked particles called amino acids. This protein is naturally made in a part of your brain called the hypothalamus. Improving the function of this protein, certain parts of it are locked in place by a pyroglutamyl residue at the start and an amide group at the end, thereby ensuring it stays intact longer and can do its job better. Now, scientists have come up with even better GnRH medicines, like leuprolide, triptorelin, and buserelin. What's special about these is that they have certain changes at the 6, 9, or 10 spots of their structure. This makes them last longer and be more effective. These changes can be done by swapping a glycine amino acid with D-amino acids or using something called ethylamide. This helps make them resist being broken down too soon. That means these updated medicines can be given only once a month, or even once a year as an injection, and still work really well.

GnRH antagonists are like a special recipe created in a lab. They attach to GnRH receptors (like keyholes ready to open with the right key), but don't activate them. Ironically, while GnRH agonists increase hormone release first before reducing it, these antagonists block these receptors immediately. Some common GnRH antagonists are cetrorelix, ganirelix, and degarelix. Their makeup includes D-amino acid swaps and modifications making them "water-fearing". This boosts how well they grasp onto these receptors and prevents them breaking down easily. These tweaks ensure they start working quickly and can cling to these receptors for a long time. This makes them perfect for injections and depot formulations that release slowly over a sustained period. Also, there have been strides made in drug design that now allow us to have non-peptide GnRH antagonists in an easy-to-take pill form. What sets these apart is their small-molecule scaffold design, which heightens how well your body can absorb and administer them (Tzoupis et al., 2020).

15.3.7.6 Gonadotropins and the Anti-Gonadotropic Class of Drugs

Gonadotropin drugs come in two types: recombinant and urinary-derived. They're basically hormones that include follicle-stimulating hormone (FSH), luteinizing hormone (LH), and human chorionic gonadotropin (hCG). These hormones are shaped like glycoproteins and are made up of two parts, labelled as α and β chains. (It's the β part that helps them attach to specific receptors.) To keep their shape stable, they have disulfide bonds and go through a process called post-translational glycosylation. This process affects how they bind to receptors and how long they exist in the body. Recombinant gonadotropins like follitropin alfa (rFSH) and lutropin alfa (rLH) are created using a process called genetic engineering. This is done in mammalian cells to ensure that they're very similar to natural hormones. On the other hand, urinary-derived gonadotropins, such as menotropins (hMG) and urofollitropin (uFSH), are taken from urine collected after menopause. They have varying levels of FSH and LH. Medical scientists have also developed modified versions of these hormones to be long-acting. This is done through processes called pegylation or using something called sustained-release carriers. The result is a type of hormone that stays available for use in the body longer, thereby reducing the need for frequent doses.

GnRH analogs, androgenic steroids, and progestins are all antigonadotropic agents. They stop gonadotropin secretion, or production. They do this by targeting something called the hypothalamic–pituitary–gonadal axis. Let's talk about steroidal antigonadotropins, like danazol and gestrinone. These are related to testosterone and have something called ethynyl or isoxazole substitutions. This means they attach better to receptors and change metabolic clearance. GnRH agonists and antagonists are types of drugs. They include things like leuprolide, degarelix, and cetrorelix. These are man-made versions of special chains of proteins, called peptide analogs, with a switch in amino acid that makes them more stable and helps them bind better to receptors. They get changed in ways like N-terminal substitutions or cyclic peptide backbones. This keeps them from breaking down too fast. Because of this, they can either block receptors immediately or do so over a longer period of time. Scientists have made advances in how drugs are designed. They've made oral non-peptide GnRH antagonists. These drugs are better because they're easier to take and your body can use them more effectively.

15.3.7.7 Progesterone and Its Antagonists' Class of Drugs

Progesterone drugs are a part of the steroid hormone family. They come from a molecule called pregnane. This molecule has a chemical structure called a cyclopentanoperhydrophenanthrene nucleus. This nucleus has a special compound, called a ketone, at two places, C-3, and C-20. Progesterone is naturally present in our bodies, but it can easily get broken down. This is because it is something called lipophilic. Scientists have created synthetic versions of progesterone, known as progestins. These synthetic versions are more stable and only react with specific receptors. These synthetic progestins have names like norethisterone, medroxyprogesterone acetate (MPA) and levonorgestrel.

Scientists make them more efficient by adding methyl or ethyl at places called C-17 and C-13. This helps them last longer and work better in the body. There are also different versions of these drugs. For example, micronized progesterone can be absorbed better by the body. Some are meant to be injected, like the kind found in IUDs and implants. These versions release the drug slowly into the body.

Antiprogestins, also called progesterone blockers, can either be steroid-based or not. They stop progesterone from working properly. They remain linked to steroids, but with changes that affect how they interact with receptors. One famous antiprogestin called mifepristone, or RU-486, has a big switch at position C-11. This stops the receptor from working right, while still clinging onto it strongly. Some other blockers, like ulipristal acetate, make changes at position 17α-acetoxy to work better with the receptor and last longer in the body. We've made leaps forward in medicine, leading to drugs that carefully manage how they interact with the progesterone receptor. Depending on the body part, these drugs, called SPRMs, can either get in the way of or support progesterone,. They figure out the best way to act.

15.3.8 Agonists

15.3.8.1 Gonadotropin Hormones

Gonadotropins are hormonal substances that have a key role in how our reproductive organs function. There are three main types: luteinizing hormone (LH); follicle-stimulating hormone (FSH); and human chorionic gonadotropin (hCG). These help produce sperm (spermatogenesis), cause eggs to mature and be released (ovulation), and stimulate the ovaries and testes (gonadal function). These hormones are vital for making other essential hormones like estrogen, progesterone, and testosterone. They're so important, in fact, that they're used in medicines to help treat infertility, issues with underactive sex organs (hypogonadism), and imbalances in our body's hormone levels. For example, FSH helps develop sperm and eggs, while LH causes eggs to be released and testosterone to be made. Meanwhile, hCG acts a lot like LH, and so can be used to treat infertility in males and issues with the production of the gonadotropin hormone (Esteves et al., 2025).

Male infertility is caused by hypogonadotropic hypogonadism, which happens due to insufficient gonadotropin production. This results in low testosterone and impaired sperm production. Hormonal medicines of hCG and FSH can help the man recover their fertility.

Doctors discovered in a study that spermatogenesis was restored by the medicines given to azoospermic men with HH who were not producing any sperm because of the gonadotropin problem. It worked and they regained their fertility.

Ladies who struggle to ovulate, a must for getting pregnant, because they suffer from PCOS can also get help from the gonadotropin medicine. A carefully managed research test found that these medicines made ovulation better, thus raising the chances of getting pregnant for women being helped to have babies (Esteves et al., 2025).

15.3.8.2 Insulin-Like Growth Factors

These types of medicines tie in with insulin-like growth factors (shortened to IGF). The medicines have strange names: recombinant IGF analogs, IGF-binding protein modulators, and IGF receptor-targeting agents. Their job is to control the activity of insulin-like growth factors (IGF-1 and IGF-2). Mostly, these medicines are proteins, and have a structure that's pretty similar to natural IGFs. They are built with one chain of things called polypeptides. This chain is made strong with three bridges made of disulfide. These bridges keep their shape just right!

Mecasermin and similar drugs belong to a group known as recombinant IGF-1. How are these drugs made? With the help of something called recombinant DNA technology. They look a lot like human IGF-1, but with some small changes. These changes help them attach to receptors better and remain stable for longer periods. Take mecasermin rinfabate, for instance. This drug is a mixture of

IGF-1 and IGF-binding protein-3 (or IGFBP-3 for short). It's specifically made to last longer and work better in the body.

Other medicines linked to IGF include monoclonal antibodies that aim at IGF-1 receptors (IGF-1R) and small compounds that tune IGF signals. These compounds often have improved binding spots to better select IGF receivers and lessen any connections with insulin receivers. Better ways to design proteins and model molecules have allowed the creation of long-lasting IGF versions and receiver changers, which enhance how the drug works in the body and its effectiveness in treatment (Li and Li, 2024).

15.3.8.3 Calcitonin

Drugs in the calcitonin class contain versions of a hormone called calcitonin. This hormone helps control how much calcium there is in our bodies. Natural calcitonin in humans, fish like salmon or eels, can be used to develop these drugs. The alterations made to this natural hormone make the drug more efficient and last longer.

Calcitonin is a teeny tiny thing made of 32 different parts called amino acids. To help keep its structure, it has a special bond between the first and seventh parts. Calcitonin from salmon is often used in treatment because it connects better to our bodies and lasts longer. This is because its sequence of amino acids is different to that of human calcitonin.

There are different ways in which one can take calcitonin medicines: by injection, a nasal spray, or orally. Some enhancements, like PEGylation and cyclization, have made the drug stay longer in our bodies and be absorbed better. One method, called recombinant DNA technology, lets us make fake versions of calcitonin that work just like the real thing, giving a steady reaction and not triggering the body's defense system as much.

15.3.8.4 Calcimimetics

Calcimimetics are a type of small drug. They work with a special receptor found mostly in two parts of the body: the glands near our throat and our kidneys. These drugs make this receptor more sensitive to a substance called calcium outside of our cells. As a result, the receptor behaves differently. It also helps lower the amount of a hormone in our body.

There are two main types of these drugs, called type I and type II. Type I drugs act like calcium ions—these are small, charged particles of calcium. Some amino acids and polyamines fall into this group. These are the types of molecules that our body produces naturally. Type II calcimimetics are a bit different. They are man-made and work in a slightly different way. An example is a drug called cinacalcet. It is a type of drug you can take orally and comes from a larger group called benzenesulfonamides. Another example is etelcalcetide. It's a drug modeled after peptides, which are small pieces of proteins. This drug is stable because of bonds called disulfide bonds.

Scientists have been trying to make these drugs even more effective. They're working on making them more selective, meaning they specifically choose which receptors to act on. They're also trying to make them last longer in the body and be easier for our body to use. These drugs can come in many forms such as tablets or injections. Scientists have made changes to these drugs to increase their solubility. That means they can be dissolved easier. This also helps them interact with the receptors for a longer time, so they don't need to be taken as often (Hou et al., 2022).

15.3.8.5 Antidiuretic Hormones

ADH-type drugs are mainly made up of something called vasopressin analogs and antidiuretic peptides. These work like a natural body chemical called arginine vasopressin (AVP). They deal with the body's balance of water. These drugs are made with nine amino acids, and it's built like a circle with a special bridge that helps keep it stable. The most used synthetic vasopressins are desmopressin, terlipressin, and felypressin. Each one is built a little differently in order to do its job more effectively. Desmopressin is made in a lab and lasts longer; terlipressin works more specifically; and felypressin helps control blood flow better.

Some medicines help our bodies hold onto water. They come in different forms: injections, sprays for your nose, or pills. Scientists have made them so that our bodies can use and dispose of them in the best way. We also have long-lasting versions—with them, you don't need to take medicine that often, but it still works just as well. These new and improved medicines give us exact control over how much water our bodies keep. They make sure our blood moves around our bodies just right. Plus, they hardly have any side effects (Wang, 2024).

15.3.8.6 Adrenal Cortical Steroids

Adrenal cortical steroids are a group of steroid hormones produced in the adrenal cortex, which are synthesized from cholesterol through a series of enzymatic reactions. They are subdivided into three functional classes—glucocorticoids, mineralocorticoids, and androgens—with the unique structural modifications of each type defining the respective function. These steroids possess a common cyclopentanoperhydrophenanthrene backbone, with variation in functional group substituents altering the compound's solubility, interaction with receptors, and metabolic stability. Synthetic alternatives such as prednisolone and dexamethasone, on the other hand, modulate chemical properties that play a role in making them more potent and longer-lasting. The biological functions and specificity of their actions are largely defined by the enzymatic production of these lipid mediators through hydroxylation, oxidation, and methylation (Kuzminac et al., 2024; Tao et al., 2024).

15.3.8.7 Corticotropin

Corticotropin, also called adrenocorticotropic hormone (ACTH), is a polypeptide hormone that is produced by the anterior pituitary gland. It is derived by enzymatic cleavage from a larger precursor protein called pro-opiomelanocortin (POMC). Corticotropin is structurally composed of 39 amino acids, with the first 24 residues—which are highly conserved across species—having the majority of its biological activity. The synthesis and secretion of corticotropin are regulated by the hypothalamic–pituitary–adrenal (HPA) axis through complex feedback loops, in addition to the interaction between corticotropin and melanocortin receptors (MC2R), which are mainly located in the adrenal cortex. Corticotropin has been synthesized and modified for use in both diagnostic and therapeutic approaches, where binding antagonists and alterations to the peptide sequence have also enhanced peptide stability and receptor binding (Harxhi et al., 2024).

15.3.8.8 Glucocorticoids

The adrenal cortex, especially the zona fasciculata, produces glucocorticoids, a kind of steroid hormone derived from cholesterol. Corticosteroids contain the same cyclopentanoperhydrophenanthrene nucleus as all other steroid molecules. Because of the hydroxyl and ketone groups in their structure, they are lipophilic and may thus pass through cell membranes and interact with intracellular receptors. Certain hydroxylation processes mediated by cytochrome P450 enzymes are involved in the synthesis of glucocorticoids ranging from pregnenolone and cortisol. Dexamethasone and prednisolone are synthetic glucocorticoids with structural modifications that improve their longevity, stability, and affinity to the receptor.

The fluorination and methylation are the modifications which occurred from naturally arise glucocorticoids that enhance the efficacy and decrease the metabolism (Jäger et al., 2024).

15.3.8.9 Mineralocorticoids

The adrenal cortex, in its zona glomerulosa, produces a class of steroid hormones, mineralocorticoids, from cholesterol. They are structurally similar to other steroid hormones, having the cyclopentanoperhydrophenanthrene nucleus. Hydroxyl (–OH) and ketone (C=O) functional groups are involved in their structure and determine their metabolic stability and receptor binding. For the production of mineralocorticoids, several enzymatic conversions are required. Cholesterol is converted to pregnenolone, which is altered by aldosterone synthase (CYP11B2) and 21-hydroxylase (CYP21A2) to produce aldosterone, the main mineralocorticoid. The structural changes made to the synthetic mineralocorticoids, such as fludrocortisone, enhance their potency and decrease their half-life, which makes them more suitable for medicinal use. Many of these changes include fluorination, which increases receptor affinity and decreases metabolic breakdown (Jäger et al., 2024).

15.4 APPLICATIONS OF HORMONAL THERAPY

Hormone replacement therapy (HRT) is often given as a treatment for symptoms that come with menopause, problems with bones being too weak, and andropause. Healing therapy with estrogen and progesterone can help ease hot flushes, mood swings, and weak bones in women who've been through menopause. Testosterone therapy is used to treat hypogonadism, which comes with aging in men. However, even though HRT can be effective, using it for a long period can lead to a higher risk of breast cancer and blood clots.

15.4.1 Hormonal Therapy in Cancer Treatment

Certain cancers, like those of the breast, prostate, and ovary, often need hormone therapy. This treatment helps stop cancer cells from growing. But there can be side effects. These include bone loss, metabolism changes, and risks to the heart. Patients need to be watched closely to stay safe (Brogna et al., 2025).

15.4.2 Reproductive Medicine and Contraception

Fertility treatments use hormones, especially when a couple is having a baby via in vitro fertilization, starting ovulation artificially, or saving fertility for later. Hormone birth control methods, like pills, patches, and IUDs, regulate periods and prevent pregnancy. Using these for a long time can cause hormone imbalances, blood clots, and change the balance of bacteria in the body (Xu et al., 2025).

15.4.3 Hormonal Therapy to Treat Metabolic and Endocrine Disorders

For people with diabetes, thyroid disorders, or adrenal insufficiency, hormone therapy can be vital. Insulin therapy is needed for type 1 diabetes, levothyroxine (T4) is the standard treatment for hypothyroidism, and glucocorticoid replacement therapy is used in Addison's disease. However, these treatments have risks like osteoporosis and metabolic dysregulation (Mei et al., 2025).

15.4.4 Gender-Affirming Hormonal Therapy (GAHT)

For transgender people, hormone therapy is key in helping the body match their gender identity. While this hormone therapy can help support mental health, ongoing check-ups are needed for heart and metabolism risks.

15.4.5 Hormonal Drugs to Remove Biofilm

Selected hormonal drugs are long being used for management of growth and biofilm formation of few selected bacteria, since dysbiosis of normal microbiota and formation of biofilm may lead to several biofilm-borne chronic diseases, and even cancers like irritable bowel disease (IBD) with colorectal cancer.

With that note, it becomes necessary to know the saga of biofilm formation, the role of hormones in development and the removal of biofilm, and, above all, the perspective of hormone therapy to manage biofilm-mediated diseases.

15.5 EFFECT OF HORMONAL DRUGS ON BACTERIAL GROWTH

Biofilms are stationary microbial colonies that colonize and develop on the surfaces of medical implants such as sutures, catheters, and dental implants with the assistance of self-produced extracellular polymeric substances (EPS). Patients are at risk for both costly medical care and mental illness as a result of these biofilms, which induce infections that can only be cured by eliminating them.

Hormonal drugs, steroids, sex hormones, metabolic hormones, and gonadotropins are known to alter metabolic pathways, quorum sensing, and immune system responses and are involved in the modulation of bacterial growth. Hormones and bacteria are interconnected in their ability to affect infection severity, antibiotic resistance, and microbial competition in both Gram-positive and Gram-negative bacteria.

15.5.1 Effect of Steroid Hormones on Bacterial Growth

Hormones, especially steroid hormones, including glucocorticoids (dexamethasone) and androgens (testosterone), exert their effects on bacteria by altering their metabolism and immune suppression. It has been established that dexamethasone, a glucocorticoid, boosted the growth of *Staphylococcus aureus* (Gram-positive) and *Pseudomonas aeruginosa* (Gram-negative) through the improvement of iron uptake and the weakening of immune clearance (Madacki-Todorović et al., 2024). In addition, testosterone has been reported to increase the growth of *E. coli*, which can lead to increased risk of infections in patients receiving testosterone therapy (Inaba and Doi, 2024).

Sex hormones like estrogen and progesterone control the bacteria that live in the urogenital and gastrointestinal tracts. Estrogen has been found to enhance the adhesion and growth of bacteria, especially for the urinary tract pathogens. A randomized study found that estrogen therapy increased the growth of *Escherichia coli* (Gram-negative) in urinary tract infections by changing the epithelial adhesion properties in response to the hormones. On the other hand, progesterone was found to suppress the growth of *Lactobacillus* (Gram-positive), which disturbs the vaginal flora and can lead to infection (Alizhan et al. 2025). Progesterone modulates *C. albicans* biofilm formation and genomic expression under acidic conditions (Gonçalves et al., 2020).

15.5.2 Effect of Thyroxine (Levothyroxine) on Bacterial Growth

Thyroid hormones, especially thyroxine (T4), are known to increase bacterial metabolism and bacterial replication rates. An experiment with microbes showed that thyroxine caused increased replication rates of *Klebsiella pneumoniae* (a Gram-negative bacterium) and *Streptococcus pneumoniae* (a Gram-positive bacterium), which suggests a possible role in bacterial virulence (Yamazaki and Seino, 2024).

15.5.3 Effect of Insulin on Bacterial Growth

Bacteria can use insulin, a vital hormone in glucose metabolism, as a nutrient, resulting in increased bacterial growth. A clinical trial on diabetic patients found that hyperinsulinemia was related to higher levels of *Pseudomonas aeruginosa* (Gram-negative) and *Streptococcus mutans* (Gram-positive) due to the higher glucose availability in the host tissues (Arevalo et al., 2024).

15.5.4 Effect of Gonadotropins on Bacterial Growth

Gonadotropins such as FSH, LH, and hCG have been seen to interact with bacterial quorum-sensing systems, affecting bacterial proliferation and toxin production. A microbial study found that gonadotropins enhanced the growth of *Klebsiella pneumoniae*, a Gram-negative bacterium, and *Listeria monocytogenes*, a Gram-positive bacterium, which increases their pathogenicity in infections of the reproductive tract (Esteves et al., 2025).

15.6 EFFECT OF HORMONAL DRUGS ON BACTERIAL BIOFILM

Bacterial biofilms are adherent communities of microorganisms encapsulated within an extrapolated polymeric substance that enhances antibiotic resistance, immune evasion, and chronic infection persistence. Hormones used for therapeutic purpose may or may not promote biofilm formation (Figure 15.2). Hormonal drugs, including steroids, sex hormones, metabolic hormones, and gonadotropins, are known to alter the biofilm formation in both Gram-positive and Gram-negative bacteria through the modification of quorum sensing, gene expression, and metabolism. These are important in the control of hormone-elicited bacterial infections and antimicrobial resistance.

15.6.1 Effect of Corticoids on Bacterial Biofilms

Hormones such as steroid hormones, the glucocorticoids, for example, dexamethasone, and the androgens, for example, testosterone, can stimulate biofilm formation in both the Gram-positive and Gram-negative bacteria through the regulation of the bacterial signaling and the host immune responses. A clinical study showed that dexamethasone boosted the biofilm formation in *Staphylococcus aureus* (Gram-positive) and *Pseudomonas aeruginosa* (Gram-negative) through the stimulation of genes that control quorum sensing and extracellular matrix production, thus making the bacteria resistant to antibiotics (Madacki-Todorović et al., 2024).

15.6.2 Effect of Estrogen and Progesterone on Bacterial Biofilms

Estrogen and progesterone-sized hormones affect urogenital biofilms by altering bacterial adhesion and metabolic activity. It has been found that estrogen increases bacterial virulence and biofilm formation in urinary tract pathogens. A randomized controlled trial found that estrogen

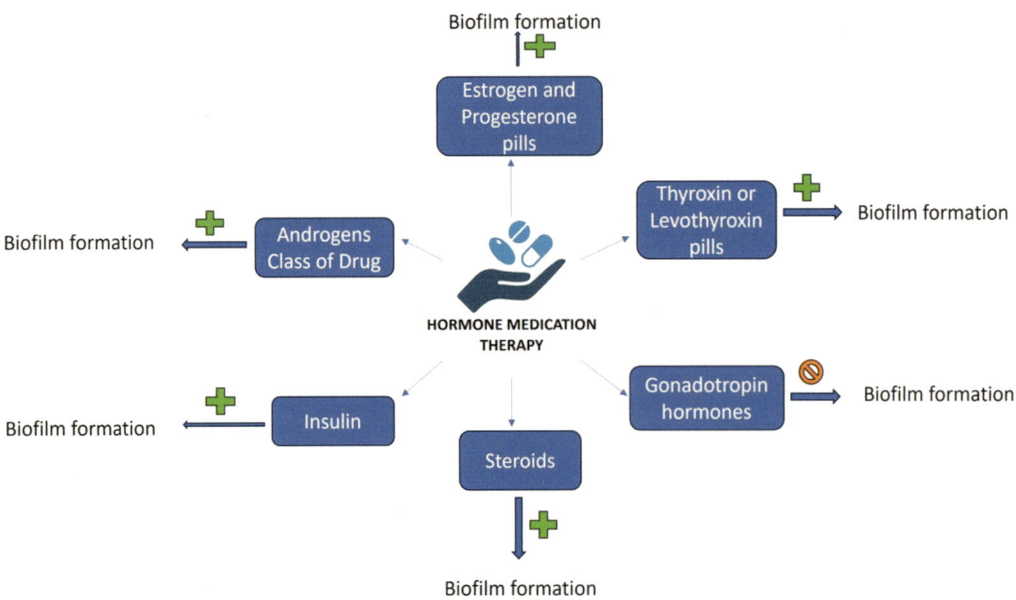

FIGURE 15.2 Effects of several hormone medication therapies on biofilm formation.

increased biofilm formation of *E. coli* (Gram-negative) and *Enterococcus faecalis* (Gram-positive) and their antibiotic resistance in urinary tract infections (UTIs). Progesterone was found to inhibit *Lactobacillus* biofilm formation, disturbing the vaginal microbiota and thus making a person prone to infections (Alizhan et al., 2025).

15.6.3 Effect of Thyroxine (Levothyroxine) on Bacterial Biofilms

Bacterial adhesion and biofilm formation are known to be influenced by thyroid hormones like thyroxine (T4). Thyroxine was found to increase biofilm formation in *Staphylococcus epidermidis* (Gram-positive) and *Klebsiella pneumoniae* (Gram-negative) in an implant-associated infection model (Yamazaki and Seino, 2024).

15.6.4 Effect of Insulin on Bacterial Biofilms

By supplying additional glucose for bacterial metabolism, insulin can increase bacterial biofilm formation, especially in diabetic patients. A clinical study of diabetic patients revealed that thicker biofilms were seen in *Pseudomonas aeruginosa* (Gram-negative) and *Streptococcus mutans* (Gram-positive) with higher insulin levels, which, in turn, increased antibiotic resistance (Arevalo et al., 2024).

15.6.5 Effect of Gonadotropins on Biofilm Development

Gonadotropins such as FSH, LH, and hCG interact with bacterial quorum-sensing pathways that may control biofilm stability and virulence. A microbial study found that gonadotropins enhanced biofilm creation by *Acinetobacter baumannii* (Gram-negative) and *Listeria monocytogenes* (Gram-positive), thus making them immune to immune clearance and antibiotic therapy (Esteves et al., 2025).

15.6.6 Effect of Hormones on Gut Bacteria

Oral hormonal drugs can affect the growth, biofilm formation, and adherence of gut bacteria at intestinal concentrations in vitro, which can explain specific microbiome signatures associated with the long-term use of these drugs. Steroid hormones increase the growth of *B. fragilis*, *E. coli*, and *L. reuteri* while reducing the growth of *B. longum*. However, thyroxine increased the growth of Gram-negative bacteria and reduced the growth of Gram-positive ones. The effect of these drugs on biofilm formation by selected bacteria was linked to their effect on growth, except *L. reuteri*, where steroid hormones reduced biofilm formation by this bacterium despite increasing bacterial growth. The effect of hormonal drugs on bacterial adherence to Caco-2/HT-28 co-culture was not solely dependent on the change in hydrophobicity; other specific binding factors might contribute to their effect. The effect of the tested drug compounds on SCFAs and lactic acid production was variable and mostly independent of their effect on bacterial growth (Hammouda et al., 2023).

15.7 CONCLUSION

Research on hormonal drugs and their relation with bacterial biofilm formation is increasing in microbiology, endocrinology, and infectious diseases. At the cellular level, hormonal drugs can impact bacterial biofilm formation by affecting cell-signaling pathways, particularly those regulated by estrogen, androgen, and thyroid hormones. Estrogen and estrogen-like compounds have been found to promote biofilm formation in various bacteria, including *Staphylococcus aureus* and *Pseudomonas aeruginosa*. These hormones can stimulate the production of biofilm-associated genes, enriching the biofilm matrix and enhancing adhesion to surfaces. Antidepressants, particularly selective serotonin reuptake inhibitors (SSRIs), have also been linked to biofilm formation in a pilot study. While the mechanism remains unclear, the authors suggested that SSRIs may influence quorum sensing and biofilm maturation through unknown mechanisms. Estrogen and tamoxifen have been reported to inhibit biofilm formation in *S. epidermidis* isolates from ventricular catheters.

This anti-biofilm effect contrasts with the pro-biofilm action observed with other hormones, indicating the complex interplay between hormonal drugs and bacterial behavior. Hormonal therapies can influence biofilm formation by altering bacterial communications. For example, the oral microbiome has been changed due to estrogen therapy, potentially increasing the prospects of biofilm-related infections. The impact of hormonal drugs on biofilm-forming bacteria intensifies the potential for antibiotic resistance. Biofilms protect the bacteria from harsh environments, allowing them to withstand preferable antibiotic treatments. Due to the exposure of healthcare workers to patients taking hormonal drugs, the combination of resistant biofilms and treatment resistance could lead to prolonged and more severe infections. This review emphasizes the need for healthcare professionals to consider the hormonal status of patients when managing device-related infections. Understanding how hormonal drugs influence biofilm formation will be crucial for developing tailored antifungal and antibacterial therapies. Biomarkers of hormonal drugs, like serum hormone levels and microbiome analyses, could be utilized to personalize anti-infective strategies. Clinical trials should be conducted to investigate the effects of specific hormonal therapies on biofilm-related infections. Additionally, research into the mechanisms through which hormonal drugs affect biofilm formation will be crucial in developing targeted therapeutic strategies. In conclusion, the relationship between hormonal drugs and bacterial biofilm formation is a significant area of research that must be understood and addressed to improve treatment outcomes for device-related infections. By fostering a greater understanding of these interactions, the field of microbiology can better contribute to precision medicine strategies aimed at optimizing therapeutic outcomes while reducing the risk of treatment failures. As for the academic contribution, this review provides a comprehensive overview of the role of hormonal drugs in influencing biofilm formation, offering valuable insights for healthcare professionals and researchers alike. By highlighting the complex interplay between hormones, bacteria, and biofilms, this review effectively captures an emerging research field that has the potential to revolutionize our understanding of medical device-related infections and develop more effective therapeutic strategies.

Among the discussed hormonal drugs, glucocorticoids such as dexamethasone were found to enhance biofilm formation in *Staphylococcus aureus* and *Pseudomonas aeruginosa*, apparently due to immune suppression and increase of bacterial stress adaptation. For instance, sex hormones such as estrogen have been reported to increase biofilm formation in *Escherichia coli* and *Klebsiella pneumoniae*, the causative agents of UTIs, while progesterone has been linked with decreased biofilm formation of the beneficial *Lactobacillus* in the vagina, thereby altering the vaginal microbiota balance. Thyroid hormones, especially thyroxine (T4), have been found to support biofilm formation in *Staphylococcus epidermidis* and *Klebsiella pneumoniae*, which makes the infections that adhere to devices hard to kill. Insulin has also been found to increase the biofilm formation in *Pseudomonas aeruginosa* and *Streptococcus mutans*, especially in diabetic patients where glucose is in abundance to support bacterial growth. Moreover, gonadotropins such as FSH and LH were found to induce biofilm formation in *Acinetobacter baumannii* and *Listeria monocytogenes*, which will make infection control in reproductive and hospital-acquired infections difficult.

These findings highlight the importance of further investigating the impact of hormonal drugs on microbial pathogenesis and biofilm development as an area that requires significant research attention. As antibiotic resistance is becoming more and more widespread, the role of hormone–microbe interactions needs to be further explored to develop new and better therapeutic strategies. Future work should also concentrate on the creation of new antimicrobial agents that act on hormones to prevent biofilm formation and the identification of biofilm inhibitors that target the hormone-induced quorum sensing. Furthermore, the impact of alternative hormone therapy on microbial pathogenicity should be assessed. In addition, there is a need to investigate the effect of prolonged treatment with unapproved hormones on biofilm formation and their impact on respiratory infections, with an emphasis on *P. aeruginosa*. By understanding these aspects, future work can focus on developing comprehensive guidelines for healthcare professionals on hormone management to prevent biofilm-related infections. Also, the fields of personalized medicine could be considered to determine how specific hormonal configurations affect bacterial infections and, therefore, distinct antimicrobial therapies.

REFERENCES

Anzar, M. A., Quamri, M. A., Sofi, G., Haider, N. Evaluation of the Add on Effect of Majoone Sarkhas with Levothyroxine in Primary Hypothyroidism: A Randomized Standard Control Adjuvant Clinical Study. *Drug Metabolism and Personalized Therapy* 2025;40(2):121–130. https://doi.org/10.1515/dmpt-2024-0096

Alizhan D, Ukybassova T, Bapayeva G, Aimagambetova G, Kongrtay K, Kamzayeva N, Terzic M. Cervicovaginal Microbiome: Physiology, Age-Related Changes, and Protective Role Against Human Papillomavirus Infection. *Journal of Clinical Medicine*. 2025 January;14(5):1521.

Arevalo G, Menendez SG, Soliman D, Galindo RJ. 6567 Novel Indication of GLP-1 Receptor Agonists Beyond Weight Loss: Addressing Post-Gastric Bypass Late Dumping Syndrome. *Journal of the Endocrine Society*. 2024 October;1:8.

Arya GC, Rathee A, Mehla S, Bisht P, Sharma R. A Review of the Azasteroid-type 5-alpha Reductase Inhibitors for the Management of Benign Prostatic Hyperplasia. *Letters in Drug Design & Discovery*. 2024 September 1;21(12):2271–2287.

Basu R, Nahar K, Kulkarni P, Kerekes O, Sattler M, Hall Z, Neggers S, Holub JM, Kopchick JJ. A Novel Peptide Antagonist of the Human Growth Hormone Receptor. *Journal of Biological Chemistry*. 2021 January 1;296:100588. doi: 10.1016/j.jbc.2021.100588. Epub 2021 Mar 24. PMID: 33774052; PMCID: PMC8086144.

Brogna MR, Ferrara G, Varone V, Montone A, Schiano M, DelSesto M, Collina F. Evaluation and Comparison of Prognostic Multigene Tests in Early-Stage Breast Cancer: Which Is the Most Effective? A Literature Review Exploring Clinical Utility to Enhance Therapeutic Management in Luminal Patients. *Molecular Carcinogenesis*. 2025;64(5):789–800.

Che X, Zhao Y, Xu Z, Hu Y, Ren A, Wu C, Yang J. Unlocking the Potential of l-α-Glycerylphosphorylcholine: From Metabolic Pathways to Therapeutic Applications. *Nutrition Reviews*. 2025 February 28;83(8): 1594–1620. doi: 10.1093/nutrit/nuaf008. PMID: 40036805.

Dai Q, Yan M, Wang C, Shen C, Jiang C, Yang B, Zhao H, Zhu Z. Cyclocarya paliurus Polysaccharide Inhibits Benign Prostatic Hyperplasia by Reducing 5α-Reductase 2. *Chinese Journal of Experimental Traditional Medical Formulae*. 2025 February 5;31(3):107–114.

Esteves SC, Viana MC, Achermann AP, Santi D. Human Chorionic Gonadotropin-Based Clinical Treatments for Infertile Men with Non-Obstructive Azoospermia. *Andrology*. 2025.

Ghasemi S, Habibi Z, Ansari F. Biotransformation of Hydrocortisone and Nandrolone Decanoate by Three Fungal Species. *Nashrieh Shimi va Mohandesi Shimi Iran*. 2024 November 21;43(3):195–208.

Gonçalves B, Bernardo R, Wang C, Schröder MS, Pedro NA, Butler G, Azeredo J, Henriques M, Pereira Mira N, Silva S. Effect of Progesterone on Candida Albicans Biofilm Formation Under Acidic Conditions: A Transcriptomic Analysis. *International journal of medical microbiology: IJMM*. 2020;310(3):151414. https://doi.org/10.1016/j.ijmm.2020.151414

Goodarzi-Khoigani M, Aminorroaya A. *Effect of Levothyroxine Therapy on Gestational Hypertension and Pre-Eclampsia in Pregnant Women with Subclinical Hypothyroidism, Hypothyroidism, and Thyroid Autoimmunity*. PubMed. 2025.

Gupta AK, Vaishnav Y, Jain SK, Annadurai S, Kumar N. Exploring novel Apalutamide Analogues as Potential Therapeutics for Prostate Cancer: Design, Molecular Docking Investigations and Molecular Dynamics Simulation. *Frontiers in Chemistry*. 2024 August 6;12:1418975.

Hahamyan HA, Basaria S. Selective Androgen Receptor Modulators—Transformative Drugs or Heralds of the Next Drug Epidemic? *JAMA*. 2024 April 23;331(16):1359–1360.

Hall JM. Endocrine Disruptors in Molecular and Structural Endocrinology. *Frontiers in Endocrinology*. 2025 February 10;16:1561253.

Hammouda ZK, Wasfi R, Abdeltawab NF. Hormonal Drugs: Influence on Growth, Biofilm Formation, and Adherence of Selected Gut Microbiota. *Frontiers in Cellular and Infection Microbiology*. 2023 March 13;13:1147585.

Harxhi R, Valentin Mihailescu D, Pilati S. 7469 A Case of Severe Transient Hypercortisolism in a Patient With Bacterial Meningitis. *Journal of the Endocrine Society*. 2024 October;8(Supplement_1):bvae163-068.

Hou YC, Zheng CM, Chiu HW, Liu WC, Lu KC, Lu CL. Role of Calcimimetics in Treating Bone and Mineral Disorders Related to Chronic Kidney Disease. *Pharmaceuticals*. 2022 July 31;15(8):952.

Hulubă IP, Crecan-Suciu BD, Păunescu R, Micluţia IV. The Link Between Sex Hormones and Depression Over a Woman's Lifespan. *Biomedical Reports*. 2025 April 1;22(4):1.

Inaba M, Doi Y. Host Stress Hormone Norepinephrine Reduces in Vitro Activity of Aminoglycoside Against Carbapenemase-Producing Enterobacterales. *Journal of Antimicrobial Chemotherapy*. 2024 June;79(6):1468–1470.

Jäger MC, González-Ruiz V, Joos FL, Winter DV, Boccard J, Degenhardt T, Brand S, Rudaz S, Thompson GR, Odermatt A. 7963 Analysis of the Inhibition of Adrenal Synthesis and Peripheral Metabolism of Corticosteroids by Oteseconazole and Two Preclinical Tetrazole Antifungals. *Journal of the Endocrine Society*. 2024 October;8(Supplement_1):bvae163–1788.

Javan MJ, Youseftabar-Miri L. Total Synthetic Studies on Potent Compounds as Anti-Prostate Cancer Agents. In *Discovery and Development of Anti-Prostate Cancerous Agents from Natural Products*. 2025 January 1 (pp. 321–375). Elsevier.

Kiyama R, Wada-Kiyama Y. Estrogenic Actions of Alkaloids: Structural Characteristics and Molecular Mechanisms. *Biochemical Pharmacology*. 2024 November 20:232:116645. doi: 10.1016/j.bcp.2024.116645. Epub 2024 Nov 20. PMID: 39577707

Kuzminac IZ, Nikolić AR, Savić MP, Ajduković JJ. Abiraterone and Galeterone, Powerful Tools Against Prostate Cancer: Present and Perspective. *Pharmaceutics*. 2024 October 30;16(11):1401.

Li C, Li G. The Role of Sphingolipid Metabolism in Endocrine Diseases. *Frontiers in Endocrinology*. 2024;15:1506971.

Lv H, Xu X, Wu Z, Lin Y, Liu Y, Liu M, Xu L, Wang X, Sun N, Abdel-Shafy H, Abdelrahman M. Yes-Associated Protein 1 is Essential for Maintaining Lactation Via Regulating Mammary Epithelial Cell Dynamics and Secretion Capacity. *International Journal of Biological Macromolecules*. 2025 March 1;293:139290.

Madacki-Todorović K, Eminović I, Mehmedinović NI, Ibrišimović M. Corticosteroid-Induced Expression of Microbial Virulence can Enhance the Development of Host Infectious Disease. *Medis–International Journal of Medical Sciences and Research*. 2024 May 19;3(2):19–24.

Marcu B, Sardari V, Munteanu R. Biochemical Aspects of Adrenocortical Insufficiency in Patients with Liver Cirrhosis. *Moldovan Journal of Health Sciences*. 2024 October;28(2):51.

Mei Y, Li W, Wang B, Chen Z, Wu X, Lin Y, Wang M. Gut Microbiota: An Emerging Target Connecting Polycystic Ovarian Syndrome and Insulin Resistance. *Frontiers in Cellular and Infection Microbiology*. 2025;15:1508893.

Mondal S, Raja K, Schweizer U, Mugesh G. Chemistry and Biology in the Biosynthesis and Action of Thyroid Hormones. *Angewandte Chemie International Edition*. 2016 June 27;55(27):7606–7630.

Mukhtar S, Haris M. Vamorolone: Revolutionizing Duchenne Muscular Dystrophy Treatment. *International Journal*. 2024 May;10(5):1.

Onyeso G, David J, Egwu M. Role of Progestins in Reproduction. *Sex Steroid Hormones-Impact on Reproductive Physiology: Impact on Reproductive Physiology*. 2025 February 5:63. https://doi.org/10.5772/intechopen.1007931

Ornetti P, Bouillet B, Denimal D. New Guidelines on Glucocorticoid-Induced Adrenal Insufficiency: The End of Short Synacthen Test in Rheumatology? *RMD Open*. 2025 January 1;11(1):e005251.

Pu S, Zhao B, Jiang Y, Cui X. Hypothyroidism/Subclinical Hypothyroidism and Metabolic Dysfunction-Associated Steatotic Liver Disease: Advances in Mechanism and Treatment. *Lipids in Health and Disease*. 2025 December;24(1):1–6.

Qin KR, Mertens E, Tempo JA, Perera M, Brennan J. Could Menopause Drug Fezolinetant Show Promise for Vasomotor Symptoms Associated with Androgen Deprivation Therapy? *Translational Andrology and Urology*. 2024 May 15;13(5):920.

Ramya GS, Bahl S, Stedman M, Heald A, Premawardhana L, Okosieme O, Taylor P, Dayan C. Thyroid Clinic Audit of Patients on Non-standard Thyroid Hormone Replacement: Review of Practice at the University Hospital of Wales (UHW). In *Endocrine Abstracts*. 2025 February 19 (Vol. 109). Bioscientifica. https://doi.org/10.1530/endoabs.109.p277

Raposo MC, Feiteira C, Ribeiro MH. Regulatory and Clinical Aspects in Biosimilar Medicines: Comparability, Extrapolation, Interchangeability, and Safety. *Drugs & Therapy Perspectives*. 2025 February 27:41(3):111–125. https://doi.org/10.1007/s40267-025-01145-5

Ru Q, Huang K, Yu R, Wu X, Shen J. Effects of Camellia Oleifera Seed Shell Polyphenols and 1, 3, 6-tri-O-galloylglucose on Androgenic Alopecia Via Inhibiting 5a-reductase and Regulating Wnt/β-catenin Pathway. *Fitoterapia*. 2024 September 1;177:106116.

Ruan X, Mueck AO. Criteria for the Choice and Monitoring of Menopausal Hormone Therapy. *Current Medicine*. 2024 December;3(1):1–9.

Saputra MFN, Nabila A, Maulida F. *Synthesis and Characterization of Insulin-MOF-5 Loaded into Cellulose Nanocrystalline-Based Dissolving Microneedle as Anti-Diabetic Delivery System*. ResearchGate. 2025. https://doi.org/10.26655/JMCHEMSCI.2025.2.5

Saretzki G. Chronic Stress, Telomeres and Aging. *Frontiers in Endocrinology*. 2024 November 18;15:1504405.

Stanczyk FZ, McGough A, Chagam L, Sitruk-Ware R. Metabolism of Progestogens Used for Contraception and Menopausal Hormone Therapy. *Steroids*. 2024 April 24:207:109427. doi: 10.1016/j.steroids.2024.109427

Suryandari DA, Yunaini L, Sunaryo H, Istiadi KA, Pratomo IP. Molecular Insights into Propylthiouracil as a Thyroid Peroxidase Inhibitor: A Computational Study Approach. *Indonesian Journal of Medical Chemistry and Bioinformatics*. 2024;3(1):2.

Szepesi Á, Soares C, Fidalgo F. Working Towards Sustainable Agriculture with Biostimulants and Biocides: From Chemical Structure to Application Methods. *Frontiers in Plant Science*. 2024 July 31;15:1443658.

Tao Y, Li Z, Yuan J, Wu H, Shi H, Wu X, Huang F. Dried Bear Bile Exerts Its Antidepressant Effect by Modulating Adrenal FXR to Reduce Peripheral Glucocorticoid Levels. *Brain, Behavior, & Immunity-Health*. 2024 November 1; 41:100856.

Todorovic A, Ericson MD, Gridley K, Sorenson NB, Haskell-Luevano C. Synthesis and Pharmacological Characterization of Adrenocorticotropic Hormone (ACTH 1–24) and C-Terminal Truncated Analogues Identifies the Minimal ACTH N-Terminal Fragment Required for Melanocorton-2 Receptor Activation. *International Journal of Peptide Research and Therapeutics*. 2024 October 30;31(1):1.

Tzoupis H, Nteli A, Androutsou ME, Tselios T. Gonadotropin-Releasing Hormone and GnRH Receptor: Structure, Function and Drug Development. *Current Medicinal Chemistry*. 2020 November 1;27(36):6136–6158.

Wang, L. Radiolabeling and Imaging of a Novel 18F-labeled Probe Targeting the Arginine Vasopressin Receptor V1a. *Journal of Nuclear Medicine*. 2024;65(supplement 2):241705.

Xu X, Lo CW, Martin LJ, Han L, Xu D. Recent Advances in Causes, Diagnosis, and Therapeutics for Congenital Heart Defects. *Frontiers in Genetics*. 2025 February 18;16:1564492.

Yamazaki Y, Seino Y. Which is the Real Nature of Glucose-Dependent Insulinotropic Peptide?: Endogenous vs Pharmacological. *Journal of Diabetes Investigation*. 2024 November 14;16(2):173–175.

Zhao K, Li Z, Sun T, Liu Q, Cheng Y, Barreto G, Li Z, Liu R. Novel Therapeutic Target and Drug Discovery for Neurological Diseases, Volume II. *Frontiers in Pharmacology*. 2025;16:1566950.

Index

Pages in *italics* refer to figures and pages in **bold** refer to tables.

Made in the USA
Columbia, SC
17 September 2026

d626dbb5-36d3-492f-b002-0db25a51b6eaR01